Algorithms for Intelligent Systems

D1796945

Series Editors

Jagdish Chand Bansal, Department of Mathematics, South Asian University,
New Delhi, Delhi, India
Kusum Deep, Department of Mathematics, Indian Institute of Technology Roorkee,
Roorkee, Uttarakhand, India
Atulya K. Nagar, Department of Mathematics and Computer Science,
Liverpool Hope University, Liverpool, UK

This book series publishes research on the analysis and development of algorithms for intelligent systems with their applications to various real world problems. It covers research related to autonomous agents, multi-agent systems, behavioral modeling, reinforcement learning, game theory, mechanism design, machine learning, meta-heuristic search, optimization, planning and scheduling, artificial neural networks, evolutionary computation, swarm intelligence and other algorithms for intelligent systems.

The book series includes recent advancements, modification and applications of the artificial neural networks, evolutionary computation, swarm intelligence, artificial immune systems, fuzzy system, autonomous and multi agent systems, machine learning and other intelligent systems related areas. The material will be beneficial for the graduate students, post-graduate students as well as the researchers who want a broader view of advances in algorithms for intelligent systems. The contents will also be useful to the researchers from other fields who have no knowledge of the power of intelligent systems, e.g. the researchers in the field of bioinformatics, biochemists, mechanical and chemical engineers, economists, musicians and medical practitioners.

The series publishes monographs, edited volumes, advanced textbooks and selected proceedings.

More information about this series at http://www.springer.com/series/16171

Basant Agarwal · Richi Nayak ·
Namita Mittal · Srikanta Patnaik
Editors

Deep Learning-Based Approaches for Sentiment Analysis

 Springer

Editors
Basant Agarwal
Indian Institute of Information Technology
Kota (IIIT-Kota)
Jaipur, Rajasthan, India

Namita Mittal
Department of Computer Science
and Engineering
Malaviya National Institute of Technology
Jaipur, Rajasthan, India

Richi Nayak
Faculty of Science and Engineering
School of Electrical Engineering
and Computer Science
Queensland University of Technology
Brisbane, QLD, Australia

Srikanta Patnaik
Department of Computer Science
and Engineering
SOA University
Bhubaneswar, Odisha, India

ISSN 2524-7565   ISSN 2524-7573 (electronic)
Algorithms for Intelligent Systems
ISBN 978-981-15-1218-6   ISBN 978-981-15-1216-2   (eBook)
https://doi.org/10.1007/978-981-15-1216-2

This Springer imprint is published by the registered company Springer Nature Singapore Pte Ltd.
The registered company address is: 152 Beach Road, #21-01/04 Gateway East, Singapore 189721, Singapore

Preface

With the exponential growth in the use of social media networks such as Twitter, Facebook, Flickr, and many others, an astronomical amount of big data has been generated. This data is present in heterogeneous forms such as text, images, videos, audio, and graphics. A substantial amount of this user-generated data is in the form of text such as reviews, tweets, and blogs that provide numerous challenges as well as opportunities to natural language processing (NLP) researchers for discovering meaningful information used in various applications. The textual information available is of two types: facts and opinion statements. Facts are objective sentences about the entities. On the other hand, opinions are subjective in nature and generally describe people's sentiments toward entities and events. Research on processing the opinionated sentences is one of the active and popular research areas due to a large number of challenges involved. It is extensively studied in data mining, text mining, web mining, and social media analytics.

Opinion mining and sentiment analysis as a research discipline have emerged during the past 15 years due to the wide range of business and social applications. They provide an automated computational approach to process and discover the sentiments and opinions from the unstructured text. Sentiment analysis is the study that analyzes people's opinion and sentiment toward entities such as products, services, person, and organisations present in the text. The automated analysis of online content to extract the opinion requires machines to build a deep understating of natural text. Customers express their opinion, feelings, and experiences about the products or services on the forums, blogs, microblogging sites, and social networks. Others, who wish to know about the product or the service, can learn from those who have already experienced it. Others' opinions and sentiments assist them in making purchasing decisions. E-commerce companies can improve their products or services on the basis of customers' opinions. They help in shaping up the future and current trends of the market.

In recent years, deep Learning approaches have emerged as powerful computational models and have shown significant success to deal with a massive amount of data in unsupervised settings. Deep learning is revolutionizing because it offers an effective way of learning representation and allows the system to learn features automatically from data without the need of explicitly designing them. Deep learning algorithms such as deep autoencoders, convolutional neural network (CNN) and recurrent neural networks (RNN), and long short-term memory (LSTM) have been reported providing significantly improved results in various natural language processing tasks including sentiment analysis.

This book focuses on recent advances in the field of sentiment analysis using deep learning-based approaches. This book is organized into 12 chapters.

The first chapter of this book "Application of Deep Learning Approaches for Sentiment Analysis" presents an introduction to the sentiment analysis in general and introduces various deep learning-based approaches for the sentiment analysis.

The chapter "Recent Trends and Advances in Deep Learning-Based Sentiment Analysis" provides the detailed discussion on recent advances in the field of deep learning in the context of sentiment analysis problem.

The chapter "Deep Learning Adaptation with Word Embeddings for Sentiment Analysis o n Online Course Reviews" presents a deep learning-based approach for sentiment analysis of textual reviews for online courses utilizing the word embedding representation which captures sentiment polarity. This chapter demonstrates how specific word embedding performs better as compared to general-purpose trained embeddings.

The chapter "Toxic Comment Detection in Online Discussions" presents various deep learning-based approaches of sentiment analysis in online discussions. The comment sections of online news platforms are an important space to express opinions and discuss political topics. This chapter discusses real-world applications such as semi-automated comment moderation and troll detection.

The chapter "Aspect-Based Sentiment Analysis of Financial Headlines and Microblogs" focuses a specific sentiment analysis problem called as aspect-based sentiment analysis in which the opinions with respect to specific aspects are analyzed. This chapter describes a novel approach which is a combination of neural network models, hand-crafted features, and attention mechanism for the aspect-based sentiment analysis for financial headlines and microblogs.

The chapter "Deep Learning-Based Frameworks for Aspect-Based Sentiment Analysis" discusses the aspect-based sentiment analysis problem with deep learning-based approaches. This chapter presents different state-of-the-art approaches for aspect-level sentiment analysis.

The chapter "Transfer Learning for Detecting Hateful Sentiments in Code Switched Language" focuses on transfer learning-based techniques to analyze code-mixed language. This chapter presents a convolutional neural network-based model which is trained on a large dataset of hateful Hindi and English mixed tweet.

The chapter "Multilingual Sentiment Analysis" focuses on sentiment analysis of various low resource languages having limited sentiment analysis resources such as annotated datasets, word embeddings, and sentiment lexicons, along with English. This chapter describes various techniques to improve word embeddings for low resource languages. This chapter discusses the challenges focused on multilingual sentiment analysis, along with how these challenges are tackled by deep learning-based solutions.

The chapter "Sarcasm Detection Using Deep Learning-Based Techniques" presents a discussion on various sarcasm detection techniques and presents a deep learning-based technique to deal with sarcasm in the text.

The chapter "Deep Learning Approaches for Speech Emotion Recognition" presents an approach for extracting the sentiments from the speech. This chapter discusses the effect of various deep learning-based techniques and feature extraction methods for analyzing sentiment from speech.

The chapter "Bidirectional Long Short-Term Memory-Based Spatio-Temporal in Community Question Answering" presents a novel deep learning approach, namely spatio-temporal bidirectional long short-term memory (ST-BiLSTM), for better semantic representation for the question answering problem.

The final chapter "Comparing Deep Neural Networks to Traditional Models for Sentiment Analysis in Turkish Language" presents a comparative study between various traditional machine learning-based approaches and a deep learning-based approach such as recurrent neural network technique for sentiment analysis of Turkish language text.

The editors are thankful to all the members of Springer (India) Private Limited, especially Aninda Bose, for the given opportunity to edit this book. Editors appreciate the efforts of authors in making this book a reality.

Jaipur, India	Basant Agarwal
Brisbane, Australia	Richi Nayak
Jaipur, India	Namita Mittal
Bhubaneswar, India	Srikanta Patnaik

Contents

About the Editors

Dr. Basant Agarwal is an Assistant Professor at the Indian Institute of Information Technology Kota (IIIT-Kota), India. He holds a Ph.D. from MNIT Jaipur, and worked as a Postdoc Research Fellow at the Norwegian University of Science and Technology (NTNU), Norway, under the prestigious ERCIM (European Research Consortium for Informatics and Mathematics) fellowship in 2016. He has also worked as a Research Scientist at Temasek Laboratories, National University of Singapore (NUS), Singapore.

Dr. Richi Nayak holds an M.E. degree from the Indian Institute of Technology, Roorkee, India, and received her Ph.D. in Computer Science from the Queensland University of Technology (QUT), Brisbane, Australia, in 2001. She is currently an Associate Professor of Computer Science at QUT, where she is also Head of Data Science. She has been successful in attaining over $4 million in external research funding in the area of text mining over the past ten years. She is a consultant to a number of government agencies in the area of data, text, and social media analytics projects. She is member of the steering committee of Australasian Data Mining in Australia (AusDM). She is the founder and leader of the Applied Data Mining Research Group at QUT. She has received a number of awards and nominations for teaching, research, and other activities.

Dr. Namita Mittal is an Associate Professor at the Department of Computer Science and Engineering, MNIT Jaipur, India. She is a recipient of the Career Award for Young Teachers (CAYT) by AICTE. She has published numerous research papers in respected international conferences and journals, and has also authored a book on the topic of sentiment analysis in the Springer book series "Socio-Affective Computing". She is an SMIEEE, and a member of ACM, CCICI, and SCRS. She has been involved in various FDPs/conferences/workshops, like the Ph.D. Colloquium FIRE 2017, and International Workshop on Text Analytics and Retrieval (WI 2018) in conjunction with Web Intelligence (WI), USA, to name a few.

Dr. Srikanta Patnaik is a Professor at the Department of Computer Science and Engineering, Faculty of Engineering and Technology, SOA University, Bhubaneswar, India. He received his Ph.D. in Computational Intelligence from Jadavpur University, India, in 1999. Dr. Patnaik was the Principal Investigator of the AICTE-sponsored TAPTEC project "Building Cognition for Intelligent Robot" and the UGC-sponsored Major Research Project "Machine Learning and Perception using Cognition Methods". He is the Editor-in-Chief of the International Journal of Information and Communication Technology and the International Journal of Computational Vision and Robotics. Dr. Patnaik is also the Editor of the Journal of Information and Communication Convergence Engineering, published by the Korean Institute of Information and Communication Engineering. He is also the Editor-in-Chief of Springer book series "Modeling and Optimization in Science and Technology".

Application of Deep Learning Approaches for Sentiment Analysis

Ajeet Ram Pathak⦿, **Basant Agarwal**⦿, **Manjusha Pandey** and **Siddharth Rautaray**

Abstract Social media platforms, forums, blogs, and opinion sites generate vast amount of data. Such data in the form of opinions, emotions, and views about services, politics, and products are characterized by unstructured format. End users, business industries, and politicians are highly influenced by sentiments of the people expressed on social media platforms. Therefore, extracting, analyzing, summarizing, and predicting the sentiments from large unstructured data needs automated sentiment analysis. Sentiment analysis is an automated process of extracting the opinionated from data and classifying the sentiments as positive, negative, and neutral. Lack of enough labeled data for sentiment analysis is one of the crucial challenges in natural language processing. Deep learning has emanated as one of the highly sought-after solutions to address this challenge due to automated and hierarchical learning capability inherently supported by deep learning models. Considering the application of deep learning approaches for sentiment analysis, this chapter aims to put forth taxonomy of traits to be considered for deep learning-based sentiment analysis and demystify the role of deep learning approaches for sentiment analysis.

Keywords Deep learning · Deep neural networks · Opinion mining · Sentiment analysis · Social media data · Taxonomy · Text representation

A. R. Pathak (✉) · M. Pandey · S. Rautaray
School of Computer Engineering, Kalinga Institute of Industrial Technology University (KIIT), Bhubaneswar 751024, India

B. Agarwal
Department of Computer Science and Engineering, Indian Institute of Information Technology (IIIT), Kota 302017, India
e-mail: basant.cse@iiitkota.ac.in

B. Agarwal et al. (eds.), *Deep Learning-Based Approaches for Sentiment Analysis*, Algorithms for Intelligent Systems, https://doi.org/10.1007/978-981-15-1216-2_1

1

1   Introduction

The drastic shifts from read-only to read-write access to the Web lead the people to interact with each other through social media networks like wikis, blogs, online forums, communities, etc. Due to this, user-generated content through social media platforms is increasing tremendously. Specifically, Web-based data of the form—opinionated text, reviews of products, and services has been one of the most contributing factors in social big data [1].

Analyzing the sentiments of people from such opinionated data helps both end users and business industries in decision-making for purchasing products, launching new products, assessing the industry reputation among the customers, etc. Sentiment analysis, also termed as opinion mining, is an automated process of extracting the polarity of the opinionated text. Alongside polarity, subject and opinion holders can also be identified using sentiment analysis. Sentiment analysis is one of the most active research areas in natural language processing since 2000 and continues to be highly sought-after research domain. It is forecasted that by 2025, NLP market would reach $22.3 billion [2].

Because of the proliferation of diverse opinion sites, it is difficult to find and monitor all the sites and collect the information pertaining to some domain and perform sentiment analysis. Moreover, it is difficult for human personnel to segregate the opinionated data from long blogs and forums and summarize the opinions. This arise the need of automated sentiment analysis systems.

Numerous techniques have been put forth till date to perform sentiment analysis based on supervised and unsupervised learning. In supervised learning, early literature focused on applying supervised machine learning techniques like naïve Bayes, support vector machines, and feature learning algorithms [3]. Unsupervised learning methods include the use of sentiment lexicons, grammatical analysis, etc.

Deep learning has emanated as a powerful technique to solve multitude of problems in the domains of computer vision [4–8], topic modeling [9–11], natural language processing [12–14], speech recognition [15], social media analytics [16–18], etc. Inspired by the same, applying deep learning-based sentiment analysis achieved great popularity in the recent lustrum. This book chapter sheds light upon the progress made in deep learning-based sentiment analysis by giving an overview of deep learning-based sentiment analysis models. Figure 1 gives a glimpse of main topics covered to demystify the application of deep learning for sentiment analysis.

2   Taxonomy of Sentiment Analysis

Figure 2 shows the taxonomy of the traits to be considered while designing the sentiment analysis models.

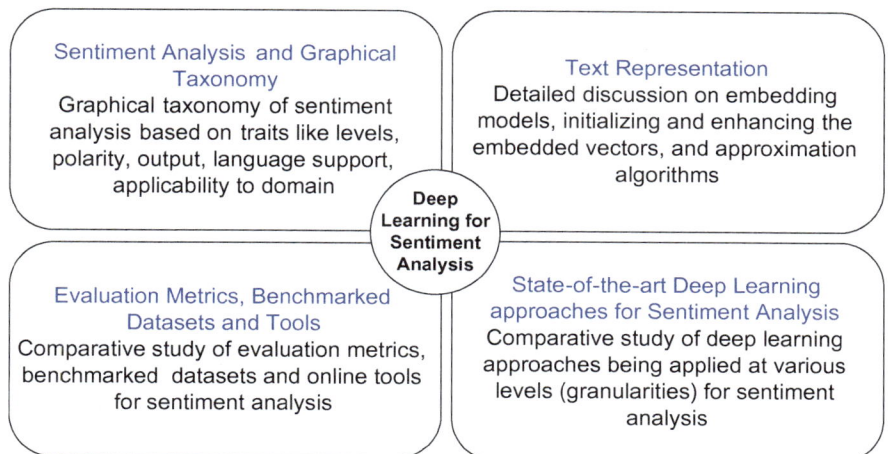

Fig. 1 Demystified overview of application of deep learning for sentiment analysis

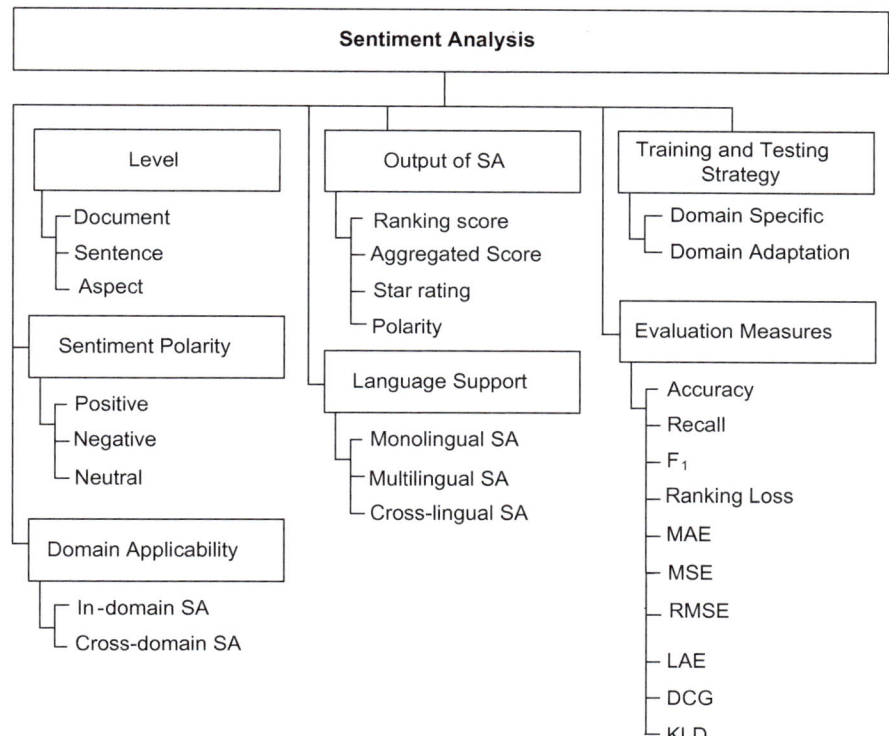

Fig. 2 Taxonomy of the traits for sentiment analysis models

2.1 Sentiment Analysis, Polarity, and Output

Sentiment analysis is an automated process, which predicts the polarity of the opinionated text in terms of positive, negative, and neutral [19]. Fine-grained sentiment
analysis involves the following categories, viz. very positive, positive, neutral, negative, and very negative. These categories can be mapped to a rating score, for example,
"very positive" can be mapped to 5 stars, whereas "very negative" to 1 star. For multiple documents, the individual polarities obtained for each document can be mapped
to the ratings and then aggregated to give aggregated score.

2.2 Levels of Sentiment Analysis

Sentiment analysis is performed at various levels of granularities such as document,
sentence, and aspect-based. These levels have been discussed in this sub-section.

Document level

This level determines the sentiment of a complete paragraph or a document. The
sentiment analysis model assumes that document contains opinionated text about
the single entity. This level does not support documents comparing the multiple
entities. The problem of determining whether the document has positive or negative
polarity is portrayed as a binary classification problem. It can also be handled as
a regression problem, for instance, assigning the rating score in the range of 1–5
stars for movie reviews. This task can also be modeled as a five-class classification
problem.

Sentence level

This level of sentiment classification aims to determine the sentiment from a single sentence. Subjectivity classification and polarity classification can be used for
inferring the sentiment from a sentence. Subjectivity classification focuses on finding whether a sentence is subjective or objective. On the other hand, the polarity
classification determines whether a given subjective sentence is positive or negative. Existing deep learning techniques focuses on predicting polarity of a sentence
as positive, negative, and neutral. As sentences are shorter compared to the document, semantic, and syntactic features obtained via POS tagger, parse trees, and
lexicons can be used for sentence-level sentiment classification. Similar to document-
level assumption, sentence-level sentiment classification assumes that each sentence
contains sentiment about single entity.

Aspect-based sentiment analysis (ABSA)

In this level, sentiments of the users expressed toward aspects (features) of the entities
(objects) such as movie and restaurant are extracted. It aims to find the aspect and
polarity pairs from a given text. This level assumes that a single entity is present per
document. As mentioned in [20], aspect-level sentiment analysis can be divided into
four tasks as aspect term extraction, aspect term polarity, aspect category detection,
and aspect category polarity. Aspect term extraction involves identifying the aspect

Table 1 Phase-wise examples in ABSA and output labels

Phase	Example	Labels
Aspect term extraction	I am happy with fast boot time, speedy Wi-Fi, and the long battery life	Aspects: "boot time", "Wi-Fi connection", and "battery life"
Aspect term polarity	I am happy with fast boot time, speedy Wi-Fi, and the long battery life	Polarities: {boot time: positive}, {Wi-Fi connection: positive}, and {battery life: positive}
Aspect category detection	It is wonderful and affordable	Categories: {General, price}
Aspect category polarity	It is wonderful and affordable	Polarities: {General: positive}, {price: positive}

terms from a set of sentences with pre-defined entities (e.g., laptops) and returning the list of distinct aspect terms. The second sub-task, namely, aspect term polarity focuses on determining the polarity of the aspect term detected in the first sub-task. Aspect category detection identifies the aspect categories in each sentence based on pre-defined set of aspect categories (e.g., general, price). The fourth sub-task aspect category polarity focuses on determining the polarity of each aspect category from a given set of sentences. Table 1 gives an example and output of each sub-task in ABSA.

Targeted ABSA is an extension of aspect-based sentiment analysis. ABSA assumes the occurrence of single entity per document, whereas targeted ABSA assumes a single sentiment toward each aspect of one or more entities. Targeted ABSA extracts the target entities, different aspects and their corresponding sentiments. For example, *"The ambience is good in Viceroy but the service is bad, on the other hand, the staff in Novotel is very prompt and the food is tasty as usual."* This instance talks about aspects of two different hotels. Targeted ABSA recognizes "Viceroy" and "Novotel" as two target entities and output the labels as {Viceroy, ambience, positive}, {Viceroy, service, positive}, {Novotel, service, positive}, {Novotel, food, positive}.

2.3 Domain Applicability, Training, and Testing Strategy

Domain applicability states weather the sentiment analysis model performs in-domain or cross-domain sentiment analysis. For in-domain sentiment analysis, training and testing are done on the same target domain, i.e., domain-specific training and testing strategy are applied. Sometimes, the target domain on which sentiment analysis is to be performed lacks or possesses very less labeled data associated with sentiment classes, and therefore it is difficult to train the model with such data.

Therefore, domain adaptation [21] (transfer learning) technique is applied for cross-domain sentiment analysis in which a model is trained on the domain with labeled data and tested on target domain with no or very less labeled data.

2.4 Language Support

Sentiment analysis models can be categorized into monolingual, multi-lingual, and cross-lingual sentiment models based on the support for the language. Cross-lingual sentiment analysis models train the model on resource-rich language and then test on resource-poor language.

2.5 Evaluation Measures

Common evaluation metrics commonly used for sentiment analysis are accuracy, F_1 score, average recall (AvgRec), macro-average F1 score, ranking loss, macro-averaged mean absolute error, least absolute error (LAE), mean squared error (MSE), Pearson correlation coefficient, KullbackLeibler divergence (KLD), and area under the ROC curve (AUC). These metrics have been discussed in this section in Sect. 5.

3 Text Representation for Sentiment Analysis

Figure 3 depicts various traits to be considered to represent the text for sentiment analysis using deep learning. Each trait has been discussed in sub-sequent sections.

3.1 Embedded Vectors

For most machine learning algorithms, which map input to output using approximation require numerical representation of input data. Embedding methods (also named as vectorizing or encoding) convert input data (i.e., words, sentences, paragraphs, document, date, emoji, graph, etc.) into real numbers capturing the hidden semantic relation between input data. Embedding models are one of the successful applications of unsupervised learning and have been popularly used in deep learning-based NLP tasks. Bengio et al. [22] introduced the concept of word embeddings. Some noteworthy models which can be used for representing the input text have been discussed.

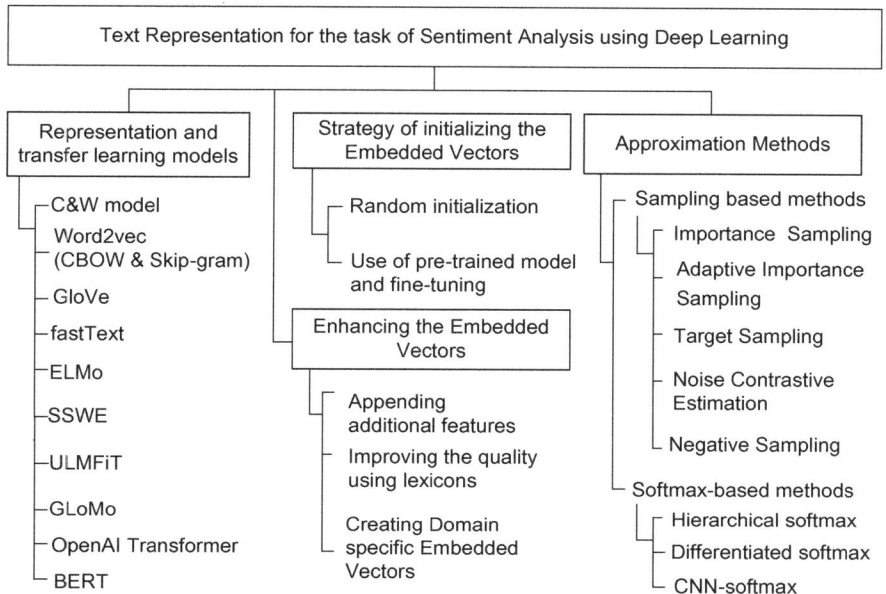

Fig. 3 Traits to be considered to represent the text for sentiment analysis using deep learning

Collobert and Weston (C&W) model

C&W model proposed in [23] has been designed using multi-layered neural network architecture, trained on large dataset and carries syntactic and semantic meaning. This model is designed agnostic to any task-specific feature engineering and therefore serves as useful word representation model for wide variety of NLP tasks.

Word2vec

The vectors used for representing the words are neural word embeddings. Word2vec [24] is used to obtain the distributed representation of words, i.e., word embeddings. Word2vec trains the words against the other words that are neighbors of each other in the input corpus. This training can be done using any of the two models such as continuous bag-of-words (CBOW)or skip-gram model. CBOW model emits a target word according to surrounding context. Skip-gram model emits words in a surrounding context provided that central word is given.

fastText

Facebook's AI research laboratory came up with fastText library [25]. It efficiently learns word representation. By making use of character-level information, fastText can be used to get the representation for rear words also.

Global Vectors for Word Representation (GloVe)

GloVemodel [26] gives vector representations for words in an unsupervised manner. It uses both global matrix factorization and local context window to get representation of the word.

Embeddings from Language Models (ELMo)

Traditional word embedding models like word2vec and GloVe can not handle the contextual meaning of the words and therefore provide the same vector representation for the word with different meanings. For instance, meaning of the word stick is different "stick" in the following example.

Sentence 1: *This stick is made up of wooden material*

Sentence 2: *Let's stick to one goal at a time*

   ELMo model [27] cleverly handles the multiple meanings of the words as mentioned in above sentences based on context by representing the embedded vector as a function of the entire sentence containing that word. ELMo representation can model syntactical and semantical characteristics of the word, handles words with multiple meanings based on context (polysemy modeling). Word vectors obtained from ELMo model are learned functions of the hidden states of a bi-directional language model. As ELMo vectors are character-based, the ELMo model can represent out-of-vocabulary words unseen in training phase by making use of morphological clues.

Sentiment-Specific Word Embeddings (SSWE)

Tang et al. [28] proposed SSWE model by incorporating sentiment knowledge in continuous representation of the words. For this, three neural network-based models have been designed, viz. $SSWE_h$, $SSWE_r$, and $SSWE_u$. $SSWE_h$ is trained with very strict constraint to predict the positive and negative n-gram in the range [1,0] and [0,1], respectively. In $SSWE_r$, the strict constraint of softmax has been removed. Both $SSWE_h$, and $SSWE_r$ prohibit generation of corrupted n-grams. Being unified, $SSWE_u$ captures both the sentiments o f sntences and syntactical contexts of the words.

Graphs from LOw-level unit Modeling (GLoMo)

Graphs from low-level unit modeling (GLoMo) framework is based on unsupervised latent graph learning [29]. It is also a transfer learning framework developed to improve the performance of NLP tasks like sentiment analysis, natural language inference, question answering, and image classification.

Universal Language Model Fine-tuning (ULMFiT)

ULMFiT [30] is transfer learning model which can be used for any natural language processing task. The pre-trained models of ULMFiT can be leveraged for sentiment analysis. In this, a language model is pre-trained on general domain and then fine-tuned on target domain. Its working is invariant to document size, number, and label and therefore claims to be universal. It follows a single architecture and training for carrying out diverse tasks and does not need domain-specific documents and labels.

OpenAITransformer

OpenAITransformer [31] first trains a transformer model on large carpus in an unsupervised manner using language model as a training signal. After this, fine-tuning the model on small supervised dataset enables to solve the specific task.

Bi-directional Encoder Representations from Transformers (BERT)

BERT [32] pretrains bi-directional representations of unlabeled data in all layers by jointly handling both left and right context. Due to this, it can be fine-tuned to solve any task of NLP by just adding one output layer to the pre-trained model.

3.2 Strategy of Initializing the Embedded Vectors

Table 2 gives details of pre-trained models which can be leveraged for sentiment analysis. Word embeddings can be initialized by setting the vector representations with random values (random initialization). Another way i s to use pre-trained word embeddings and then fine-tune these embeddings for initializing the model.

Pre-trained models based on various corpora such as Wikipedia (C&W), Google News (Google), Twitter with emoticons (SSWE), Amazon corpus (Amazon), Wikipedia and Twitter (Glove) have been developed. Applying word2vec to a specific corpus yields customized embeddings [37, 38]. As mentioned in [33], random initialization may result in getting local minima with stochastic gradient descent (SGD) and if the pre-trained embeddings are not fine-tuned then automatic feature learning capacity of deep neural networks can not be leveraged. Therefore, use of pre-trained embeddings as initializer and then fine-tuning them helps to make the model efficient [39].

3.3 Enhancing the Embedded Vectors

For enhancing the effectiveness of the embedded vector, additional feature (from a word, sentence, and document) can be extracted and appended to a pre-trained embedded vector. For example, word vector can be appended with sentiment, parts-of-speech (POS) tag, word subjectivity, total count of syllables, number of characters with or without punctuation, etc.

The words which are out-of-vocabulary to the embedding model lack vector representation. For such OOV words, vector representation is obtained by approximation

Table 2 Pre-trained word embedding models and corpora

Word embedding model	Corpus
C&W [23]	Wikipedia
Word2vec CBOW [34]	Google News
Word2vec CBOW [33]	Amazon
SSWE$_h$, SSWE$_r$, and SSWE$_u$ [28]	5 million tweets having positive emoticons and 5 million tweets having negative emoticons
fastText [25]	Trained for 157 languages
ELMo small, medium, and original [35]	1 billion word benchmark
ELMo original (5.5B) [35]	Dataset with 5.5 billion tokens + Wikipedia with 1.9 billion tokens
ULMFiT [30]	Wikitext-103 having 28,595 Wikipedia articles and 103 million words
BERT [32]	BooksCorpus [36]

based on OOV word's context. The following are some solutions to handle OOV words. (1) Specifically, given a sentence and corresponding OOV word, language modeling performs sequencing of words in sentence and then predicts the meaning of word by comparing it with similar sentences. (2) Another solution is to use character or n-gram-level embeddings obtained from fastText. (3) Embeddings can be trained from scratch on the text. However, it suffers from overfitting and can not handle sentences having complex structure. Tang et al. [40] handled the problem of OOV words for the domain of users and products by averaging the representation of available data related to users and products. Creating a domain-specific word embedding model also helps to improve the performance [28, 41, 42].

3.4 Approximation Methods

Reducing the computational complexity of final softmax layer is one of the crucial challenges to be handled while designing the better word embedding model. Therefore, approximation algorithms based on sampling and softmax-based approaches have been devised by the research community. These approaches have been discussed in this sub-section.

3.5 Sampling-Based Approaches

Sampling-based approaches approximate the normalization term present in the denominator of the softmax with other computationally inexpensive loss function. Sampling-based methods are useful only for training. During testing, the full softmax needs to be computed to get a normalized probability.

- Importance sampling: Traditional importance sampling is based on Monte-Carlo sampling. It approximates a target distribution via unigram distribution.
- Adaptive importance sampling: Approximation using importance sampling works better for large samples [43]. Bengio and Senécal proposed an Adaptive importance sampling [44] which works on n-gram distribution.
- Target sampling: Jean et al.'s [45] approximation training algorithm is based on biased importance sampling, namely target sampling, which allows training neural machine translation model with a much large target vocabulary. Once the model is trained, they limit the target words being sampled by forming a subset of the vocabulary obtained by partitioning and selecting pre-defined sample words in each partition.
- Noise contrastive estimation (NCE): NCE [46] is more stable compared to importance sampling. Importance sampling has the risk of proposal distribution getting divergent from target distribution. Compared to importance sampling, NCE

does not find the probability of the word directly. NCE uses an auxiliary loss for maximizing the probability of correct words using optimization.

- Negative sampling: It minimizes the negative log-likelihood of words in training set using logistic loss function and focuses on learning word-representations of high quality.

3.6 Softmax-Based Approaches

- Hierarchical softmax (H-Softmax): Approximation based on hierarchical softmax [47] replaces the softmax layer with hierarchical tree in which leaves correspond to the words. Hierarchical layer decomposes the process of probability calculation. This alleviates the need of calculating the expensive normalization over the words. Therefore, it achieves a speed-up for word prediction tasks.
- Differentiated softmax: Differentiated softmax [48] is a variant of traditional softmax layer. It is based on the philosophy that a number of parameters required by words are different and varies according to the occurrence of the words. Due to this principle, D-softmax works faster during testing. However, the assignment of a smaller number of parameters to rarely occurring words does not help the model to handle rare words efficiently.
- CNN-softmax: Kim et al.'s [49] work focuses on modifying the traditional softmax layer using character-level convolutional neural network (CNN). Character-level CNN has been used for producing the input word embeddings. Jozefowicz et al. [50] designed softmax loss based on character-level CNN, named as CNN-softmax. However, character-based models can not handle the same words with different meanings. This is because continuous space representation is used for the characters and the model prone to learn mapping from characters to word embeddings using smooth function. Therefore, a correction factor can be introduced which is learned per word.

4 Deep Learning Approaches for Sentiment Analysis

In this section, highly significant deep learning approaches for sentiment analysis at document, sentence, and aspect-level have been discussed. Table 3 compares these approaches based on text representation, neural network model, dataset, and crux of each approach.

Table 3 Comparative study of deep learning-based sentiment analysis approaches

Text representation	Neural network model	Dataset	Crux
Bag-of-words (BoW) [34]	Denoising autoencoder	IMDB, Amazon product review (books, DVDs, music, electronics, and kitchenware)	Learns task-oriented representation using Bregman divergence loss
Bilingual embeddings [52]	Denoising autoencoder	NLP and CC 2013	Incorporates sentiment information into embeddings
Word embeddings (Word2vec) [51]	CNN/LSTM for sentence representation, GRU for document representation	IMDB, Yelp 2013, Yelp 2014, Yelp 2013	Represents documents by considering the relation among semantics of the sentences
Dense vector representation at sentence, paragraph, and document [54]	Distributed memory model of paragraph vector	Stanford Sentiment Treebank, IMDB	Learns vector representation by predicting the surrounding words in the context
Word embeddings [40]	CNN-based user-product neural network	IMDB, Yelp 2013, Yelp 2014	Incorporates user- and product-level information
Word embeddings [53]	LSTM and deep memory network with content-based attention mechanism	IMDB, Yelp 2013, Yelp 2014	Incorporates user- and product-level information via attention mechanism imbibed in computational layers of deep memory network
Word embeddings [76]	Cached LSTM	IMDB, Yelp 2013, Yelp 2014	Captures local and global sentiment information using cached
Word embeddings [55]	GRU-based sequence encoder	Yelp 2013, Yelp 2014, and Yelp 2015 dataset, IMDB review, Yahoo Answer, Amazon review	Hierarchical attention applied at word and sentence level
Word embeddings [56]	Bi-directional LSTM model	TripAdvisor, BeerAdvocate dataset	Iterative attention module with multiple hop mechanism modeling interaction between documents and aspect questions

(continued)

Table 3 (continued)

Text representation	Neural network model	Dataset	Crux
Word embeddings [57]	Bilingual bi-directional LSTMs	NLP and CC 2013	Hierarchical attention mechanism
Word embeddings [58]	Adversarial memory network	Amazon reviews	Alleviates need of manually identifying pivot for cross-domain SA
Word embeddings [59]	LSTM	IMDB, Yelp 2014, Yelp 2015	Use of varied sentence representation
Word vectors [60]	Semi-supervised recursive autoencoders	EP, Movie reviews, and MPQA opinion	Greedy approximation for tree construction
Word and matrix vectors [61]	Matrix-vector recursive neural network	Movie reviews	Compositional vector representation enables
Tag and word embeddings [62]	Tag-guided recursive neural network, tag-embedded recursive neural network/recursive neural tenser network	Stanford Sentiment Treebank	Tag-guided compositional functions and use of integrated embeddings to capture syntactic information with less parameters and complexity
Word embeddings [63]	Dynamic CNN	1.6 million tweets comprising of positive and negative labels (training) and 400 hand-annotated tweets (testing)	Captures word relations
Character embeddings, word embeddings, sentence representation [64]	Character to sentence CNN	Stanford Sentiment Treebank, Stanford Twitter Sentiment corpus with Twitter messages	Uses character- to sentence-level information
Modified and fine-tuned word embeddings [65]	LSTM network	Stanford Twitter Sentiment corpus	Use of flexible compositional function, explores task-distinctive function words
Word vectors [66]	Regional CNN-LSTM	Stanford Sentiment Treebank, Chinese valence-arousal texts	Hierarchical learning by combination of deep learning models

(continued)

Table 3 (continued)

Text representation	Neural network model	Dataset	Crux
Word embeddings [41]	CNN or RNN	Movie review, camera review, laptop, and restaurant dataset from SemEval 2015 Task 12	Jointing learning of sentence embeddings and sentiment classifier
Pre-trained GloVe embeddings [67]	Hierarchical bi-directional long short-term memory	SemEval-2016 Task 5 dataset for ABSA with five domains and eight languages	Modeling review structure and sentential context in a hierarchical manner
Word embeddings (pre-trained word2vec) [68]	Recursive neural network and conditional random field	SemEval-2014 Task 4 (laptop), SemEval-16 Task 5 (restaurant)	Dual propagation of label information from parameter learning in CRF to representation learning in recursive neural network
General (GloVe) and domain-specific embeddings (fastText) [69]	CNN	SemEval-2014 laptop, SemEval-2016 restaurant	Double embedding mechanism
Word embeddings (GloVe) [70]	Bi-directional LSTM with mutual attention	SemEval-2014 Task 4	Mutual generation of attention from aspect to text and text to aspect
Word embeddings (pre-trained GloVe) [71]	CNN and bi-directional LSTM	SemEval-2014 Task 4, ACL14-target dataset	Context preserving and positive relevance mechanism
Word embeddings (pre-trained GloVe) [72]	LSTM	SemEval-2014 Task 4	Attention mechanism
Word embeddings (pre-trained GloVe) [73]	LSTM	SemEval-2014 Task 4	Interactive attention mechanism
Concatenated embeddings (SSWE + word2vec) [74]	bi-directional and three-way gated neural network	ACL14-target dataset, MPQA, Mitchell et al.'s corpus	Models syntax and semantics of the text and also interaction among target entity and its surrounding context

(continued)

Table 3 (continued)

Text representation	Neural network model	Dataset	Crux
Word embeddings (word2vec) [75]	Bi-directional LSTM	SentiHood	Identifies sentiments toward multiple target entities
Word embeddings (pre-trained word2vec Skip-gram) [77]	LSTM with hierarchical attention	SentiHood, SemEval-2015 (partial)	Incorporates commonsense knowledge of sentiment concepts
Word embeddings (pre-trained GloVe) [78]	Memory augmented model	SentiHood	Delayed memory update mechanism
Pre-trained BERT language model [79]	Fine-tuned and pre-trained BERT model	SentiHood, SemEval-2014 Task 4	Construction of auxiliary sentence

Document-level sentiment analysis approaches

Zhai and Zhang [34] proposed a semi-supervised denoising autoencoder model for document-level sentiment analysis. It considers sentiment information during learning phase for getting good representation of document vectors. It learns a task-oriented data representation by using Bregman divergence function as a loss in the autoencoder and obtaining discriminative loss function from class labels.

Zhou et al. [52] proposed bilingual sentiment embeddings for cross-lingual sentiment classification. In this, denoising autoencoder is used to learn bilingual embeddings in unsupervised way. Then via supervised learning, sentiment information is incorporated into bilingual embeddings from sentiment labels of documents to get bilingual sentiment word embeddings.

For learning the document representation, Tang et al. [51] utilized the sentence relationships. For this, they first used CNN or long short-term memory (LSTM) for sentence representation learning and then applied gated recurrent unit (GRU) for adaptively encoding the semantics of sentences and their relation in document representation for sentiment analysis.

For overcoming the shortcomings of bag-of-words model, Le and Mikolov proposed unsupervised algorithm, namely paragraph vector [54] which learns fixed-length representation of text data from variable-sized text such as sentence, paragraphs, and documents. It learns representation by predicting the surrounding words based on contextual information from the text. After learning the vector representation, logistic classifier is applied to learn to predict the sentiments. During testing, the network for vector representation freezes and representation for test data (sentence, paragraph, or document) is learnt using gradient descent. The leant vector representation is then fed to logistic regression for predicting the sentient.

Tang et al. [40] proposed supervised learning framework which incorporates user- and product-level information in a neural network model to perform document-level sentiment classification. Incorporation of user-level and product-level information facilitates to capture the individual choices of users and overall qualities of products, respectively, to provide better representation of the text.

Like [51], Chen et al. [52] incorporated user- and product-level information in a hierarchical LSTM model via word and sentence-level attention mechanism. Based on the principle of compositionality [80], they modeled document semantics in a hierarchical manner at word, sentence, and document level. They used word-level user-product attention to get sentence representation and sentence-level user-product attention to get document representation.

Dou [53] also proposed user-product deep memory network (UPDMN) for capturing user and product information. Initially, a document is represented using LSTM and then deep memory network having computational layers with content-based attention mechanism is applied for predicting review rating. For handling semantic knowledge in long text, Xu et al. [76] put forth cached LSTM model. Cache mechanism divides the memory in different groups with varying forgetting patterns and enable to capture emotional information locally and globally for improved sentiment classification. Compared to standard LSTM, this model converges faster. Hierarchical attention network based on GRU-based sequence encoder proposed in [55] applies attention mechanism at word- and sentence-level for document-level sentiment classification. It incrementally constructs a document vector by aggregating significant words into sentence vectors and in turn significant sentence vectors into document vectors via aggregation. Song et al. [56] proposed hierarchical iterative attention model using bi-directional LSTM which captures interaction between documents and aspects at word- and sentence-level to learn the document representation in aspect-specific fashion. This model performs multi-aspect sentiment classification. Zhou et al. [57] proposed to use bi-directional LSTM with sentence-level attention mechanism for cross-lingual sentiment classification. Initially, machine translation tool translated training data into target language. They used bi-directional LSTM for modeling the document representation in source and target language. To remove the noise effect introduced due to machine translation, hierarchical attention mechanism is introduced which jointly trains with the LSTM network. Li et al. [58] addressed the issue of selecting the pivots for cross-domain sentiment analysis in transfer learning mode. They used adversarial memory network and jointly trained two networks for sentiment and domain classification. Huang et al. [59] proposed two variants of representations to be used with LSTM for document-level sentiment classification. In the first variant, document is represented by capturing the semantics of sentences from sentence vectors. In the second variant, document is represented using sorted sentence vectors. For getting sorted sentence representation, dataset is pre-processed to remove irrelevant sentences, which does not carry sentiment information.

Sentence-level sentiment analysis approaches

Socher et al. [60] first put forth recursive autoencoder network working in semi-supervised manner for sentiment classification at sentence level. This approach retrieves vector representation with reduced dimensions for multi-word phrases.

As this method is based on single-vector space model, it can not capture the compositional meaning of long phrases.

Socher et al. [61] put forth recursive matrix-vector model which additionally associates matrix representation with a word in a tree structure. This approach alleviates the problem of capturing the compositional meaning of long sentences with arbitrary syntax and length by representing the word and phrase using both the vector and matrix. Word vector captures inherent meaning and change in meaning of neighboring words is captured by matrix representation. An external parser has been used for building a tree structure.

To perform supervised training and evaluate sentiment compositional models, Socher et al. [81] developed Stanford Sentiment Treebank dataset [82]. They proposed recursive neural tensor network based on tensor-oriented compositional features for efficiently capturing the interaction among the words in a sentence. The model was tested on movie reviews dataset where sentiment polarities varied from very negative to very positive as five-sentiment classes.

Qian et al. [62] proposed two models based on compositional functions, namely, tag-guided recursive neural network (TG-RNN), tag-embedded recursive neural network/recursive neural tenser network (TE-RNN/RNTN). The former model selects a composition function based on POS tags of a phrase, whereas the later model combines tag and word embeddings. They tested the performance on Sentiment Treebank corpus and the models achieved significant performance over baseline models.

Dynamic CNN proposed by Kalchbrenner et al. [63] uses dynamic K-max pooling operator to capture semantics of sentences. They experimented on DCNN by varying the initialization parameters of word embeddings such as CNN with random initialization, CNN with pre-trained and fine-tuned embeddings, and CNN with multiple sets of word embeddings. Character to sentence CNN model proposed in [64] uses two layers of CNN for extracting word- and sentence-level features with varying length of input sentences for sentiment analysis. Wang et al. [65] utilized gates and constant error carousels in the memory structure of LSTM for handling the interaction among words for via compositional function. A regional CNN-LSTM model [66] performs dimensional sentiment analysis in which regional CNN captures sentence-level information locally and LSTM captures long-distance dependency.

Motivated by structural correspondence learning method preferably used for domain adaptation [83], Yu and Jiang [41] proposed the idea of learning generalized sentence embeddings for cross-domain sentence-level sentiment analysis and designed CNN models to joint learning of hidden feature representations of labeled and unlabeled data.

Aspect-based sentiment analysis approaches

Ruder et al. [67] captured intra- and inter-sentence relation using hierarchical bi-directional LSTM for aspect-based sentiment analysis. The complete reliance on sentence and its structure made their approach language-independent, and thus supports multi-lingual ABSA.

Wang et al. [68] proposed integrated recursive neural networks with conditional random field for jointly extracting the explicit aspect terms and opinion terms as the first step toward ABSA. Xu et al. [69] applied double embedding mechanism with CNN model for aspect extraction. This approach uses both general embeddings (GloVe-CNN) and domain-specific embeddings (DE-CNN) without any extra supervision for aspect extraction.

Attention-over-attention mechanism proposed in [70] jointly models representation of aspects and sentences to capture interaction among aspects and the context of the sentences. It used two bi-directional LSTM networks for learning the hidden semantics of the words in sentence and target. Target-specific transformation networks (TNet) [71] adapts convolutional neural network for handling target-level sentiment classification. For integrating target information into word representation, target-specific transformation network is proposed.

Wang et al. [72] proposed attention-based LSTM for ABSA. They proposed two ways of considering the aspect information while applying attention mechanism. Interactive attention network [73] (IAN) leverages target and context information for computing the attention vector and learns target and context representations. By concatenating target representation with context representation, IAN predicts polarity of the target. Zhang et al. [74] proposed to use gated recurrent neural networks for targeted sentiment analysis. First, for better representation of target and context by applying pooling layer over hidden layer instead of words, bi-directional gated neural network is used. A three-way gated neural network has been used to model interaction between surrounding context and the target. Saeidi et al. [84] proposed SentiHood dataset for targeted ABSA. They proposed to use the bi-directional LSTM model and logistic regression model to learn a classifier for each aspect.

Ma et al. [77] proposed a solution for handling targeted ABSA by applying attention mechanism in two-step model at target- and sentence-level and extending LSTM to incorporate commonsense knowledge associated with sentiments. Inspired by the use of memory augmented models in machine reading, Liu et al. [78] proposed to use external memory chains with a delayed memory update mechanism, enabling to track multiple target entities for targeted ABSA. Sun et al. [79] utilized pre-trained BERT language model for targeted ABSA. Specifically, they represented single sentence and a pair of sentences using pre-trained BERT language model and constructed the auxiliary sentences. After this, the task of targeted ABSA has been transformed into sentence-pair classification task. By fine-tuning the pre-trained BERT model, sentiment analysis has been performed.

5   Evaluation Metrics for Sentiment Analysis

Evaluation metrics commonly used for sentiment analysis have been discussed in this section.

- Accuracy: Accuracy (precision) relates to how often the sentiment rating predicted by the model is correct. Higher is the accuracy, better is the model. Accuracy is calculated as

$$\text{Acc} = \frac{\text{TP} + \text{TN}}{\text{TP} + \text{TN} + \text{FP} + \text{FN}} \tag{1}$$

where TP, TN, FP, and FN denote true positive, true negative, false positive, and false negative, respectively.

- F_1 score: It uses both precision and recall of test data for finding its score. It is calculated as follows.

$$F_1 = \frac{2(Precision \times Recall)}{Precision + Recall} \tag{2}$$

- Average recall (AvgRec): For the models, which find the overall sentiment of a document or text, average recall is used. Average recall is calculated by averaging the recall across the sentiment classes such as positive, negative, and neutral.

$$AvgRec = \frac{1}{2}\left(R^P + R^N + R^U\right) \tag{3}$$

where R^P, R^N, and R^U refer to recall associated with positive, negative, and neutral class, respectively. The value of AvgRec varies in the range [0, 1]. Average recall is more robust to class imbalance as compared to standard accuracy. Higher the value of AvgRec, better is the model.

- Macro-average F_1 score: Macro-average F_1 score is calculated with respect to positive and negative classes as

$$F_1^{PN} = \frac{1}{2}\left(F_1^P + F_1^N\right) \tag{4}$$

where F_1^P and F_1^N denote F_1 score with respect to positive and negative class, respectively.

- Ranking loss: It averages the distance between actual and predicted rank [85, 86]. It is calculated as follows.

$$Ranking\,loss = \sum_{i=1}^{n} \frac{|t_i - \hat{t}_i|}{k \times n} \tag{5}$$

where t_i and $\hat{t}_i$ denote values associated with actual sentiment and predicted sentiment, respectively, k is number of sentiment classes, and n is instances used for testing.

- Macro-averaged mean absolute error: It is robust for imbalanced datasets [87]

$$MAE^M(t, \hat{t}) = \frac{1}{k} \sum_{j=1}^{k} \frac{1}{|t_j|} \sum_{t_i \in t_j} |t_i - \hat{t}_i| \tag{6}$$

where t and $\hat{t}$ denote vector of actual and predicted sentiment values, respectively, $t_j = \{t_i : t_i \in t, t_i = j\}$ and k denotes sentiment classes in t.

- Least absolute error (LAE) [88]: It is widely used evaluation measure to calculate the error of sentiment classification. It is given as

$$LAE = \sum_{i=1}^{n} |\hat{t}_i - t_i| \tag{7}$$

where $\hat{t}_i$ and t_i denote vector of predicted sentiment values and actual sentiment values.

- Mean squared error (MSE) [89]: It is used for evaluating the sentiment prediction error. It is specifically used for regression. MSE and Root MSE are computed as follows.

$$MSE = \frac{1}{n} \sum_{i=1}^{n} (\hat{t}_i - t_i)^2 \tag{8}$$

$$RMSE = \sqrt{\frac{1}{n} \sum_{i=1}^{n} (\hat{t}_i - t_i)^2} \tag{9}$$

where n denotes number of test instances, $\hat{t}_i$ and t_i denote vector of predicted sentiment values and actual sentiment values. It can be noted that lower values of MSE and RMSE indicates better performance of prediction model.

- Pearson correlation coefficient: It is calculated as

$$r = \frac{1}{n-1} \sum_{i=1}^{n} \left(\frac{t_i - \bar{t}}{\sigma_t}\right) \left(\frac{\hat{t}_i - \bar{\hat{t}}}{\sigma_{\hat{t}}}\right) \tag{10}$$

where n denotes number of test instances, $\hat{t}_i$ and t_i denote value of predicted and actual sentiments, $\bar{\hat{t}}$ and $\bar{t}$ denote arithmetic means of predicted and actual values, and σ represents standard deviation. Higher the value of r indicates better prediction accuracy of the model.

- Distributed cumulative grain (DCG): While performing sentiment analysis using topic modeling technique, first topics (aspects) are detected and then the sentiments associated with detected topics (aspects) are predicted. Therefore, for the sake of evaluating the relevance of returned topics (aspects), normalized Discounted Cumulative Gain (nDCG) is used [90]. The regular DCG is computed as follows.

$$\text{DCG}_m = \sum_{i=1}^{m} \frac{2^{rel(i)} - 1}{\log_2(i + 1)} \tag{11}$$

where m represents top m topics (aspects), rel(i) denotes relevance score of topics (aspect) i. For the models which produce the rankings of the detected topics (aspects), normalized DCG summarizes the quality of the rankings.

- KullbackLeibler divergence (KLD): KLD [91] is used for measuring error in estimating actual distribution t over a set $\mathcal{K}$ of sentiment classes by means of a predicted distribution $\hat{t}$. Like MAE^M, lower the values of KLD, better is the model. KLS is calculated as follows.

$$KLD(\hat{t}, t, K) = \sum_{k_j \in K} t(k_j) \log_e \frac{t(k_j)}{\hat{t}(k_j)} \tag{12}$$

- Area under the ROC curve (AUC): Saeidi et al. [84] proposed to use the AUC metric for tasks of aspect and sentiment detection. AUC helps to measure the quality of ranking the output scores without relying on the threshold.

6 Benchmarked Datasets and Tools

Table 4 gives the glimpse of standard benchmarked datasets used for sentiment analysis at document, sentence, aspect, and targeted aspect-level.

These are numerous tools available which offer sentiment analysis as one of its services. The details of the tools providing sentiment analysis as a service have been mentioned in Table 5.

With reference to popularity of sentiment analysis, dedicated search engines have been developed such as Social Mention [116], Social Searcher [117], Talkwalker's Quick Search [118]. Social Mention [116] combines the user-generated data across the Web and gives the sentiments of a given keyword based on how many times the positive, negative, and neutral mentions of the keyword are present in the collected data. Social Searcher [117] is a real-time search engine for quickly pulling recent mentions from popular social networks and displays analytics in the form of mentions, users, and sentiments for the topic entered in the search box. It also offers sentiment filters to get a set of mentions.

Table 4 Benchmarked datasets for sentiment analysis

Dataset	Description	Level
Yelp 2013, 2014, 2015 [92]	These are restaurant review datasets from Yelp dataset challenge. Each review is rated from 1 star to 5 stars	Document
IMDB [93]	It consists of 84,919 movie reviews ranging from 1 to 10. The average length of each review is about 394.6 words	Document
Amazon review [94]	This dataset is developed from Stanford Network Analysis Project spanning 18 years with 34,686,770 reviews on 2,441,053 products with review ratings from 1 star to 5 stars	Document
NLP and CC 2013 [95]	It contains reviews from three domains as book, DVD, and music. For each domain, training data contains the reviews in English (2000 positive reviews and 2000 negative reviews), whereas test data contains the 4000 reviews in Chinese	Document
Experience project (EP) [60]	It contains 31,676 confession entries, and 74,859 votes for the five-sentiment labels	Sentence
Movie reviews [96]	Movie review documents are labeled with polarity (positive or negative) or rating based on subjectivity. Sentences are labeled with subjectivity status (subjective or objective) or polarity	Document and sentence
MPQA opinion [97]	It contains news articles from varied news sources with manual annotation for opinions and states. There are 22 average number of words per sentence	Sentence
Stanford Sentiment Treebank with sentences from movie reviews [81]	It includes sentiment labels for 215,154 phrases in the parse trees of 11,855 sentences	Sentence
Stanford Twitter Sentiment corpus [98]	It contains 1.6 million tweets with automatically labeled positive or negative sentiments and manually annotated test set	Sentence

(continued)

Table 4 (continued)

Dataset	Description	Level
Chinese Valence-Arousal Texts [99]	It consists of 2009 texts from social media with valence and arousal dimensions of the range from 1 to 9 annotated manually	Sentence
Camera review dataset [100]	It contains reviews related to digital products like MP3 players and cameras	Sentence
SemEval-2016 Task 5 dataset [101]	It contains dataset from five domains in eight languages	ABSA
ACL14-target dataset [102]	It consists of the target entities like celebrities, products, and companies obtained from Twitter platform	Targeted ABSA
Mitchell et al.'s corpus [75]	It is a corpus with 3288 entities for targeted ABSA	Targeted ABSA
SentiHood [84]	It consists of 5215 sentences in which 3862 sentences contain single target and remaining sentences have multiple targets	Targeted ABSA
SemEval-2014 Task 4 [20]	It consists of laptop and restaurant review dataset with more than 6000 sentences having fine-grained aspect-level human annotations	ABSA and targeted ABSA

7 Conclusion

This chapter gives a demystified overview of state-of-the-art approaches for sentiment analysis. The proposed graphical taxonomy gives traits to be considered for designing the sentiment analysis systems. Providing suitable input to the deep learning models plays crucial role in achieving the good performance. Therefore, parameters associated with text representation techniques such as use of embedded vectors, language models, ways of improving the functionality of embedded vectors, and approximating the computationally expensive softmax function in embedding models have been thoroughly discussed.

A comparative overview of the noteworthy research papers focusing on sentiment analysis at document, sentence, and aspect level using deep learning approaches has been given in the chapter. We also shed light upon state-of-the-art benchmarked datasets and the tools and services available for sentiment analysis.

Table 5 Comparative study of existing tools for sentiment analysis

Tool	Level	Features
Brand24 [103]	Document	Its *Mentions tab* feature allows to track the sentiments over time depicted via graph. The spikes and dips regarding the sentiments enable to identify the reason behind the change in the sentiment. It also provides *analysis* and *summary tab* to analyze the sentiments thoroughly and summarize the results, respectively
Clarabridge [104]	Document, topic	It gives the sentiment analysis score on the 11-point scale by using lexical and grammatical methods to perform the sentiment analysis at clause level. It also provides emotion analysis service by analyzing the customer feedbacks
Repustate [105]	Document, topic	It offers sentiment analysis for 23 languages and handles data from surveys, social media, blogs, news, forums, or business data. It uses parts-of-speech tagging methods, lemmatization, and concept of prior polarity for sentiment analysis
OpenText [106]	Document, sentence, and topic	It offers support for English, German–French, Spanish, and Portuguese language. It uses machine learning and NLP-based techniques. OpenText analyzes the sentiments based on positive, negative, mixed, or neutral polarities, and supports subjective pattern evaluation
ParallelDots [107]	Document	It works for 14 different languages and uses LSTM-based algorithms for analyzing the sentiments based on the positive, negative, and neutral polarities

(continued)

Table 5 (continued)

Tool	Level	Features
Lexalytics [108]	Document	It offers user-centric sentiment analysis (customer-centric and employee-centric). It follows hybrid sentiment analysis approach encompassing machine learning and rule-based approaches
Hi-Tech BPO [109]	Document, images, speech, emoji, and visuals)	It performs sentiment analysis of data in the form of text, speech, emoji, images, and visuals by combining NLP and ML algorithms. The problem of sentiment analysis has been divided into opinion mining, text mining, social sentiment, and social listening
Sentiment Analyzer [110]	Document	It serves as general-purpose sentiment analysis tool developed using computational linguistics and text mining approaches and works for the text in English language. It gives sentiment score in the range of -100 to $+100$ where former stands for very negative sentiment and later for very positive sentiment
SentiStrength [111]	Document	For this tool, both downloaded software and Web version are available. It performs sentiment analysis at rate of 16,000 texts per second. It measures sentiment strength in terms of two scores as -1 (not negative) to -5 (extremely negative) and 1 (not positive) to 5 (extremely positive)

(continued)

Table 5 (continued)

Tool	Level	Features
Meaning Cloud [112]	Document, sentence, and attribute	This tool provides global sentiment as a general opinion, performs attribute-level sentiment analysis. It also offers irony detection feature. It can distinguish between very positive, very negative, and sentence without sentiment. This tool identifies conflicting and contradictory messages
Tweet Sentiment Visualization [113]	Topic (keyword-based query)	This tool accepts the keyword(s) from the user and then pulls Tweets from Twitter and visualizes them in various ways, viz. sentiment, topics, heatmap, tag cloud, timeline, map, affinity, narrative, and Tweets' date, author, pleasure, arousal, and text
Rapidminer [114]	Document	It is a data science platform and offers sentiment analysis as one of its services
Brandwatch [115]	Topic	As one of its services, Brandwatch supports sentiment analysis for 44 languages and performs sentiment analysis and key topic detection
Sentigem [119]	Document	It performs sentiment analysis of a document or a text block. API is also provided through input text can be provided for sentiment analysis

References

1. Pathak, A.R., M. Pandey, and S. Rautaray. 2018. Construing the big data based on taxonomy, analytics and approaches. *Iran Journal of Computer Science* 1: 237–259.
2. NLP market. https://www.tractica.com/newsroom/press-releases/natural-language-processing-market-to-reach-22–3-billion-by-2025/.
3. Agarwal, B., and N. Mittal. 2016. Prominent feature extraction for sentiment analysis. In *Springer Book Series: Socio-Affective Computing series*, 1–115. Springer International Publishing, ISBN: 978-3-319-25343-5. https://doi.org/10.1007/978-3-319-25343-5.

4. Pathak, A.R., M. Pandey, and S. Rautaray. 2018. Application of deep learning for object detection. *Procedia Computer Science* 132: 1706–1717.
5. Pathak, A.R., M. Pandey, and S. Rautaray. 2018. Deep learning approaches for detecting objects from images: A review. In *Progress in Computing, Analytics and Networking*, ed. Pattnaik, P. K., S.S. Rautaray, H. Das, J. Nayak, J., 491–499. Springer Singapore.
6. Pathak, A.R., M. Pandey, S. Rautaray, and K. Pawar. 2018. Assessment of object detection using deep convolutional neural networks. In *Advances in Intelligent Systems and Computing*, 673.
7. Pawar, K., and V. Attar. 2019. Deep learning approaches for video-based anomalous activity detection. *World Wide Web* 22: 571–601.
8. Pawar, K., and V. Attar. 2019. Deep Learning approach for detection of anomalous activities from surveillance videos. In *CCIS*. Springer, In Press.
9. Pathak, A.R., M. Pandey, and S. Rautaray. 2019. Adaptive model for dynamic and temporal topic modeling from big data using deep learning architecture. *Internationl Journal of Intelligent Systems and Applications* 11: 13–27. https://doi.org/10.5815/ijisa.2019.06.02.
10. Bhat, M.R., M.A. Kundroo, T.A. Tarray, and B. Agarwal. 2019. Deep LDA: A new way to topic model. *Journal of Information and Optimization Sciences* 1–12 (2019).
11. Pathak, A.R., M. Pandey, and S. Rautaray. 2019. Adaptive framework for deep learning based dynamic and temporal topic modeling from big data. *Recent Patents on Engineering, Bentham Science* 13: 1. https://doi.org/10.2174/1872212113666190329234812.
12. Pathak, A.R., M. Pandey, and S. Rautaray. 2019. Empirical evaluation of deep learning models for sentiment analysis. *Journal of Statistics and Management Systems* 22: 741–752.
13. Pathak, A.R., M. Pandey, and S. Rautaray. 2019. Adaptive model for sentiment analysis of social media data using deep learning. In *International Conference on Intelligent Computing and Communication Technologies*, 416–423.
14. Ram, S., S. Gupta, and B. Agarwal. 2018. Devanagri character recognition model using deep convolution neural network. *Journal of Statistics and Management Systems,* 21: 593–599.
15. Hinton, G., and et al. 2012. Deep neural networks for acoustic modeling in speech recognition. *IEEE Signal Processing Magazine*, 29.
16. Jain, G., M. Sharma, and B. Agarwal. 2019. Spam detection in social media using convolutional and long short term memory neural network. *Annals of Mathematics and Artificial Intelligence* 85: 21–44.
17. Agarwal, B., H. Ramampiaro, H. Langseth, and M. Ruocco. 2018. A deep network model for paraphrase detection in short text messages. *Information Processing & Management* 54: 922–937.
18. Jain, G., M. Sharma, and B. Agarwal. 2019. Optimizing semantic LSTM for spam detection. *International Journal of Information Technology* 11: 239–250.
19. Liu, B. 2012. Sentiment Analysis and Opinion Mining, 1–108. https://doi.org/10.2200/s00416ed1v01y201204hlt016.
20. SemEval-2014. http://alt.qcri.org/semeval2014/task4/.
21. Glorot, X., A. Bordes, and Y. Bengio. 2011. Domain adaptation for large-scale sentiment classification: A deep learning approach. In *Proceedings of the 28th International Conference on Machine Learning (ICML-11)*, 513–520.
22. Bengio, Y., R. Ducharme, P. Vincent, and C. Jauvin. 2003. A neural probabilistic language model. *Journal of Machine Learning Research* 3: 1137–1155.
23. Collobert, R., et al. 2011. Natural language processing (almost) from scratch. *Journal of Machine Learning Research* 12: 2493–2537.
24. Mikolov, T., K. Chen, G. Corrado, and J. Dean. 2013. Efficient estimation of word representations in vector space. arXiv Prepr. arXiv1301.3781.
25. Bojanowski, P., E. Grave, A. Joulin, and T. Mikolov. 2017. Enriching word vectors with subword information. *Transactions of the Association for Computational Linguistics* 5: 135–146.
26. Pennington, J., R. Socher, and C. Manning. 2014. Glove: Global vectors for word representation. In *Proceedings of the 2014 Conference on Empirical Methods in Natural Language Processing (EMNLP)*, 1532–1543.

27. Peters, M.E., and et al. 2018. Deep contextualized word representations. In *Proceedings of NAACL.*
28. Tang, D., and et al. 2014. Learning sentiment-specific word embedding for twitter sentiment classification. In *Proceedings of the 52nd Annual Meeting of the Association for Computational Linguistics (Volume 1: Long Papers)*, 1555–1565.
29. Yang, Z., and et al. 2018. Glomo: Unsupervisedly learned relational graphs as transferable representations. arXiv Prepr. arXiv1806.05662.
30. Howard, J., and S. Ruder. Universal language model fine-tuning for text classification. arXiv Prepr. arXiv1801.06146.
31. Radford, A., K. Narasimhan, T. Salimans, I. Sutskever. 2018. Improving language understanding by generative pre-training. URL https://s3-us-west-2.amazonaws.com/openai-assets/research-covers/language-unsupervised/language_understanding_paper.pdf.
32. Devlin, J., M.-W. Chang, K. Lee, and K. Toutanova. 2018. Bert: Pre-training of deep bidirectional transformers for language understanding. arXiv Prepr. arXiv1810.04805.
33. Liu, P., S. Joty, and H. Meng. 2015. Fine-grained opinion mining with recurrent neural networks and word embeddings. In *Proceedings of the 2015 Conference on Empirical Methods in Natural Language Processing*, 1433–1443.
34. Zhai, S., and Z.M. Zhang. 2016. Semisupervised autoencoder for sentiment analysis. In *Thirtieth AAAI Conference on Artificial Intelligence.*
35. EMLo. https://allennlp.org/elmo.
36. Zhu, Y., and et al. 2015. Aligning books and movies: Towards story-like visual explanations by watching movies and reading books. In *Proceedings of the IEEE International Conference on Computer Vision*, 19–27.
37. Poria, S., E. Cambria, and A. Gelbukh. 2016. Aspect extraction for opinion mining with a deep convolutional neural network. *Knowledge-Based Systems* 108: 42–49.
38. Wang, W., S.J. Pan, D. Dahlmeier, and X. Xiao. 2016. Recursive neural conditional random fields for aspect-based sentiment analysis. arXiv Prepr. arXiv1603.06679.
39. Jebbara, S., and P. Cimiano. 2016. Aspect-based relational sentiment analysis using a stacked neural network architecture. In *Proceedings of the Twenty-second European Conference on Artificial Intelligence*, 1123–1131.
40. Tang, D., B. Qin, and T. Liu. 2015. Learning semantic representations of users and products for document level sentiment classification. In *Proceedings of the 53rd Annual Meeting of the Association for Computational Linguistics and the 7th International Joint Conference on Natural Language Processing (Volume 1: Long Papers)*, 1014–1023.
41. Yu, J., and J. Jiang. 2016. Learning sentence embeddings with auxiliary tasks for cross-domain sentiment classification. In *Proceedings of the 2016 Conference on Empirical Methods in Natural Language Processing*, 236–246.
42. Sarma, P.K., Y. Liang, and W.A. Sethares. 2018. Domain adapted word embeddings for improved sentiment classification. arXiv Prepr. arXiv1805.04576.
43. Bengio, Y., J.-S. Senécal, and Others. 2003. Quick training of probabilistic neural nets by importance sampling. In *AISTATS*, 1–9.
44. Bengio, Y., and J.-S. Senécal. 2008. Adaptive importance sampling to accelerate training of a neural probabilistic language model. *IEEE Transactions on Neural Networks* 19: 713–722.
45. Jean, S., K. Cho, R. Memisevic, and Y. Bengio. 2014. On using very large target vocabulary for neural machine translation. arXiv Prepr. arXiv1412.2007.
46. Mnih, A., and Y.W. Teh. 2012. A fast and simple algorithm for training neural probabilistic language models. arXiv Prepr. arXiv1206.6426.
47. Morin, F., and Y. Bengio. 2005. Hierarchical probabilistic neural network language model. *Aistats* 5: 246–252.
48. Chen, W., D. Grangier, and M. Auli. 2015. Strategies for training large vocabulary neural language models. arXiv Prepr. arXiv1512.04906.
49. Kim, Y., Y. Jernite, D. Sontag, and A.M. Rush. 2016. Character-aware neural language models. In *Thirtieth AAAI Conference on Artificial Intelligence.*

50. Jozefowicz, R., O. Vinyals, M. Schuster, N. Shazeer, and Y. Wu. 2016. Exploring the limits of language modeling. arXiv Prepr. arXiv1602.02410.
51. Tang, D., B. Qin, and T. Liu. 2015. Document modeling with gated recurrent neural network for sentiment classification. In *Proceedings of the 2015 Conference on Empirical Methods in Natural Language Processing*, 1422–1432.
52. Zhou, H., L. Chen, F. Shi, and D. Huang. 2015. Learning bilingual sentiment word embeddings for cross-language sentiment classification. In *Proceedings of the 53rd Annual Meeting of the Association for Computational Linguistics and the 7th International Joint Conference on Natural Language Processing (Volume 1: Long Papers)*, 430–440.
53. Dou, Z.-Y. 2017. Capturing user and product information for document level sentiment analysis with deep memory network. In *Proceedings of the 2017 Conference on Empirical Methods in Natural Language Processing*, 521–526.
54. Le, Q., and T. Mikolov. 2014. Distributed representations of sentences and documents. In *International Conference on Machine Learning*, 1188–1196.
55. Yang, Z., and et al. 2016. Hierarchical attention networks for document classification. In *Proceedings of the 2016 Conference of the North American Chapter of the Association for Computational Linguistics: Human Language Technologies*, 1480–1489.
56. Yin, Y., Y. Song, and M. Zhang. 2017. Document-level multi-aspect sentiment classification as machine comprehension. In *Proceedings of the 2017 Conference on Empirical Methods in Natural Language Processing*, 2044–2054.
57. Zhou, X., X. Wan, and J. Xiao. 2016. Attention-based LSTM network for cross-lingual sentiment classification. In *Proceedings of the 2016 Conference on Empirical Methods in Natural Language Processing*, 247–256.
58. Li, Z., Y. Zhang, Y. Wei, Y. Wu, and Q. Yang. 2017. End-to-end adversarial memory network for cross-domain sentiment classification. In *IJCAI*, 2237–2243.
59. Rao, G., W. Huang, Z. Feng, and Q. Cong. 2018. LSTM with sentence representations for document-level sentiment classification. *Neurocomputing* 308: 49–57.
60. Socher, R., J. Pennington, E.H. Huang, A.Y. Ng, and C.D. Manning. semi-supervised recursive autoencoders for predicting sentiment distributions. In *Proceedings of the Conference on Empirical Methods in Natural Language Processing*, 151–161.
61. Socher, R., B. Huval, C.D. Manning, and A.Y. Ng. 2012. Semantic compositionality through recursive matrix-vector spaces. In *Proceedings of the 2012 Joint Conference on Empirical Methods in Natural Language Processing and Computational Natural Language Learning*, 1201–1211.
62. Qian, Q., and et al. 2015. Learning tag embeddings and tag-specific composition functions in recursive neural network. In *Proceedings of the 53rd Annual Meeting of the Association for Computational Linguistics and the 7th International Joint Conference on Natural Language Processing (Volume 1: Long Papers)*, 1365–1374.
63. Kalchbrenner, N., E. Grefenstette, and P. Blunsom. 2014. A convolutional neural network for modelling sentences. arXiv Prepr. arXiv1404.2188.
64. dos Santos, C., and M. Gatti. 2014. Deep convolutional neural networks for sentiment analysis of short texts. In *Proceedings of COLING 2014, the 25th International Conference on Computational Linguistics: Technical Papers*, 69–78.
65. Wang, X., Y. Liu, S.U.N. Chengjie, B. Wang, and X. Wang. 2015. Predicting polarities of tweets by composing word embeddings with long short-term memory. In *Proceedings of the 53rd Annual Meeting of the Association for Computational Linguistics and the 7th International Joint Conference on Natural Language Processing (Volume 1: Long Papers)*, vol. 1, 1343–1353.
66. Wang, J., L.-C. Yu, K. Lai, and X. Zhang. 2016. Dimensional sentiment analysis using a regional CNN-LSTM model. In *Proceedings of the 54th Annual Meeting of the Association for Computational Linguistics (Volume 2: Short Papers)*, 225–230.
67. Ruder, S., P. Ghaffari, and J.G. Breslin. 2016. A hierarchical model of reviews for aspect-based sentiment analysis. arXiv Prepr. arXiv1609.02745.

68. Wang, W., S.J. Pan, and D. Dahlmeier, and X. Xiao. 2016. Recursive neural conditional random fields for aspect-based sentiment analysis. arXiv Prepr. arXiv1603.06679.
69. Xu, H., B. Liu, L. Shu, and P.S. Yu. 2018. Double embeddings and cnn-based sequence labeling for aspect extraction. arXiv Prepr. arXiv1805.04601.
70. Huang, B., Y. Ou, and K.M. Carley. 2018. Aspect level sentiment classification with attention-over-attention neural networks. In *International Conference on Social Computing, Behavioral-Cultural Modeling and Prediction and Behavior Representation in Modeling and Simulation*, 197–206.
71. Li, X., L. Bing, W. Lam, and B. Shi. 2018. Transformation networks for target-oriented sentiment classification. In *Proceedings of the 56th Annual Meeting of the Association for Computational Linguistics (Volume 1: Long Papers)*, 946–956.
72. Wang, Y., M. Huang, L. Zhao and Others. 2016. Attention-based lstm for aspect-level sentiment classification. In *Proceedings of the 2016 Conference on Empirical Methods in Natural Language Processing*, 606–615.
73. Ma, D., S. Li, X. Zhang, and H. Wang. 2017. Interactive attention networks for aspect-level sentiment classification. arXiv Prepr. arXiv1709.00893.
74. Zhang, M., Y. Zhang, and D.-T. Vo. 2016. Gated neural networks for targeted sentiment analysis. In *Thirtieth AAAI Conference on Artificial Intelligence*.
75. Mitchell et al. Corpus. http://www.m-mitchell.com/code/index.html.
76. Xu, J., D. Chen, X. Qiu, and X. Huang. 2016. Cached long short-term memory neural networks for document-level sentiment classification. arXiv Prepr. arXiv1610.04989.
77. Ma, Y., H. Peng, and E. Cambria. 2018. Targeted aspect-based sentiment analysis via embedding commonsense knowledge into an attentive LSTM. In *Thirty-Second AAAI Conference on Artificial Intelligence*.
78. Liu, F., T. Cohn, and T. Baldwin. 2018. Recurrent entity networks with delayed memory update for targeted aspect-based sentiment analysis. In *Proceedings of the 2018 Conference of the North American Chapter of the Association for Computational Linguistics: Human Language Technologies, Volume 2 (Short Papers)*, 278–283.
79. Sun, C., L. Huang, and X. Qiu. 2019. Utilizing BERT for aspect-based sentiment analysis via constructing auxiliary sentence. In *Proceedings of the 2019 Conference of the North American Chapter of the Association for Computational Linguistics: Human Language Technologies, Volume 1 (Long and Short Papers)*, 380–385.
80. Pelletier, F.J. 1994. The principle of semantic compositionality. *Topoi* 13: 11–24.
81. Socher, R., and et al. 2013. Recursive deep models for semantic compositionality over a sentiment treebank. In *Proceedings of the 2013 Conference on Empirical Methods in Natural Language Processing*, 1631–1642.
82. Sentiment Treebank. https://nlp.stanford.edu/sentiment/treebank.html.
83. Blitzer, J., M. Dredze, and F. Pereira. 2007. Biographies, Bollywood, boom-boxes and blenders: Domain adaptation for sentiment classification. In *Proceedings of the 45th Annual Meeting of the Association of Computational Linguistics*, 440–447.
84. Saeidi, M., G. Bouchard, M. Liakata, and S. Riedel. 2016. SentiHood: targeted aspect based sentiment analysis dataset for urban neighbourhoods. In *Proceeding COLING 2016, 26th International Conference Computational Linguistics: Technical Papers*, 1546–1556.
85. Crammer, K., and Y. Singer. 2002. Pranking with ranking. In *Advances in Neural Information Processing Systems*, 641–647.
86. Moghaddam, S., and M. Ester. 2010. Opinion digger: an unsupervised opinion miner from unstructured product reviews. In *Proceedings of the 19th ACM International Conference on Information and Knowledge Management*, 1825–1828.
87. Marcheggiani, D., O. Täckström, A. Esuli, and F. Sebastiani. 2014. Hierarchical multi-label conditional random fields for aspect-oriented opinion mining. In *European Conference on Information Retrieval*, 273–285.
88. Lu, B., M. Ott, C. Cardie, and B.K. Tsou. 2011. Multi-aspect sentiment analysis with topic models. In *2011 IEEE 11th International Conference on Data Mining Workshops (ICDMW)*, 81–88.

89. Wang, H., Y. Lu, and C. Zhai. 2011. Latent aspect rating analysis without aspect keyword supervision. In *Proceedings of the 17th ACM SIGKDD International Conference on Knowledge Discovery and Data Mining*, 618–626.
90. Wang, Q., J. Xu, H. Li, and N. Craswell. 2013. Regularized latent semantic indexing: A new approach to large-scale topic modeling. *ACM Transactions on Information Systems (TOIS)* 31: 5.
91. Kullback, S., and R.A. Leibler. 1951. On information and sufficiency. *The Annals of Mathematical Statistics* 22: 79–86.
92. Yelp Dataset. https://www.yelp.com/dataset/challenge.
93. Diao, Q. and et al. 2014. Jointly modeling aspects, ratings and sentiments for movie recommendation (JMARS). In *Proceedings of the 20th ACM SIGKDD International Conference on Knowledge Discovery and Data Mining*, 193–202.
94. Zhang, X., J. Zhao and Y. LeCun. 2015. Character-level convolutional networks for text classification. In *Advances in Neural Information Processing Systems*, 649–657.
95. NLP and CC 2013. http://tcci.ccf.org.cn/conference/2013/index.html.
96. Movie Reviews. http://www.cs.cornell.edu/people/pabo/movie-review-data/.
97. MPQA Opinion. http://mpqa.cs.pitt.edu.
98. Go, A., R. Bhayani, and L. Huang. 2009. Twitter sentiment classification using distant supervision. In *CS224 N Project Report, Stanford*, 1.
99. Yu, L.-C., and et al. 2016. Building Chinese affective resources in valence-arousal dimensions. In *Proceedings of the 2016 Conference of the North American Chapter of the Association for Computational Linguistics: Human Language Technologies*, 540–545.
100. Camera Review. https://www.cs.uic.edu/~liub/FBS/sentiment-analysis.html.
101. Pontiki, M., and et al. 2016. SemEval-2016 task 5: Aspect based sentiment analysis. In *Proceedings of the 10th International Workshop on Semantic Evaluation (SemEval-2016)*, 19–30.
102. Dong, L., and et al. 2014. Adaptive recursive neural network for target-dependent twitter sentiment classification. In *Proceedings of the 52nd annual meeting of the association for computational linguistics (volume 2: Short papers)*, 49–54.
103. Brand24. https://brand24.com.
104. Clarabridge. https://www.clarabridge.com/platform/analytics/.
105. Repustate. https://www.repustate.com.
106. OpenText. https://www.opentext.com/products-and-solutions/products/discovery/information-access-platform/sentiment-analysis.
107. ParallelDots. https://www.paralleldots.com/sentiment-analysis.
108. Lexalytics. https://www.lexalytics.com/technology/sentiment-analysis.
109. Hi-Tech BPO. https://www.hitechbpo.com/sentiment-analysis.php.
110. Sentiment Analyzer. https://www.danielsoper.com/sentimentanalysis/.
111. SentiStrength. http://sentistrength.wlv.ac.uk.
112. Meaning Cloud. https://www.meaningcloud.com/products/sentiment-analysis.
113. Tweet Sentiment Visualization. https://www.csc2.ncsu.edu/faculty/healey/tweet_viz/tweet_app/.
114. Rapidminer. https://rapidminer.com/solutions/text-mining/.
115. Brandwatch. https://www.brandwatch.com/products/analytics/.
116. Social Mention. http://www.socialmention.com.
117. Social Searcher. https://www.social-searcher.com/social-buzz/.
118. Talkwalker's Quick Search. https://www.talkwalker.com/quick-search-form.
119. Sentigem. https://sentigem.com.

Recent Trends and Advances in Deep Learning-Based Sentiment Analysis

Ahmed Ahmet and Tariq Abdullah

Abstract Sentiment analysis is a fundamental branch of natural language processing. It is an essential task of identifying and extracting sentiment in opinionated data from sources such as social media, product feedback or blogs. Deep learning-based approaches have exceeded human-level performance in areas such as computer vision and speech recognition. Deep learning is widely accepted as the most promising in machine learning. In this chapter, we survey and analyse the current trends and advances in deep learning-based sentiment analysis approaches for document-level, sentence-level and aspect-based sentiment analysis for short and long text. A detailed discussion of deep learning architectures for sentiment analysis is provided. The studied approaches are classified into coarse-grain (including document and sentence level), fine-grain (includes target and aspect level) and cross-domain. Lastly, we provide a summary and in-depth analysis of the surveyed studies, for each of the aforementioned categories. The overwhelming number of studies explored convolutional neural networks (CNNs), long short-term memory (LSTM), gated recurrent unit (GRU) and attention mechanism. For coarse-grain sentiment analysis, LSTM and CNN-based models compete on performance, but it is CNNs that offer reduced model complexity and training overhead. Fine-grain sentiment analysis requires a model to learn complex interactions between target/aspect words and opinion words. Bi-directional LSTM and attention mechanisms offer the most promise, although CNN-based models have been adept at aspect extraction. The efforts in cross-domain sentiment analysis are dominated by LSTM and attention models. Our survey of cross-domain approaches revealed the use of multitask learning, adversarial training and joint training for domain adaptation.

Keywords Deep learning · Sentiment analysis · Artificial neural networks · Machine learning · Sentiment classification · Document level · Sentence level · Aspect-level opinion mining · Cross-domain sentiment analysis

A. Ahmet · T. Abdullah (✉)
University of Derby, Derby, UK
e-mail: t.abdullah@derby.ac.uk

A. Ahmet
e-mail: a.ahmet@derby.ac.uk

© Springer Nature Singapore Pte Ltd. 2020
B. Agarwal et al. (eds.), *Deep Learning-Based Approaches for Sentiment Analysis*, Algorithms for Intelligent Systems, https://doi.org/10.1007/978-981-15-1216-2_2

1 Introduction

Sentiment analysis is a fundamental branch in natural language processing (NLP). It is the process of understanding sentiment in user-generated opinionated data in social media, product feedback or blogs. Sentiment analysis is increasingly important in business intelligence, empowering industry in key areas such as brand monitoring, product analytics and market research and analysis. Organizations can classify sentiment towards a product or service, helping to improve customer experience by positively impacting product life cycle [1]. Sentiment analysis can be applied to capture sentiment tendencies of an individual towards products, enabling an organization to predict and recommend products to an individual [2]. With social media becoming omnipresent in modern society, sentiment analysis approaches are also used to gauge public opinion on issues in real time, providing valuable intelligence on government policies, marketing campaigns or live events [3]. The value of sentiment analysis goes beyond business intelligence, and work on sentiment analysis can be extended to areas such as detecting social media content that violates terms and conditions, including hate speech [4] or cyber-bullying [5].

With the maturation of big data and the advent of IoT, extracting value from unstructured opinionated data is a key challenge for both industry and academia. It has made sentiment analysis an active area within natural language processing (NLP) domain.

Efficient machine learning models capable of accurate sentiment analysis have many advantages over existing approaches, and these include scalability, real-time analysis and a consistent criterion. With the tremendous data generated from digitization, scalability is key to capturing accurate sentiment across the generated data. Ability to execute real-time analysis on vast data gives an unparalleled ability to gain valuable feedback on topic at hand.

Deep learning is a prominent branch of machine learning. It has gone through a surge of activity in recent years due to the alignment of several factors, which includes advancements in high-end machines that make the training of deeper networks viable, the availability of large training data sets and key advancements in artificial neural networks. These have led to remarkable performance for some deep learning models that have not only matched human-level performance but surpassed it in areas like computer vision, voice generation and recognition and art and style imitation [6]. In this survey of current deep learning trends in sentiment analysis, we categorize sentiment analysis into the coarse-grain and fine-grain.

The remainder of the chapter is arranged as follows: the 'Related Work' section looks at related surveys into the application of deep learning in sentiment analysis. This section is followed by the study rationale that explains the challenges and motivations for conducting this study. Most commonly used deep learning architectures are summarized in 'Deep Learning Architectures' section followed by the research methodology of this study. A summary and detailed discussion of the deep learning-based studies for 'coarse-grain', 'fine-grain' and 'cross-domain' sentiment analysis are provided in the subsequent sections. These are followed by our survey highlights and conclusion.

2 Related Work

In this section, we present a summary of existing deep learning- and machine learning-based sentiment analysis studies. We conclude this section by identifying research gaps from existing studies and provide a study rationale for this chapter.

Zhang et al. [7] take a detailed investigation into deep learning and sentiment analysis, examining state-of-the-art approaches. A weakness of this study is the limited selection of models surveyed. It does contain sufficient summaries for the selected studies; however, summaries for document, sentence and aspect level are lacking.

Comparative studies on traditional machine learning algorithms and deep learning models show a clear preference for deep learning [8]. This edge is not limited to sentiment analysis, and deep learning approaches have been shown to outperform on predictive analysis [9].

Broad surveys into sentiment analysis split the topic into document, sentence and aspect level [7, 10–12], while more focused studies lay out sentiment analysis approaches methodologically [13].

Prior to the surge in interest surrounding deep learning, researchers focused their attention on a selection of machine learning algorithms which included probabilistic classifiers [such as Naive Bayes (NB), Bayesian network (BN), maximum entropy (MaxEnt), multinomial Naive Bayes (MNB), conditional random field (CRF)], linear classifiers such as logistic regression (LR), support vector machines (SVM) [14–16] and decision trees (such as random forest (RF) [8, 12, 13]). SVM and NB classifiers are most dominant approaches followed by decision trees [8, 9, 12, 13]. Studies evaluating the performance of these algorithms show a clear preference for SVM classifiers [8, 12, 13].

3 Machine Learning Approaches for Sentiment Analysis

This part of the literature review is added for completeness of this section. Most of the literature review on deep learning architectures refers to machine learning-based sentiment analysis approaches that are trained on a set of features. Most of the work involves determining a set of features to train the algorithm [17]. The performance of these approaches is strongly influenced by the set of features. Feature selection plays an important role in improving classification accuracy [1]. It is the process of ranking features on how informative or noisy they are, with lower rank features being less informative and noisier. Feature selection techniques are categorized into four main classes: (1) statistical, (2) hybrid and heuristic (3) clustering based, or

(4) NLP based. Statistical methods include hybrid, univariate and multivariate. Hybrid methods include a combination of multivariate and univariate and other methods. Examples of univariate techniques are information gain, occurrence frequency, chi-square, minimum frequency thresholds and log-likelihood. While univariate methods are computational efficient, attribute interactions are ignored [1]. Some examples of multivariate methods are recursive feature eliminates and decision tree models. They consider a collection of attributes for attribute selection using a wrapper model. Multivariate methods are more computationally expensive compared to univariate as they assess attribute interactions. Clustering approaches involve a small number of parameters to tune but come with a key drawback of minor features being more difficult to extract [2]. An example of a hybrid method is POS tagging with WordNet dictionary. While NLP-based methods have high accuracy, they have low recall with dependency on accuracy part of speech tagging, they work on the following principles: adjectives, nouns, noun phrases and adverbs typically express features [3], and subjective expressions with terms appearing near can signify features.

Ly et al. [16] utilized sentence-level syntactic information by incorporating them to isolate product features from redundant features. Stanford Dependency Parser was used to achieve this. Somprasertsri and Lalitrojwong [18] merged syntactic and lexical features with a maximum entropy model to extract product features. Zhang et al. [2] merged point-wise mutual information and association rules for distinguishing features while employing HowNet sentiment dictionary. Wang and Wang [19] used bootstrapping iterative learning strategy with supplementary linguistic rules to classify product features. Zhang et al. [20] employed two steps for feature extraction. The first step included using part–whole relation patterns and 'a' pattern for extracting features and also to deliver performance improvements. The second step involved the ranking of candidate features based on their importance. The features in this step were extracted using conditional random fields and maximum entropy model.

Lexicon-based approaches rely on sentiment lexicons. This approach relies on static lists of words, and words not present in the lexicon are ignored. This requires frequent updates when working with social media data set like Twitter.

A 'bag-of-words' approach in combination with one nominal and seven binary features divide tweets into a predefined set of generic classes for retrieving new relevant tweets with better accuracy [21]. This approach was used to pick out 'slangs', 'signs', 'phrases' and 'opinion words. The noise was found in new tweets.

An alternative system for topic classification uses Lexicon-based approach to sum the amount of positive and negative words appearing in the text [22]. However, this approach failed poor identification of words when a negative text was used for testing. Kamps [23] used a simple technique based on lexical relations to perform text classification. Andrea [24] used WordNet to classify text using an assumption that words with similar polarity have a similar orientation.

Ting-Chun [25] used an algorithm based on part of speech (POS) pattern. A text phrase was used as a query for a search engine, and the results were used to classify text.

Naive Bayes-based approaches [26] have been a common technique in text classification due to their simplicity as well as efficiency. These approaches exploit the likelihood of two events to forecast the probability of one activity given the happening of the other activity is straightforward. However, these approaches are not suitable for text categorization as topics are not necessarily dependent on one another.

Support vector machine (SMV)-based method is regarded as one of the greatest text classification approach [13–15] and employs a statistical classification technique which is built on the intensification of the events as well as the hyperplane separation.

K-nearest neighbour (KNN)-based methods [27] are not efficient and failed to extract any aspect from the data sets which were used for the training purpose and instead could only specify the similarities in the documents to that of its neighbours.

Jain and Katkar [28] compared nearest neighbour (NN), Naive Bayes (NB), subspace classifier (SS) and decision tree (DT) on two data sets.

Sentiment analysis for text containing spelling variation, informal words, short length status message, word shortening, emoticons and negation word can be more problematic, where each feature could lead to a reversal of the polarity. Turney [29] used semantic orientation on user reviews to identify the underlying sentiments. Sentiment analysis while handling negation and intensifying words was reported in [30, 31]. Keyword spotting-based technique relied heavily on words clearly indicating the sentiment being expressed [32]. However, this approach fails when the regular meaning of such keywords is negated in the context of the entire statement. Lee and Vaithyanathan [33] proposed a support vector-based binary model (positive or negative) of sentiment analysis.

A comparison [34] of lexicon-based and machine learning techniques for sentiment analysis was performed. They trained their model on existing data and used it for making predictions on new data received based on the closest topic match. A cross-disciplinary approach for validating prediction results was proposed in [35].

A summary of the challenges identified from the surveyed study for both deep learning and machine learning approaches and rationale for conducting our survey for this chapter is explained in the next section.

4 Study Rationale

Contrary to machine learning approaches, deep learning models do not involve manually engineered features created using expert knowledge and available linguistic resources. Deep learning models learn through a method called backpropagation, enabling end-to-end training without human input. This innovation combined with the ability to stack neural networks into deep models enables unparalleled performance compared to traditional machine learning algorithms. While the performance of machine learning approaches tends to plateau with increasing training data, deep learning models perform better with more data.

Due to the sheer success of deep learning, research activity is continually increasing year upon year. Separating important studies from the deluge of activity is important in making sense of which deep learning approaches are trending for sentiment analysis.

In our survey of related work, we did not find studies that captured a broad survey of important studies with summaries and analysis for key categories of sentiment analysis such as coarse-grain, fine-grain and cross-domain sentiment analyses.

In this chapter, we explore the most important deep learning studies in sentiment analysis with a focus on the following areas: recent trends and advances in deep learning-based sentiment analysis, coarse-grain (document and sentence level), fine-grain (target and aspect level) and cross-domain sentiment analysis.

5 Deep Learning Architectures

In this section, we explain important deep learning architectures commonly utilized in sentiment analysis. We provide a description of each architecture and explain their inner working for better understanding of our analysis sections discussing the surveyed models.

A detailed discussion for different variants of these architectures below is beyond the scope of this chapter. Interested readers are encouraged to refer to the respective studies.

5.1 Convolutional Neural Networks

Convolutional neural networks (CNNs) are a class of artificial neural networks (ANNs) that soared to prominence with machine vision tasks, significantly outperforming traditional artificial neural networks [36]. ANNs do not scale well with large input matrices while CNNs scale very well with larger input. Like traditional ANNs, CNNs contain input, fully connected and output layers but with additional layers that are key to CNN's effectiveness, these include convolutional and pooling layers. These layers involve obtaining feature maps using filters. Feature maps are feature representations of input data. The filters used in convolutional and pooling layers are learned during backpropagation.

In Fig. 1, we see a 5×5 filter being applied to the input layer and a 2×2 filter being applied to the feature maps in convolutional layer 1. Pooling layers serve to drastically reduce feature map size while also extracting the position of a feature relative to other objects in a feature map.

The effectiveness of CNNs in classification tasks stem from their ability to exploit feature locality. This is done at different granularities, enabling CNNs to model hierarchical-modular decomposition of input data.

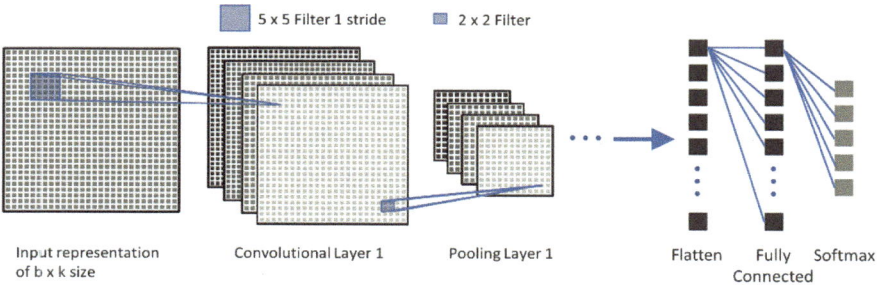

Fig. 1 CNN architecture

Fig. 2 Unfolded RNN

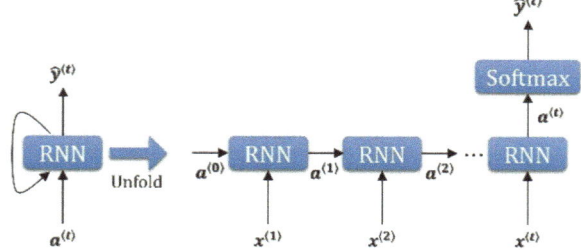

5.2 Recurrent Neural Networks

Recurrent neural networks (RNN) are a class of artificial neural networks designed to excel with a sequence of inputs, where information is captured from the sequence. This is critical in time series tasks such as speech recognition, music generation, natural language processing (NLP), DNA sequence analysis, machine translation, video activity recognition and name entity recognition. RNN architecture processes sequence by iterating through the sequence elements and maintaining a state containing information relative to what it has seen, capturing the correlation between the current time step and previous ones. Standard RNN cell architecture, shown in Fig. 3, contains a single tanh layer. Each input element is fed to the network one by one (see Fig. 2).

5.3 Bi-directional Recurrent Neural Network

The output of a unidirectional RNN (Fig. 2) at any given time step is dependent on the information encoded from the previous time steps. Some tasks, particularly in NLP domain, require information from future outputs to be incorporated. Bi-directional RNN enables a model to encode information from both the beginning and end of a

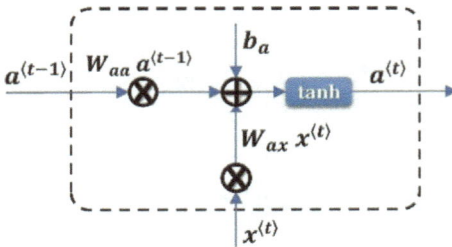

Fig. 3 Standard recurrent neural net

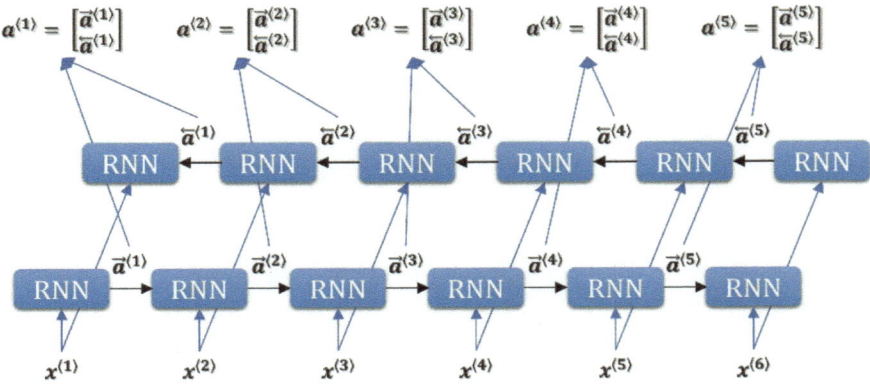

Fig. 4 Bi-directional RNN

sequence of inputs. The architecture of a bi-directional RNN comprises two RNNs placed on top of each other, with each RNN starting sequence of input from opposite ends. Figure 4 shows a bi-directional RNN. At each time step, the output of forward and backward RNN is often concatenated and passed to the next layer in the model.

6   Long Short-Term Memory (LSTMs)

For longer series of inputs, RNNs architecture is unable to train gradients, and this creates the vanishing and exploding gradient problem. These arise during backpropagation where lots of vanishing gradients (less than 1) or exploding gradient (more than 1) are multiplied together. This multiplication leads to gradients converging to either 0 or infinity and essentially prevents the network from learning. This overarching drawback to RNNs was addressed with LSTMs in 1997 by Hochreiter and Schmidhuber [37]. LSTMs have proven to be very effective with a longer series of input. LSTMs are the most widely used RNN approach followed by GRUs. A generic architecture of LSTM is represented in Fig. 5.

Fig. 5 Long short-term
memory cell architecture

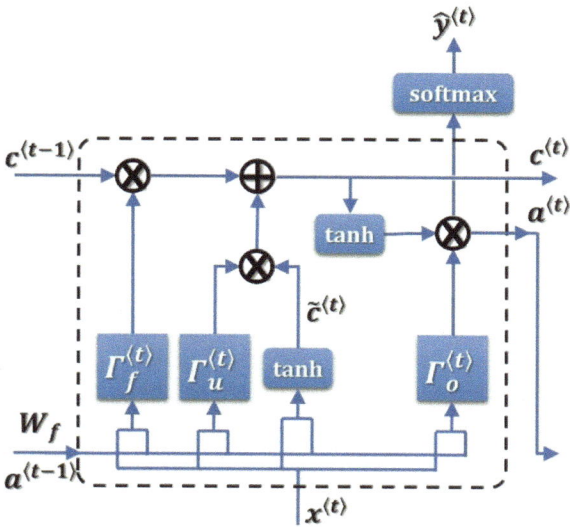

LSTMs also have a chain-like structure. Unlike RNNs, the repeating cell in LSTM has a different architecture. Instead of having a single gating mechanism, there are three, interacting in a very special way:

- **Forget gate**: sigmoid layer (2) takes concatenated input $x^{\langle t \rangle}$ and $a^{\langle t-1 \rangle}$ and decides which parts from old output should be removed.
- **Update gate**: sigmoid layer (1) decides which new information should be updated or ignored from the concatenated input $x^{\langle t \rangle}$ in the cell state $c^{\langle t \rangle}$. Tan*h* layer (4) creates a vector of all the possible values from the new input. Two are multiplied to update the new cell sate. This is added to old memory $c^{\langle t-1 \rangle}$ to give $c^{\langle t \rangle}$.
- **Output gate**: sigmoid layer (3) decides which parts of cell state to output. Cell state puts through a tan*h* layer generating all the possible values and multiplies it by the output of the sigmoid gate so that we only output the parts we decided to. The hidden state is obtained by element-wise multiplicated of sigmoid layer and tan*h* layer (6) gives the hidden state.

$$\Gamma_u^{\langle t \rangle} = \sigma \left(W_u \big[a^{<t-1>}, x^{<t>} \big] + b_u \right) \tag{1}$$

$$\Gamma_f^{\langle t \rangle} = \sigma \left(W_f \big[a^{<t-1>}, x^{<t>} \big] + b_f \right) \tag{2}$$

$$\Gamma_o^{\langle t \rangle} = \sigma \left(W_o \big[a^{<t-1>}, x^{<t>} \big] + b_o \right) \tag{3}$$

$$\widetilde{c}^{<t>} = \tanh \left(W_c \big[a^{<t-1>}, x^{<t>} \big] + b_c \right) \tag{4}$$

$$c^{<t>} = c^{<t-1>} \times \Gamma_f^{\langle t \rangle} + \tilde{c}^{<t>} \times \Gamma_u^{\langle t \rangle} \tag{5}$$

$$a^{<t>} = \Gamma_o^{\langle t \rangle} \times \tanh c^{<t>} \tag{6}$$

RNNs overwrite their memory at each time step in a rather unrestrained fashion, whereas LSTM transforms its memory in a more careful manner using a specific gating mechanism deciding which bits of information to remember, update and pay attention to. This enables LSTM to learn from longer sequences of input while avoiding vanishing and exploding gradients.

7 Gated Recurrent Units (GRUs)

A variant of LSTMs, gated recurrent unit (GRU), was proposed in 2014 by Cho et al. [38]. GRUs are also equipped to avoid vanishing and exploding gradient problem but with a simpler internal architecture. Unlike the three gates present in LSTMs, GRUs consists of two gates: an update and reset gate. Update gated is designed to help learn long-term dependencies from a series of input, while reset gate is intended to learn short-term dependencies. GRUs does not maintain an internal memory state different from their hidden state. Figure 6 shows a generic gated recurrent unit cell architecture.

GRU cell contains two gates: an update gate which decides what new information to add and what to throw away using sigmoid layer (7), taking in concatenated $x^{\langle t \rangle}$ and $h^{\langle t-1 \rangle}$, and a reset gate which also uses a sigmoid layer (8) to decide how much of past memory to forget, taking concatenated input $x^{\langle t \rangle}$ and $h^{\langle t-1 \rangle}$. Tanh layer (9) calculates proposed $\tilde{h}^{<t>}$. Final memory is calculated (10) by determining what to collect from the current memory and previous steps.

Fig. 6 Gated recurrent unit architecture

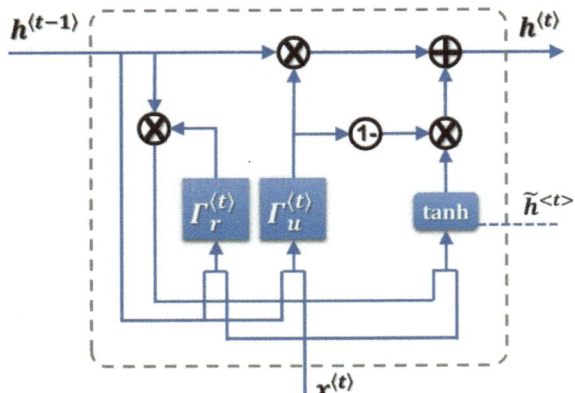

$$\Gamma_u^{\langle t \rangle} = \sigma \left(W_u \left[a^{<t-1>}, x^{<t>} \right] + b_u \right) \tag{7}$$

$$\Gamma_r^{\langle t \rangle} = \sigma \left(W_r \left[a^{<t-1>}, x^{<t>} \right] + b_r \right) \tag{8}$$

$$\widetilde{h}^{<t>} = \tanh \left(x^{<t>} W_h + \left(\Gamma_r^{\langle t \rangle} \right) W_h + b_h \right) \tag{9}$$

$$h^{<t>} = \Gamma_r^{\langle t \rangle} \cdot c^{<t-1>} + \left(1 - \Gamma_r^{\langle t \rangle} \right) \cdot h^{<t>} \tag{10}$$

As GRUs have fewer parameters, they are more ideal for deeper networks due to less resource utilization during training while also requiring fewer data to generalize. Empirical evaluations [39] comparing LSTMs and GRUs are inconclusive on the question of performance, but rather, they emphasize hyper-parameterization as the most important factor.

8 Attention Mechanism

Attention mechanisms rose to prominence in NLP with an influential paper on machine translation [40] and represented major progress in NLP domain. To gain an intuition for attention mechanism, we can think of it as a single hidden layer neural network. The purpose of attention mechanism network is to determine the importance of each hidden state and provide a weighted sum of all the features fed as input.

Figure 7 shows a 'sequence to sequence model', a model used in machine translation tasks. A single attention layer is placed between an encoder Bi-LSTM and

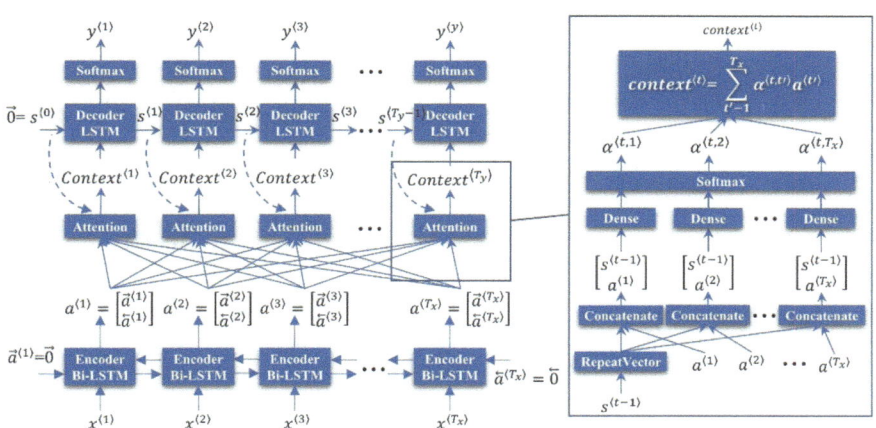

Fig. 7 Attention mechanism

decoder LSTM layers. All feature vectors $a^{(t)}$ from encoder Bi-LSTM are fed to attention. The output of attention mechanism is the computed context vector $c^{(t)}$ which is fed to decoder LSTM layer. The context vector is a weighted sum of the features from the encoder layer weighted using attention weights. Contribution of feature vector is computed using a soft max where the attention weights a^j sum to 1, and all feature vectors contribute to decoder at each time step. The attention weights, trained during backpropagation, are key to determining how much attention $y^{(t)}$ pays to each feature vector.

Attention mechanisms allow models to replicate how the human brain is able to focus on specific parts of the input, determining the importance of each part for the desired output.

9   Research Methodology

Inclusion Criteria

We restricted the survey to studies published between January 2013 and January 2019. Only studies exploring deep learning approaches for sentiment analysis of short text from social media, product reviews and different sentiment analysis competitions. Coarse-grain, fine-grain and cross-domain sentiment analysis were considered.

Exclusion Criteria

We excluded publications that were simple in their approach or had very poor performance. Many papers taking part in sentiment classification competitions fall into this category. We did, however, include many papers from these competitions that either performed well or were innovative in their approach.

Search Method

We used digital libraries to search for the studies. These libraries included ACM digital, IEEEXplorer, Scopus and Google Scholar which we found returned more results for each key words.

10   Approach to Sentiment Analysis Task Categorization

In this survey we categorize SA into the coarse-grain and fine-grain granularities. Coarse-grain can be broken down to include document, sentence-level and cross-domain tasks, while fine-grain is used to describe aspect level and cross-domain tasks. Cross-domain is not restricted to a granularity. Document- and Sentence-level subjectivity classification involves establishing if a sentence contains emotions, opinions, beliefs or evaluations. Aspect-level SA consists of aspect term extraction, aspect term categorization, aspect term polarity classification and target term polarity classification. We provide summary and analysis for granularities and cross-domain, laying out an overview of architectures and their performance on the most widely used data sets. Lastly, a detailed discussion summarizing the trends in SA is provided in the next sections.

11 Coarse-Grain Sentiment Analysis

Coarse-grain sentiment analysis is performed on the document or sentence level. More specifically, coarse-grain sentiment analysis involves sentiment classification and subjectivity classification for sentences and documents. Our survey of coarse-grain sentiment analysis is focused on studies involving sentiment classification. The goal of document or sentence sentiment classification is to predict the sentiment polarity of a document or sentence of words. We have categorized the fine-grain approaches by neural network architecture (see Table 1).

RNNs have often been associated with tasks in the natural language processing domain due to their innate ability to learn from the order of input. Our survey of coarse-grain sentiment analysis yielded a varied selection of studies clustered around three neural network approaches like CNNs [41–44], RNNs [45–49] and attention-RNN [50–52]-based approaches.

Table 1 Coarse-grain sentiment analysis neural network types

Architecture	Description	Study
CNN	Convolutional neural network (CNN) architecture consists of convolutional, pooling, fully connected and classifier layers	[41–44, 72–76]
LSTM	A variant of RNNs, long short-term memory networks, differs from generic RNN with the addition of another memory cell with forget, update and output gates	[45, 47–49, 77–80]
GRU	GRU, an LSTM variant, differs from LSTM with gating mechanism, containing an update and reset gate	[54, 81]
Attention-LSTM	Attention-LSTM models consist of one or more layers of long short-term memory and attention networks	[46, 50, 52]
Attention-GRU	Attention-GRU models consist of one or more layers of gated recurrent units and attention networks	[51]
Ensemble	Ensemble models consist of multiple homogeneous deep learning networks, jointly working in the same model	[56, 82, 83]
Hybrid	Hybrid models consist of multiple heterogeneous deep learning networks, jointly working in the same model	[53, 55]

CNN's ability to extract local and position-invariant features from imagery data crosses over to textual data, with strikingly good results [43, 44, 53]. Although almost all the studies involve textual data exclusively, recent work into multimodal sentiment analysis [53] show deep learning approaches can perform close to state-of-the-art models designed for textual data.

Within the RNN family, LSTMs and GRUs are the most prevalent. Both approaches use gating mechanism to control the flow of information. GRUs has a simpler internal architecture with one less gate and the absence of memory unit present in LSTMs. As a result, they train faster but due to the less complex gating mechanism, LSTMs are better equipped to capture more information from longer sequences of input.

All RNN-based approaches use LSTMs and GRUs to learn hidden vectors from input [45, 47–49, 54]. Bi-directional LSTMs are used to capture more structural information and perform better with longer sequences. Regularization is ubiquitous with L2 regularization [45, 47] and dropout [46, 48, 49] being the technique of choice.

In recent years, the attention mechanism has seen increasing usage with LSTM and GRU networks [43, 46, 50, 51]. LSTMs and GRUs are used to learn high-level presentations, capturing compositional semantics from a sequence of inputs. Attention layers follow RNN layers, enabling models to learn the contribution of each sentence or word towards the polarity.

Attention mechanism, LSTM and GRU have also been utilized to model the hierarchical structure of text [50, 51]. Sentence-level attention is used to capture important sentences contributing to polarity while word-level attention is used to capture keywords contributing to the polarity. In combination with LSTM/GRU encoders, hierarchical attention has proven to be very effective in gaining new insights into hierarchical dependencies of text contributing to sentiment polarity.

Bi-directional LSTMs [45, 46, 48–50, 55, 56] and GRUs [51] are heavily utilized in coarse-grain approaches for both LSTM/GRU and attention-LSTM/GRU models. Bi-directional layers can learn forward and backward features from the sequence of input. Resulting hidden vectors for each time step are usually concatenated.

Bi-directional networks enable a model to capture dependencies on both sides of input rather than capturing information from just the previous input as is the case with unidirectional networks.

In Table 2, we surveyed the top three deep learning models for coarse-grain data sets, displaying rank and identifying the model architecture. Stanford Sentiment Treebank 2 and Yelp Binary classification are both dominated by CNN and LSTM approaches, and IMDB has top three spots with LSTM models. CNN also does very well with SemEval Twitter tasks, occupying first two spots on SemEval-2016 Task 4. Ensemble approaches which utilized a CNN and LSTM or GRU appeared in top 3 for SemEval 2016 and 2017.

CNN, LSTM, GRU and attention-LSTM/GRU models compete for performance, with no distinct performance edge from any of the approaches. CNN's ability to model hierarchical-modular decomposition of input data for computer vision tasks is also well suited to textual data in sentiment analysis. The advantage of CNNs

Table 2 Coarse-gain data sets and top performing architectures

Data set	DL architecture	Model/variant	Rank[a]	Study
SST-2 Binary classification	DNN	MT-DNN	1	[84]
	CNN	CNN large	2	[72]
	LSTM	Block-sparse LSTM	3	[77]
Yelp Binary classification	LSTM	ULMFiT	1	[78]
	CNN	DPCNN	2	[73]
	GRU	DRNN	3	[54]
IMDB	LSTM	ULMFiT	1	[78]
		Block-sparse LSTM	2	[77]
		oh-LSTM	3	[79]
Amazon review polarity	GRU	DRNN	1	[54]
		SRNN	3	[81]
SemEval-2014 task 9[b]	ANN	Coooolll	2	[85]
	CNN	ThinkPositive	10	[74]
SemEval-2016 task 4: sentiment analysis in Twitter subtask A	CNN	SwissCheese	1	[75]
		SENSEI-LIF	2	[76]
	Ensemble	UNIMELB: CNN, LSTM	3	[82]
SemEval-2017 task 4[c]	LSTM	DataStories	1	[80]
	Ensemble	LIA: CNN, LSTM	3	[83]
SemEval-2017 task 5[d]	Hybrid	RiTUAL-UH: CNN, Bi-GRU	2	[86]

[a]Denotes ranking after completion of workshop ranking
[b]Twitter sentiment analysis
[c]Sentiment analysis in Twitter subtask A
[d]Fine-grained sentiment analysis on financial microblogs and news

becomes more pronounced when considering the edge CNNs have over LSTMs, GRUs and attention-LSTM/GRU on model complexity.

CNN models have fewer parameters, requiring less tuning and computational resources to train. LSTM and attention-LSTM/GRU are much costly to train with more parameters. Although RNN-based models are ideal for sentiment analysis due to their inherent ability to model the semantic dependencies of sequential data, they do not provide a strong performance edge over CNNs.

12 Fine-Grain Sentiment Analysis

Fine-grain sentiment analysis encompasses tasks on a subsentence level, often involving multiple targets in a sentence. Aspect-based sentiment analysis involves several subtasks such as aspect extraction, aspect categorization, entity extraction and

aspect sentiment classification. Most common fine-grain task is aspect-level sentiment classification which involves determining the sentiment polarity towards a pre-
defined aspect. In the remainder of this section, we discuss our survey into fine-grain
sentiment analysis studies.

Table 3 summarizes our survey into deep learning architectures for fine-grain
sentiment analysis. We have categorized the studies into deep learning architectures.

Contrary to coarse-grain sentiment analysis, our survey revealed significantly
more activity for LSTMs, GRUs and attention compared to CNN, and we discuss
the reason behind this later in this section. Models that utilize LSTMs and GRUs
use them on top of an embedding layer to learn high level features, capturing the
interaction between aspect/target with opinion words. The resulting hidden vectors
are either fed to other LSTMs/GRUs layers or a classifier for predicting sentiment
polarity.

Attention-LSTM models are the most popular approach we surveyed for fine-
grain sentiment analysis. LSTM layers are used to learn semantic information from
a series of word input. Standard LSTM model cannot efficiently capture key parts of

Table 3 Fine-grain sentiment analysis model architectures

Architectures	Description	Study
CNN	CNN architecture consists of convolutional, pooling, fully connected and classifier layers	[65, 73, 87]
Recursive neural networks	RNNs employ a shared weight matrix in conjunction with a binary tree structure (often used as parser). It enables RNNs to perform well on variable size inputs	[88, 89]
LSTM	A variant of RNNs, long short-term memory networks, differs from generic RNN with the addition of another memory cell with forget, update and output gates	[59, 78, 90–94]
GRU	GRU, an LSTM variant, differs from LSTM with gating mechanism, containing an update and reset gate	[54, 64]
Attention	Attention mechanism is a simple one-layer ANN designed to pay more or less attention to individual inputs	[95]
Attention-LSTM	Attention-LSTM models consist of one or more layers of long short-term memory and attention networks	[57, 60–63, 96–102]
Attention-GRU	Attention-GRU models consist of one or more layers of gated recurrent units and attention networks	[58, 64]
Hybrid	Hybrid models consist of multiple heterogeneous deep learning networks, jointly working in the same model	[47, 86, 103]

a sentence contributing to the polarity of an aspect or target. Attention mechanism can be used to capture the contribution of each word towards an aspect or target by mapping their semantic association. Adding attention mechanism to LSTM or GRU models substantially improves the accuracy of aspect-level sentiment classification.

Hierarchical attention to capturing hierarchical dependencies is not limited to coarse-grain approaches. This approach has been successfully utilized to learn aspect specific representations [57, 58].

Bi-directional LSTMs [59–63] and GRUs [57, 58, 64] are heavily utilized in fine-grain approaches for both LSTM/GRU and attention-LSTM/GRU models. Bi-directional layers can encode forward and backward information from a sequence of input. Resulting hidden vectors for each time step are usually concatenated. Bi-directional networks enable a model to capture dependencies on both sides of input rather than capturing information from just the previous input as is the case with unidirectional networks.

In Table 4, we surveyed the top three deep learning models for fine-grain data sets, displaying their rank and identifying the model architecture. Both Stanford Sentiment Treebank 5 and Yelp fine-grain data set use a 5-point scale to classify the polarity of text. LSTM models dominate both SST-5 and Yelp data sets, with SST-5 having top three ranks with LSTM models. Ranking for SemEval data sets is much

Table 4 Survey of fine-grain data sets and top performing models

Data set	DL architecture	Model/variant	Rank[a]	Study
SST-5 fine-grained classification	LSTM	EDD-LG (shared)	1	[91]
		Suffix Bi-LSTM	2	[92]
		BCN + ELMo	3	[93]
Yelp fine-grained classification	LSTM	ULMFiT	1	[78]
	CNN	DPCNN	2	[73]
	GRU	DRNN	3	[54]
SemEval 2014 task 4: aspect-based sentiment analysis	Attention-GRU	HAPN	1	[58]
	Attention-LSTM	SA-LSTM-P	2	[101]
		MGAN	3	[100]
SemEval-2016 task 5[b]	CNN	NileTMRG	3	[87]
SemEval-2017 task 5[c]	Hybrid	RiTUAL-UH: CNN, Bi-GRU	2	[86]
SentiHood	LSTM	REN with DMU[d]	1	[90]
		LSTM-LOC	3	[94]
	Attention-LSTM	Sentic LSTM + TA + SA	2	[102]

[a]Denotes ranking after completion of workshop ranking
[b]Aspect-based sentiment analysis
[c]Fine-grained sentiment analysis on financial microblogs and news
[d]Recurrent entity networks with delayed memory update

more mixed due to the difficulty of obtaining sources which track the performance of models long after the SemEval task has expired. Some tasks, like SemEval 2014 Task 4, are tracked and we can see attention-GRU and attention-LSTM models occupying top three rankings. SentiHood data set is another popular data set used to benchmark models. LSTM and attention-LSTM models fill top three rankings for this data set.

While CNNs have shown very promising performance and strong presence on coarse-grain sentiment analysis, it is much less prominent on fine-grain tasks. CNN's ability to learn sentiment features on a less granular level does not transfer over to finer granularities, although CNNs have shown to be very effective at aspect detection [65]. We found on more granular level LSTMs, GRUs and attention showed the most promise in capturing syntactic information on a subsentence level.

13   Cross-Domain Sentiment Analysis

To apply machine learning approaches in sentiment analysis, previously labelled data is required for a model to learn patterns and be able to classify sentiment polarity. This process where the model is trained using labelled data is called supervised learning. The performance of a model is directly related to the quality of the training data. Thus, machine learning approaches, including deep learning, use same data to train a model and test it. Certain domains have abundant training data while other domains are lacking. The costly process of labelling new training sets by humans can be avoided with approaches that can perform efficiently over different domains.

Cross-domain sentiment analysis involves predicting the sentiment polarity of a target domain using a model trained on the source domain. It aims to overcome constraints around the availably of training data across different domains. The domain is defined as a semantic concept or a class of objects. Using commercial products as an example, different types of products such as toys, DVDs or books are considered domains. The central challenge of cross-domain sentiment analysis is to extract domain-invariant sentiment features from the source domain and utilize this knowledge in predicting sentiment polarity of the target domain. Although cross-domain sentiment analysis lags coarse and fine-grain sentiment analysis on research activity, recent trends suggest an uptick in interest.

Table 5 shows the survey into cross-domain sentiment analysis approaches. Research is focused on two popular deep learning architectures: LSTM and attention, both dominating approaches with exception of Moro et al. [66] using a differential neural computer and Chen and Cardie [67] utilizing CNN for a minor role. The proposed approaches have been heavily augmented to achieve the task of cross-domain sentiment analysis. The models we surveyed are more complex than coarse and fine-grain sentiment analysis models, where there is strong use of multitask learning [67–69], adversarial training [67, 70] and joint simultaneous training of multiple networks with shared layers [67, 71].

A naive method for domain-agnostic approach involves merging all training data. This approach risks losing valuable domain knowledge. To avoid this, domain-aware

Table 5 Cross-domain models and deep learning architecture

Architecture	Description	Study
LSTM	A variant of RNNs, LSTM, differs from generic RNN with the addition of another memory cell with forget, update and output gates	[17, 66–68]
Attention	Attention mechanism is a simple one-layer ANN designed to pay more or less attention to individual inputs	[69, 70]
Differential neural computer	Differential neural computer is an RNN with auto-associative memory	[103]

models can retain domain knowledge via multitask learning [17, 49]. Adversarial training enables a model to learn domain-invariant features with an adversarial domain classifier. Adversarial training mechanism is designed to assist a model to discover shared parameters which are not restricted to a domain. The joint training of networks simultaneously can assist a network domain classifier to be less discriminative towards the source and target representations [67, 70].

14 Conclusion and Survey Highlights

Our survey of deep learning in coarse-grain, fine-grain and cross-domain sentiment analysis revealed four deep learning architectures dominating the approaches: CNNs, LSTMs, GRUs and attention. For coarse-grain sentiment analysis, CNNs and LSTMs compete on performance, but CNNs offer reduced model complexity and training overhead. Fine-grained sentiment analysis requires a model to learn the complex interactions between aspect and opinion words on a subsentence level. Bi-directional LSTMs and attention mechanism offer the most promise, although CNN-based models have proven efficient at aspect extraction. Cross-domain sentiment analysis is also dominated by approaches using LSTM, GRU and attention. The addition of multitask learning, adversarial training and joint training is being the key distinguishing factor between fine-grain and cross-domain sentiment analysis approaches.

We also note that the performance of deep learning approaches for fine-grained sentiment analysis lags coarse-grain and requires more attention from the research community.

References

1. Koncz, P., and J. Paralic. 2011. An approach to feature selection for sentiment analysis. In *2011 15th IEEE International Conference on Intelligent Engineering Systems*, 357–362. IEEE.
2. Zhang, H., Z. Yu, M. Xu, and Y. Shi. 2011. Feature-level sentiment analysis for Chinese product reviews. In *2011 3rd International Conference on Computer Research and Development*, vol. 2, 135–140. IEEE.

3. Hu, M., and B. Liu. 2004. Mining opinion features in customer reviews. *AAAI* 4 (4): 755–760.
4. Biere, S. 2019. Hate speech detection using natural language processing techniques (online) Beta.vu.nl. Available at: https://beta.vu.nl/nl/Images/werkstuk-biere_tcm235-893877.pdf. Accessed 3 Jun 2019.
5. Nahar, V., S. Unankard, X. Li, and C. Pang. 2012. Sentiment analysis for effective detection of cyber bullying. In *Asia-Pacific Web Conference*, 767–774. Berlin, Heidelberg: Springer.
6. Roman Steinberg, u. 2019. 6 areas where artificial neural networks outperform humans (online) VentureBeat. Available at: https://venturebeat.com/2017/12/08/6-areas-where-artificial-neural-networks-outperform-humans/. Accessed 10 Apr 2019.
7. Zhang, L., Wang, S. and Liu, B. 2018. Deep learning for sentiment analysis: A survey. Wiley Interdisciplinary Reviews: Data Mining and Knowledge Discovery, e1253.
8. Llombart, O.R. 2017. Using machine learning techniques for sentiment analysis. Published in June of 2017.
9. Kursuncu, U., M. Gaur, U. Lokala, K. Thirunarayan, A. Sheth, and I.B. Arpinar. 2019. Predictive analysis on Twitter: Techniques and applications. In *Emerging Research Challenges and Opportunities in Computational Social Network Analysis and Mining*, 67–104. Cham: Springer.
10. Giachanou, A., and F. Crestani. 2016. Like it or not: A survey of twitter sentiment analysis methods. *ACM Computing Surveys (CSUR)* 49 (2): 28.
11. Mariel, W.C.F., S. Mariyah, and S. Pramana. 2018. Sentiment analysis: A comparison of deep learning neural network algorithm with SVM and naïve Bayes for Indonesian text. In *Journal of Physics: Conference Series*, vol. 971(1), 012049. IOP Publishing.
12. Pradhan, V.M., J. Vala, and P. Balani. 2016. A survey on sentiment analysis algorithms for opinion mining. *International Journal of Computer Applications* 133 (9): 7–11.
13. Xia, R., C. Zong, and S. Li. 2011. Ensemble of feature sets and classification algorithms for sentiment classification. *Information Sciences* 181 (6): 1138–1152.
14. Joachims, T. 1998. Text categorization with support vector machines: learning with many relevant features. In *10th European Conference on Machine Learning (ECML'98)*.
15. Rennie, J. and R. Rifkin. 2001. Improving multiclass text classification with the support vector machine. MIT, Technical Report. AIM-2001-026.2001.
16. Ly, D.K., K. Sugiyama, Z. Lin, and M.Y. Kan. 2011. June. Product review summarization from a deeper perspective. In *Proceedings of the 11th Annual International ACM/IEEE Joint Conference on Digital Libraries*, 311–314. ACM.
17. Agarwal, B, N. Mittal. 2016. Prominent feature extraction for sentiment analysis. In *Springer Book Series: Socio-Affective Computing series*, 1–115. Springer International Publishing. ISBN: 978-3-319-25343-5, https://doi.org/10.1007/978-3-319-25343-5.
18. Somprasertsri, G. and P. Lalitrojwong. 2008. Automatic product feature extraction from online product reviews using maximum entropy with lexical and syntactic features. In *2008 IEEE International Conference on Information Reuse and Integration*, 250–255. IEEE.
19. Wang, B, and H. Wang. 2008. Bootstrapping both product features and opinion words from Chinese customer reviews with cross-inducing. In *Proceedings of the Third International Joint Conference on Natural Language Processing: Volume-I*.
20. Zhang, L., B. Liu, S.H. Lim, and E. O'Brien-Strain. 2010. Extracting and ranking product features in opinion documents. In *Proceedings of the 23rd International Conference on Computational Linguistics: Posters*, 1462–1470. Association for Computational Linguistics.
21. Sriram, B., D. Fuhry, E. Demir, H. Ferhatosmanoglu, and M. Demirbas. 2010. Short text classification in twitter to improve information filtering. In *Proceedings of the 33rd International ACM SIGIR Conference on Research and Development in Information Retrieval*, 841–842. ACM.
22. Cambria, E.e.a. 2013. Statistical approaches to concept-level sentiment analysis. *IEEE Intelligent Systems* 28 (3): 6–9.
23. Mokken, M.M.K.R.J, and M.D. Rijke. 2004. Using wordnet to measure semantic orientation of adjectives. In *4th International Conference on Language Resources and Evaluation*, 1115–1118.

24. Esuli, A, and F. Sebastiani. 2005. Determining the semantic orientation of terms through gloss classification. In *14th ACM International Conference on Information and Knowledge Management*, 617–624.
25. Peng, T.-C., and C.-C. Shih. 2010. An unsupervised snippet-based sentiment classification method for chinese unknown phrases without using reference word pairs. *Journal of Computing* 2 (8).
26. Melville, P., W. Gryc, and R.D. Lawrence. 2009. Sentiment analysis of blogs by combining lexical knowledge with text classification. In *15th ACM SIGKDD International Conference on Knowledge Discovery and Data Mining, ser. KDD '09* 1275–1284. ACM.
27. Han, E. 1999. Text categorisation using weight adjusted k-nearest neighbour classification. Ph.D. dissertation, University of Minnesota.
28. A.P. Jain, and V.D. Katkar. 2015. Sentiments analysis of Twitter data using data mining. In *International Conference on Information Processing (ICIP)*.
29. Turney, P., and M. Littman. 2003. Measuring praise and criticism: Inference of semantic orientation from association. *ACM Transactions on Information Systems* 21 (4): 315–346.
30. Wiegand, M., A. Balahur, B. Roth, D. Klakow, and A. es Montoyo. 2010. A survey on the role of negation in sentiment analysis. In *Proceedings of the Workshop on Negation and Speculation in Natural Language Processing*, ser., 60–68.
31. Bifet, A., and E. Frank. 2010. Sentiment knowledge discovery in Twitter streaming data. In *Proceedings of the 13th International Conference on Discovery Science*, 15, Ed.
32. Cambria, E., B. Schuller, Y. Xia, and C. Havasi. 2013. New avenues in opinion mining and sentiment analysis. *IEEE Intelligent Systems* 28 (2): 15–21.
33. Pang, B., L. Lee, and S. Vaithyanathan. 2002. Thumbs up? Sentiment classification using machine learning techniques. In *Proceedings of the ACL-02 Conference on Empirical Methods in Natural Language Processing*, 79–86.
34. Ficamos, P., and Y. Liu. 2016. A topic based approach for sentiment analysis on twitter data. *International Journal of Advanced Computer Science and Applications* 7 (12).
35. Cambria, E., and A. Hussain. 2015. *Sentic computing: A common-sense-based framework for concept-level sentiment analysis*. Springer Publishing Company, Incorporated.
36. LeCun, Y., L. Bottou, Y. Bengio, and P. Haffner. 1998. Gradient-based learning applied to document recognition. *Proceedings of the IEEE* 86 (11): 2278–2324.
37. Hochreiter, S., and J. Schmidhuber. 1997. Long short-term memory. *Neural Computation* 9 (8): 1735–1780.
38. Cho, K., B. Van Merriënboer, C. Gulcehre, D. Bahdanau, F. Bougares, H. Schwenk, and Y. Bengio. 2014. Learning phrase representations using RNN encoder-decoder for statistical machine translation. arXiv preprint arXiv:1406.1078.
39. Chung, J., C. Gulcehre, K. Cho, and Y. Bengio. 2014. Empirical evaluation of gated recurrent neural networks on sequence modeling. arXiv preprint arXiv:1412.3555.
40. Bahdanau, D., K. Cho, and Y. Bengio. 2014. Neural machine translation by jointly learning to align and translate. arXiv preprint arXiv:1409.0473.
41. Akhtar, S., A. Kumar, A. Ekbal, and P. Bhattacharyya. 2016. A hybrid deep learning architecture for sentiment analysis. In *Proceedings of COLING 2016, the 26th International Conference on Computational Linguistics: Technical Papers* 482–493. Osaka, Japan: The COLING 2016 Organizing Committee.
42. Johnson R, and T. Zhang. 2015. Effective use of word order for text categorization with convolutional neural networks. In *Proceedings of the Conference of the North American Chapter of the Association for Computational Linguistics: Human Language Technologies (NAACL-HLT 2015)*.
43. Salinca, A. 2017. Convolutional neural networks for sentiment classification on business reviews. arXiv preprint arXiv:1710.05978.
44. Rani, S., and P. Kumar. 2018. Deep learning based sentiment analysis using convolution neural network. *Arabian Journal for Science and Engineering*.
45. Xu, J., D. Chen, X. Qiu, and X. Huang. 2016. Cached long short-term memory neural networks for document-level sentiment classification. In *Proceedings of the 2016 Conference on Empirical Methods in Natural Language Processing*.

46. Wang, Y., A. Sun, J. Han, Y. Liu, and X. Zhu. 2018. Sentiment analysis by capsules. In *Proceedings of the 2018 World Wide Web Conference on World Wide Web*, 1165–1174. International World Wide Web Conferences Steering Committee.
47. Zadeh, A., M. Chen, S. Poria, E. Cambria, and L.-P. Morency. 2017. Tensor fusion network for multimodal sentiment analysis. In *Proceedings of the 2017 Conference on Empirical Methods in Natural Language Processing*.
48. Nguyen, D., K. Vo, D. Pham, M. Nguyen, and T. Quan. 2017. A deep architecture for sentiment analysis of news articles. In *International Conference on Computer Science, Applied Mathematics and Applications*, 129–140. Cham: Springer.
49. Liu, P., X. Qiu, and X. Huang. 2016. Recurrent neural network for text classification with multi-task learning. arXiv preprint arXiv:1605.05101.
50. Zhou X, X. Wan, J. Xiao. 2016. Attention-based LSTM network for cross-lingual sentiment classification. In *Proceedings of the Conference on Empirical Methods in Natural Language Processing (EMNLP 2016)*.
51. Yang, Z., D. Yang, C. Dyer, X. He, A. Smola, and E. Hovy. 2016. Hierarchical attention networks for document classification. In *Proceedings of the 2016 Conference of the North American Chapter of the Association for Computational Linguistics: Human Language Technologies*, 1480–1489.
52. Baziotis, C., N. Pelekis, and C. Doulkeridis. 2017. Datastories at semeval-2017 task 4: Deep lstm with attention for message-level and topic-based sentiment analysis. In *Proceedings of the 11th International Workshop on Semantic Evaluation (SemEval-2017)*, 747–754.
53. Poria, S., I. Chaturvedi, E. Cambria, and A. Hussain. 2016. Convolutional MKL based multimodal emotion recognition and sentiment analysis. In *2016 IEEE 16th International Conference on Data Mining (ICDM)*, 439–448. IEEE.
54. Wang, B. 2018. Disconnected recurrent neural networks for text categorization. In *Proceedings of the 56th Annual Meeting of the Association for Computational Linguistics*, vol. 1: Long Papers, 2311–2320.
55. Chen, T., R. Xu, Y. He, and X. Wang. 2017. Improving sentiment analysis via sentence type classification using BiLSTM-CRF and CNN. *Expert Systems with Applications* 72: 221–230.
56. Cliche, M. 2017. BB_twtr at SemEval-2017 Task 4: Twitter sentiment analysis with CNNs and LSTMs. arXiv preprint arXiv:1704.06125.
57. Yin Y, Y. Song, M. Zhang. 2017. Document-level multi-aspect sentiment classification as machine comprehension. In *Proceedings of the Conference on Empirical Methods in Natural Language Processing (EMNLP 2017)*.
58. Xue, W, and T. Li. 2018. Aspect based sentiment analysis with gated convolutional networks. arXiv preprint arXiv:1805.07043.
59. Zhu, P., and T. Qian. 2018. Enhanced aspect level sentiment classification with auxiliary memory. In *Proceedings of the 27th International Conference on Computational Linguistics*, 1077–1087.
60. Wang, W., S.J. Pan, and D. Dahlmeier. 2018. Memory networks for fine-grained opinion mining. *Artificial Intelligence* 265: 1–17.
61. Huang, B., Y. Ou, and K.M. Carley. 2018. Aspect level sentiment classification with attention-over-attention neural networks. arXiv preprint arXiv:1804.06536.
62. Zheng, S, and R. Xia. 2018. Left-center-right separated neural network for aspect-based sentiment analysis with rotatory attention. arXiv preprint arXiv:1802.00892.
63. Wang, B, and W. Lu. 2018. Learning latent opinions for aspect-level sentiment classification. In *Thirty-Second AAAI Conference on Artificial Intelligence*.
64. Zhang, M., Y. Zhang, and D.T. Vo. 2016. Gated neural networks for targeted sentiment analysis. In AAAI, 3087–3093.
65. Poria, S., E. Cambria, and A. Gelbukh. 2016. Aspect extraction for opinion mining with a deep convolutional neural network. *Knowledge-Based Systems* 108: 42–49.
66. Chen, X, and C. Cardie. 2018. Multinomial adversarial networks for multi-domain text classification. arXiv preprint arXiv:1802.05694.

67. Ji, J., C. Luo, X. Chen, L. Yu, and P. Li. 2018. Cross-domain sentiment classification via a bifurcated-LSTM. In *Pacific-Asia Conference on Knowledge Discovery and Data Mining*, 681–693. Cham: Springer.

68. Ruder, S, and B. Plank. 2018. Strong baselines for neural semi-supervised learning under domain shift. arXiv preprint arXiv:1804.09530.

69. Li, Z, Y. Zhang, Y. Wei, Y. Wu, and Q. Yang. 2017. End-to-end adversarial memory network for cross-domain sentiment classification. In *Proceedings of the International Joint Conference on Artificial Intelligence (IJCAI 2017)*.

70. Li, Z., Y. Wei, Y. Zhang, and Q. Yang. 2018. Hierarchical attention transfer network for cross-domain sentiment classification. In *Proceedings of the Thirty-Second AAAI Conference on Artificial Intelligence, AAAI 2018*, New Orleans, Lousiana, USA, 2–7 Feb 2018.

71. Nam, H, and B. Han. 2016. Learning multi-domain convolutional neural networks for visual tracking. In *Proceedings of the IEEE Conference on Computer Vision and Pattern Recognition*, 4293–4302.

72. Baevski, A., Edunov, S., Liu, Y., Zettlemoyer, L. and Auli, M., 2019. Cloze-driven Pretraining of Self-attention Networks. arXiv preprint arXiv:1903.07785.

73. Johnson, R. and Zhang, T. 2017. Deep pyramid convolutional neural networks for text categorization. In *Proceedings of the 55th Annual Meeting of the Association for Computational Linguistics*, vol. 1: Long Papers, 562–570.

74. dos Santos, C. 2014. Think positive: Towards Twitter sentiment analysis from scratch. In *Proceedings of the 8th International Workshop on Semantic Evaluation (SemEval 2014)*, 647–651.

75. Deriu, J., M. Gonzenbach, F. Uzdilli, A. Lucchi, V.D. Luca, and M. Jaggi. 2016. Swisscheese at semeval-2016 task 4: Sentiment classification using an ensemble of convolutional neural networks with distant supervision. In *Proceedings of the 10th International Workshop on Semantic Evaluation* (No. CONF, 1124–1128).

76. Rouvier, M., and B. Favre. 2016. SENSEI-LIF at SemEval-2016 task 4: Polarity embedding fusion for robust sentiment analysis. In *Proceedings of the 10th International Workshop on Semantic Evaluation (SemEval-2016)*, 202–208.

77. Gray, S., A. Radford, and D.P. Kingma. 2017. Gpu kernels for block-sparse weights. arXiv preprint arXiv:1711.09224.

78. Howard, J., and S. Ruder. 2018. Universal language model fine-tuning for text classification. arXiv preprint arXiv:1801.06146.

79. Johnson, R., and T. Zhang. 2016. Supervised and semi-supervised text categorization using LSTM for region embeddings. arXiv preprint arXiv:1602.02373.

80. Baziotis, C., N. Pelekis, and C. Doulkeridis. 2017. Datastories at semeval-2017 task 4: Deep lstm with attention for message-level and topic-based sentiment analysis. In *Proceedings of the 11th International Workshop on Semantic Evaluation (SemEval-2017)*, 747–754.

81. Yu, Z., and G. Liu. 2019. Sliced recurrent neural networks (online) arXiv.org. Available at: https://arxiv.org/abs/1807.02291. Accessed 11 Apr 2019.

82. Xu, S., H. Liang, and T. Baldwin. 2016. Unimelb at semeval-2016 tasks 4a and 4b: An ensemble of neural networks and a word2vec based model for sentiment classification. In *Proceedings of the 10th International Workshop on Semantic Evaluation (SemEval-2016)*, 183–189.

83. Rouvier, M. 2017. LIA at SemEval-2017 task 4: An ensemble of neural networks for sentiment classification. In *Proceedings of the 11th International Workshop on Semantic Evaluation (SemEval-2017)*, 760–765.

84. Liu, X., P. He, W. Chen, and J. Gao. 2019. Multi-task deep neural networks for natural language understanding. arXiv preprint arXiv:1901.11504.

85. Tang, D., F. Wei, B. Qin, T. Liu, and M. Zhou. 2014. Coooolll: A deep learning system for twitter sentiment classification. In *Proceedings of the 8th International Workshop on Semantic Evaluation (SemEval 2014)*, 208–212.

86. Kar, S., S. Maharjan, and T. Solorio. 2017. RiTUAL-UH at SemEval-2017 task 5: Sentiment analysis on financial data using neural networks. In *Proceedings of the 11th International Workshop on Semantic Evaluation (SemEval-2017)*, 877–882.

87. Khalil, T., and S.R. El-Beltagy. 2016. Niletmrg at semeval-2016 task 5: Deep convolutional neural networks for aspect category and sentiment extraction. In *Proceedings of the 10th International Workshop on Semantic Evaluation (SEMEVAL-2016)*, 271–276.
88. Wang, W., S.J. Pan, D. Dahlmeier, and X. Xiao. 2016. Recursive neural conditional random fields for aspect-based sentiment analysis. arXiv preprint arXiv:1603.06679.
89. Socher, R., A. Perelygin, J. Wu, J. Chuang, C.D. Manning, A. Ng, and C. Potts. 2013. Recursive deep models for semantic compositionality over a sentiment treebank. In *Proceedings of the 2013 Conference on Empirical Methods in Natural Language Processing*, 1631–1642.
90. Liu, F., T. Cohn, and T. Baldwin. 2018. Recurrent entity networks with delayed memory update for targeted aspect-based sentiment analysis. arXiv preprint arXiv:1804.11019.
91. Patro, B.N., V.K. Kurmi, S. Kumar, and V.P. Namboodiri. 2018. Learning semantic sentence embeddings using pair-wise discriminator. arXiv preprint arXiv:1806.00807.
92. Brahma, S. 2019. Improved sentence modeling using suffix bidirectional LSTM. (online) Export.arxiv.org. Available at: http://export.arxiv.org/abs/1805.07340. Accessed 11 Apr 2019.
93. Peters, M.E., M. Neumann, M. Iyyer, M. Gardner, C. Clark, K. Lee, and L. Zettlemoyer. 2018. Deep contextualized word representations. arXiv preprint arXiv:1802.05365.
94. Saeidi, M., G. Bouchard, M. Liakata, and S. Riedel. 2016. Sentihood: Targeted aspect based sentiment analysis dataset for urban neighbourhoods. arXiv preprint arXiv:1610.03771.
95. Tang, D., B. Qin, and T. Liu. 2016. Aspect level sentiment classification with deep memory network. arXiv preprint arXiv:1605.08900.
96. He, R., W.S. Lee, H.T. Ng, and D. Dahlmeier. 2018. Effective attention modeling for aspect-level sentiment classification. In *Proceedings of the 27th International Conference on Computational Linguistics*, 1121–1131.
97. Wang, Y., M. Huang, and L. Zhao. 2016. Attention-based LSTM for aspect-level sentiment classification. In *Proceedings of the 2016 Conference on Empirical Methods in Natural Language Processing*, 606–615.
98. He, R., W.S. Lee, H.T. Ng, and D. Dahlmeier. 2018. Exploiting document knowledge for aspect-level sentiment classification. arXiv preprint arXiv:1806.04346.
99. Ma, D., S. Li, X. Zhang, and H. Wang. 2017. Interactive attention networks for aspect-level sentiment classification. arXiv preprint arXiv:1709.00893.
100. Ma, Y., H. Peng, and E. Cambria. 2018. Targeted aspect-based sentiment analysis via embedding commonsense knowledge into an attentive LSTM. In *Thirty-Second AAAI Conference on Artificial Intelligence*.
101. Li, Z., Y. Wei, Y. Zhang, X. Zhang, X. Li, and Q. Yang, Q. 2018. Exploiting coarse-to-fine task transfer for aspect-level sentiment classification. arXiv preprint arXiv:1811.10999.
102. Li, L., Y. Liu, and A. Zhou. 2018. Hierarchical attention based position-aware network for aspect-level sentiment analysis. In *Proceedings of the 22nd Conference on Computational Natural Language Learning*, 181–189.
103. Moro, G., A. Pagliarani, R. Pasolini, and C. Sartori. 2018. Cross-domain & in-domain sentiment analysis with memory-based deep neural networks. In *Proceedings of the 10th International Joint Conference on Knowledge Discovery, Knowledge Engineering and Knowledge Management*.
104. Aggarwal, U., and G. Aggarwal. 2017. Sentiment Analysis: A Survey. *International Journal of Computer Sciences and Engineering* 5 (5): 222–225.
105. Chen, P., Z. Sun, L. Bing, and W. Yang. 2017. Recurrent attention network on memory for aspect sentiment analysis. In *Proceedings of the 2017 Conference on Empirical Methods in Natural Language Processing*, 452–461.
106. Liu, Q., Y. Zhang, and J. Liu. 2018. Learning domain representation for multi-domain sentiment classification. In *Proceedings of the 2018 Conference of the North American Chapter of the Association for Computational Linguistics: Human Language Technologies*, (Long Papers), vol. 1, 541–550.

Deep Learning Adaptation with Word Embeddings for Sentiment Analysis on Online Course Reviews

Danilo Dessí, Mauro Dragoni, Gianni Fenu, Mirko Marras and Diego Reforgiato Recupero

Abstract Online educational platforms are enabling learners to consume a great variety of content and share opinions on their learning experience. The analysis of the sentiment behind such a collective intelligence represents a key element for supporting both instructors and learning institutions on shaping the offered educational experience. Combining Word Embedding representations and deep learning architectures has made possible to design sentiment analysis systems able to accurately measure the text polarity on several contexts. However, the application of such representations and architectures on educational data still appears limited. Therefore, considering the over-sensitiveness of the emerging models to the context where the training data is collected, conducting adaptation processes that target the e-learning context becomes crucial to unlock the full potential of a model. In this chapter, we describe a deep learning approach that, starting from Word Embedding representations, measures the sentiment polarity of textual reviews posted by learners after attending online courses. Then, we demonstrate how Word Embeddings trained on smaller e-learning-specific resources are more effective with respect to those trained on bigger general-purpose resources. Moreover, we show the benefits achieved by combining Word Embeddings representations with deep learning architectures instead of common machine learning models. We expect that this chapter will help stakeholders to get a clear view and shape the future research on this field.

D. Dessí (✉) · G. Fenu · M. Marras · D. Reforgiato Recupero
Department of Mathematics and Computer Science, University of Cagliari, Cagliari, Italy
e-mail: danilo_dessi@unica.it

G. Fenu
e-mail: fenu@unica.it

M. Marras
e-mail: mirko.marras@unica.it

D. Reforgiato Recupero
e-mail: diego.reforgiato@unica.it

M. Dragoni
Fondazione Bruno Kessler, Trento, Italy
e-mail: dragoni@fbk.eu

© Springer Nature Singapore Pte Ltd. 2020 57
B. Agarwal et al. (eds.), *Deep Learning-Based Approaches for Sentiment Analysis*, Algorithms for Intelligent Systems, https://doi.org/10.1007/978-981-15-1216-2_3

Keywords Big data · E-learning · Deep learning · Online education · Sentiment analysis · Word Embeddings · Domain adaptation

1   Introduction

The advent of Social Web has enabled the development and the sharing of experiences among people around the world. Individuals use online social platforms to express opinions about products and/or services in a wide range of domains, influencing the point of view and the behavior of their peers. Such user-generated data, which generally come in form of text (e.g., reviews, tweets, wikis, blogs), is often characterized by a positive or negative polarity according to the satisfaction of people who write the content. Understanding individual's satisfaction is a key element for businesses, policy-makers, organizations, and social institutions to hear and act on the voice of people. The automatic investigation performed on top of person's opinions in order to detect subjective information usually relies on sentiment analysis (SA). Techniques and systems in this field aim to identify, extract, and classify emotions and sentiments by combining Natural Language Processing (NLP), Text Mining, and Computational Linguistics [18]. Exploiting this artificial intelligence makes possible to replace or complement common practices (e.g., focus groups or surveys) for uncovering opinions, but still presents challenges because of large sources and context dependencies of data.

SA approaches can be classified into supervised and unsupervised. Supervised approaches require a training dataset annotated with numerical values left by users or inferred from the content in the text (e.g., emoticons), and leverage it to build a classifier which predicts a sentiment score for unseen data. Common supervised pipelines first require the extraction of features from the input text, such as terms frequencies, parts of speech, and emotional words. Such features are then fed into an algorithm which characterizes each input text with a positive or negative polarity, i.e., sentiment detection. On the other hand, unsupervised approaches rely on lexicons associated with sentiment scores in order to model the sentiment polarity of a given text. While both types of approaches have been shown feasible on the sentiment detection task, they tend to suffer from over-sensitiveness to the context where the training data is collected. This results in lower performance when applied to other contexts. Hence, understanding how SA techniques work in emerging areas, such as online education, becomes crucial.

Online educational platforms deployed at large scale, such as Coursera[1] and Udemy,[2] are earning more and more attention as social spaces where students can discover and consume a great variety of contents about many topics, and share opinions regarding their educational experience [13]. Such collective intelligence might be useful for various stakeholders, including peers who are planning to attend a given

---

[1]https://www.coursera.org/.
[2]https://www.udemy.com/.

course, instructors who are interested in improving their teaching practices and increasing students' satisfaction, and providers who can get benefits from the feedback left by users to refine tools and services in the platform itself. With this in mind, these platforms can be envisioned as dedicated social media where discussions are limited to specific topics concerning course content quality, teachers' skills, and so on [7]. SA approaches on students' opinions have recently started to receive the attention of the involved stakeholders [19], and their design and development is still an open challenge.

The most prominent SA solutions leverage Word Embeddings, i.e., distributed representations that model words properties in vectors of real numbers which capture syntactic features and semantic word relationships. These resources have been shown useful in NLP tasks, like Part-Of-Speech (POS) tagging [25] and Word Analogy [28], and they have been also exploited for SA [15, 20, 23, 24, 40]. The generation of Word Embeddings is based on distributional co-occurrences of adjacent words able to model words meanings that are not visible from their surface. This exploits the fact that words with a similar meaning tend to be connected by a given relation. For instance, the verbs *utilize* and *use*, which are synonym although syntactically different, present similar sets of co-occurring words and can be considered similar, while a third verb, such as *play*, has different co-occurrences and should be considered different from both *utilize* and *use*. The literature acknowledges that Word Embeddings generated from general-purpose datasets, independently from any specific domain, under-perform the ones built on texts from the target context of the SA task [17]. This happens because context-trained Word Embeddings may capture specific patterns from the target context, while generic-trained ones might learn patterns acting as noise for the target context. Hence, training Word Embeddings tailored for the e-learning context is crucial to reach high SA performance.

Deep learning (DL) has emerged as subclass of the machine learning (ML) area, where various neural network approaches are combined together for pattern classification and regression tasks. It usually employs multiple layers able to learn complex data representation, increasingly higher level features, and to correctly classify or measure properties held by data. The success of DL is due to the new advances in the ML field as well as to the ever more increasing computational abilities of computers through the use of Graphical Processing Units (GPUs) [10]. DL has shown improvements in addressing SA tasks [2, 20, 23]. Using DL models powered by Word Embeddings trained on the e-learning context can enable improving the effectiveness of the dedicated SA systems.

In this chapter, we first discuss the state-of-the-art literature on DL and Word Embeddings for SA (Sect. 2). We provide a high-level description of the existing Word Embeddings generation algorithms (Sect. 3) and common DL layers and networks (Sect. 4). Then, we propose a DL model, which is an extension of our preliminary work in [11], trained on Word Embedding representations coming from the e-learning context and able to predict a sentiment score for reviews posted by learners (Sect. 5). We also report its experimental evaluation on a large-scale dataset of online course reviews (Sect. 6). We show how Word Embeddings trained on smaller context-specific textual resources are more effective with respect to those trained on bigger general-purpose resources. Moreover, we highlight the benefits derived from the combination of Word Embeddings and DL instead of common ML approaches. Finally, conclusions, open challenges, and future directions are discussed (Sect. 7). Code and models accompanying this chapter would support the development of next-generation sentiment-aware e-learning platforms.[3]

2   State of the Art

In this section, we discuss the literature on DL and SA with a particular focus on the e-learning domain. The reader notices that we discuss separately the use of DL methods and Word Embeddings although they might be adopted together.

2.1   Sentiment Analysis in E-Learning Systems

E-learning domain has recently gained the attention of SA in order to get knowledge from new dynamics that e-learning platforms allow to. By leveraging students' emotions, it might be possible contributing to increase the students' motivation and improve learning processes. More specifically, the study of sentiments in a e-learning platform can contribute to learning and teaching evaluation, investigate how technology can influence students' learning process, evaluate the use of e-learning tools, and improve learning content recommendations [35].

The measurement of sentiments and emotions in such platforms should be as less invasive as possible for not disturbing the learning process and not influencing the overall opinion. The adoption of textual reviews left by students' for analyzing sentiments and emotions is one of the less invasive techniques, because data collection and analysis are completely transparent for students. In literature, various scenarios where SA was used to study learning aspects using textual reviews can be found. For instance, one work embraces the use of sentiment analysis to mine students' interactions in collaborative tools, guaranteeing the students' communication privacy in their opinions [9]. Another relevant area that exploited SA was the teachers' assessment. For example, the authors in [19] adopted a support vector machine to evaluate the teachers' performance using 1040 comments of systems engineering students as a dataset. The evaluation of the teaching–learning process was also object of study by means of SA in [8]. Its authors adopted comments posted by both students and teachers. Similarly, the authors in [36] studied text sentiment to build an adaptive learning environment with improved recommendations. For example, they described how to choose a learning activity for a student based on his/her goals and emotional profile.

---

[3]Please find code and models at https://github.com/mirkomarras/dl-sentiment-coco.

The study of SA within the e-learning domain is still an open research area. With this paper, we aim to make a contribution in this direction by designing an effective DL model for improving sentiment detection in students' reviews.

2.2 Deep Learning for Sentiment Analysis

ML has been extensively adopted for SA tasks, using different types of algorithm and fitting various types of extracted features. For example, authors in [39] used maximum entropy (ME) and Naive Bayes (NB) algorithms, adopting syntactical and semantic patterns extracted from words on Twitter. Their method relies on the concept of contextual semantic, i.e., considering word co-occurrences in order to extract the meaning of words [46]. In the evaluation on nine Twitter datasets, they obtained better performance when the ML algorithms were trained with their method both at tweet and entity levels. More recently, authors in [45] applied NB, ME, stochastic gradient descent (SGD), and support vector machine (SVM) algorithms to classify movies reviews in a binary classification problem (i.e., positive or negative evaluation of reviews). They showed that the use of a n-gram model to represent reviews with the above algorithms obtains higher levels of accuracy when the value n was small. Moreover, they showed that combining uni-gram, bi-gram, and tri-gram features enabled to enhance the accuracy of the method against the use of a single representation at once. ML methods rely on lexical syntactical features representations which are derived from text, not considering semantic relationships that can occur between words. Hence, in spite of feature engineering advancements, there has been a growth of techniques to infer semantics as DL that has emerged as an effective paradigm to automatically learn continuously information from text. The first DL approaches were studied at the begin of years 1990, but, due to high computational costs, they lost interest among scientific communities [48]. However, in the last years, more and more powerful computers and a huge availability of big data, DL approaches became state-of-the-art solutions across various domains.

SA domain also experienced the influence of the wide spread of DL approaches. For example, in [15] a convolutional neural network (CNN) composed by two layers was designed to capture features from character to sentence level. An ensemble DL method was proposed by [1], where various sentiment classifiers trained on different sets of features were combined. They performed experiments on six different datasets coming from Twitter and movies reviews. With their experiments, they improved the state of the art against DL baselines. Another approach to combine various classifiers with DL ones was also proposed by authors in [31], where a SVM classifier was mounted on top of a seven-layer CNN in order to complement the characteristics of each other and obtain a more advanced classifier. With their variation, they were able to obtain more than 3% of improvement compared to a classic DL model. To learn continuous representations of words for SA, a combination of CNN with a long short-term memory (LSTM) was exploited by [42]. They were able to assign fixed-length vectors to sentences of varying lengths, showing how DL approaches outperform common ML algorithms. Although DL models showed improvements in many domains, they have not been deeply studied for applications in e-learning.

2.3   Word Embeddings for Sentiment Analysis

Word Embeddings have been successfully used in various domains, ranging from behavioral targeting [4] to SA. Within the latter, they have been widely employed for improving accuracy of baselines methods not using Word Embeddings. As traditional Word Embeddings methods do not usually take into account words distributions for a specific task, resulting representations might lose important information for a given task. In the context of SA, authors in [24] incorporated prior knowledge at both word and document levels with the aim to investigate how contextual sentiment was influenced by each word. On the same direction, other researchers [43] employed sentiment of text for the generation of Words Embeddings. In particular, they joined context semantics and sentiment characteristics so that in the embedding model neighboring words have both a similar meaning and sentiment. The rationale behind that depends on the fact that many words with a similar context are usually mapped on similar vector representations even if they have an opposite sentiment polarity (e.g., *bad* and *good*). Similarly, authors in [49] augmented sentiment information into semantic word representations and extended Continuous Skip-gram model (Skip-gram), coming up with two sentiment Word Embedding models. The learned sentiment Word Embeddings were able to correctly represent sentiment and semantics. Furthermore, authors in [26] presented a model that uses a mix of unsupervised and supervised techniques to learn word vector representations, including semantic term-document features. The model showed performances higher than several ML methods adopted for sentiment detection. Focusing on Twitter sentiment classification, authors in [44] trained sentiment-sensitive Words Embeddings through the adoption of three neural networks designed to detect the sentiment polarity of texts. Their methods encoded sentiment information in the continuous representation of words, and experiments on a benchmark Twitter classification dataset in SemEval 2013 showed that it outperformed the competitors. Last but not least, authors in [38] described a procedure with Word Embeddings for the estimation of levels of negativity in a sample of 56,000 plenary speeches from the Austrian parliament. They found out that the different levels of negativity shown by speakers in different roles from government or opposition parties agree with expected patterns indicated by common sense hypotheses. Their results showed that the Word Embeddings approach offers a lot of potential for SA in social sciences.

Several challenges have been created to solve SA polarity detection task and several resulting winning systems employed Word Embeddings within their core. For example, the semantic sentiment analysis challenge [16, 32–34], held within the

ESWC conference, reached its fifth edition[4] [6]. The 2018 edition included a polarity detection task where participants were asked to train their systems by using a combination of Word Embeddings already generated by the organizers. The aim was to both validate the quality of their systems (precision/recall analysis) and detect which combination of embeddings worked better. SemEval is a workshop on semantic evaluation that takes place each year and includes a set of tasks in NLP and Semantic Web (e.g., SA polarity detection). One participant of the SemEval-2018 edition targeted the task of irony detection in Twitter [47]. It employed a simple neural network architecture of multi-layer perceptron with various types of input features, including lexical, syntactic, semantic, and polarity features. The proposed system used 300-dimensional pre-trained Word Embeddings from GloVe [30] to compute a tweet embedding as the average of the embeddings of words in the tweet. By applying latent semantic indexing and extracting tweet representation through the Brown clustering algorithm, it achieved high performance in both subtasks of binary and multi-class irony detection in tweets. It ranked third using the accuracy metric and fifth using the F_1 metric. Kaggle[5] is the world's largest community of data scientists and offers ML competitions, a public data platform, and a cloud-based workbench for data science. It hosts several challenges, and some were related to SA. For instance, the *Sentiment Analysis on Movie Reviews* challenge[6] asked participants to label the movie reviews collected in the Rotten Tomatoes dataset [29] on a scale of five values: negative, somewhat negative, neutral, somewhat positive, positive. One recent challenge, namely *Bag of Words Meets Bags of Popcorn*,[7] looked for DL models combined with Word Embeddings for polarity detection of movie reviews collected by authors in [26].

3 Word Embedding Representations for Text Mining

Before using words in a model, they should be encoded as numbers. For instance, a function can be used to map words to integers or to one-hot encode words. When applying such an encoding to words, sparse vectors of high dimensionality are commonly obtained. On large datasets, this could cause performance issues. Additionally, such encoding functions do not take into account the semantics of the words. On the other hand, Word Embeddings are dense vectors with lower dimensionality, and the semantic relationships between words are reflected in the distance and direction of the vectors. Each word is positioned into a multi-dimensional space whose dimensions can be empirically chosen. The vector values for a word represent its position in this embedding space. Synonyms are found close to each other while words with opposite

---

[4]http://www.maurodragoni.com/research/opinionmining/events/challenge-2018/.

[5]https://www.kaggle.com/.

[6]https://www.kaggle.com/c/sentiment-analysis-on-movie-reviews.

[7]https://www.kaggle.com/c/word2vec-nlp-tutorial.

meanings have a large distance between them. In this section, we introduce the most representative and recent Word Embedding generator algorithms, highlighting their pros and cons.

3.1 Word2Vec

The *Word2Vec* Word Embedding generator [27] aims to detect the meaning and semantic relations between words by exploiting the co-occurrence of words in documents belonging to a given corpus. The core idea is to capture the context of words, using ML approaches such as recurrent or deep neural networks. In order to eliminate noise, *Word2Vec* operates on a corpus of sentences by constructing a vocabulary based on the words that appear in the corpus more often than a user-defined threshold. Then, it trains either the Continuous Bag-Of-Words (CBOW) or the Skip-gram algorithm on the input documents to learn the word vector representations. In this chapter, we will consider the Skip-gram algorithm since it works well with small amount of training data and represents well even rare words or phrases.

3.2 GloVe

The *GloVe* [30] Word Embedding generator is a unsupervised learning algorithm developed by Stanford. It creates Word Embeddings by aggregating global word–word co-occurrence matrices from a corpus. The resulting embeddings show interesting linear substructures of the word in the vector space. More precisely, the algorithm consists of collecting word co-occurrence statistics in a form of word co-occurrence matrix. Each element of this matrix represents how often the word i appears in context of word j. The corpus is scanned in the following manner: For each term, it looks for context terms within some area defined by a $windowsize$ before the term and a $windowsize$ after the term, and it gives less weight for more distant words.

3.3 FastText

The *FastText* [22] Word Embedding generator is an algorithm for learning word representations. It differs form the previous ones in the sense that word vectors as the ones learned in *Word2Vec* treat every single word as the smallest unit whose vector representation is to be found, while *FastText* assumes a word to be formed by n-grams of character. This new representation of a word is helpful to find the vector representation for rare words. Since rare words could still be broken into character n-grams, they could share these n-grams with the common words. This can help to manage vector representations for words not present in the dictionary since they can also be broken down into character n-grams. Character n-grams embeddings tend to perform superior to *Word2Vec* and *GloVe* on smaller datasets [3].

3.4 Intel

Intel proposes to improve the data structures in *Word2Vec* through the use of mini-batching and negative sample sharing, allowing to solve the neural Word Embedding generation problem using matrix multiply operations [21]. They explored different techniques to distribute *Word2Vec* computation across nodes in a cluster and demonstrate strong scalability. Their algorithm is suitable for modern multi-core/many-core architectures and allows scaling up the computation near linearly across cores and nodes, and processing millions of words per second. *Intel* embeddings generally differ from *Word2Vec* embeddings since the number of updates of the model is different across these two implementations, and the convergence is not equal for the same the number o f epochs.

4 Deep Learning Components for Text Mining

DL has recently emerged as a new area within ML research. It embraces information processing methods consisting of a sequence of complex nonlinear models. Each model forms a layer that independently processes data. The output of a layer is fed as an input to the subsequent layer in the sequence until the final output is obtained. In this section, we provide a high-level overview of the most popular layers and networks used or combined for mining texts and, thus, useful for conducting SA.

4.1 Feed-Forward Neural Network (FNN)

Feed-forward neural networks (FNNs) were one of the first and simplest components applied to learn from data using DL paradigms. One or more levels of nodes, often called perceptrons, are randomly joined by weighted connections in a many-to-many fashion. These networks were historically thought in order to simulate a biological model where nodes are neurons and links between them represent synapses. For this reason, they are also called multi-layer perceptron (MLP) networks. On the basis of the input values fed into the network, nodes of a certain level can be activated and their signal is broadcasted to the subsequent level. In order to activate nodes of a subsequent level, the signal generated at a given level is weighted and must be greater than a given threshold. Weights are generally initialized with random values and adjusted during training in order to minimize a predefined objective function. This family of networks has been proved to be useful for pattern classification, but

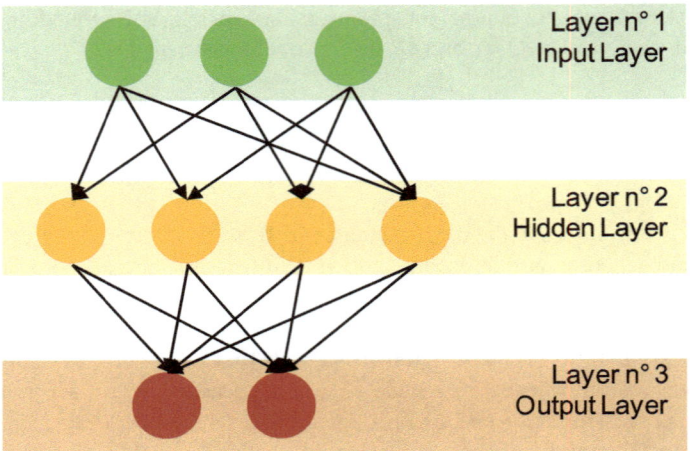

Fig. 1 An example of a feed-forward network composed by three layers

less suitable for labeling sequences since it does not take into account the sequence of input data. A simple three-layer feed-forward neural network is shown in Fig. 1.

The sample FNN as it is accepts three-dimensional inputs and returns two-dimensional outputs. Each node of a given level is connected to nodes of the subsequent layer. The input data is fed into the network by means of Layer 1, which acts as input layer, and then sent to the first hidden layer, i.e., Layer 2. The output of this layer is finally propagated to Layer 3, which represents the output layer. The action to move data from a layer to another by activating or not the corresponding nodes is generally called *forward pass* of the network.

4.2 Recurrent Neural Network (RNN)

Recurrent neural networks (RNNs) are tailored for processing data as a sequence. In contrast to FNNs which commonly pass the input data directly from input to output nodes, RNNs have cyclic or recurrent connections among nodes of distinct levels. This makes possible to model the output of the network by taking into account the entire history of the received input data. Recurrent connections connect past data with the one that is currently being processed, simulating a state memory. The forward pass is similar to the one in FNNs. The unique difference is that the activation of a node depends on both the current input and the previous status of the hidden layers. The text is fed into the network by means of vector representations that are recurrently processed. This means that RNNs view a sample text as an ordered sequence of word identifiers, differently from common FNNs working on hand-crafted inputs, e.g., Bag of Words (BOW).

In a wide range of applications, data can present patterns from the past to the future and vice versa. For instance, for classifying the sections of a given story, it could be useful to have access to both future and past sections. However, the future content of a text is ignored by common FNNs and RNNs, as they work sequentially. Bidirectional RNNs (BiRNNs) let the network, at a given point in time, to take information from both earlier and later data in the sequence, going beyond the exposed limitations. The idea behind this kind of networks consists of presenting the training data forward and backward by two hidden RNNs which are then combined into a common output layer. This strategy makes possible to find patterns that can be learned from both past and future history of data.

4.3 Long Short-Term Memory (LSTM) Network

Long short-term memory (LSTM) networks are an RNN extension designed to work on sequence problems and that has achieved state-of-the-art results on challenging prediction tasks. LSTM networks employ recurrent connections and add memory blocks in their recurrent hidden layers. These memory blocks save the current temporal state during training and make possible to learn temporal observations hidden in the input data. The fact of using connections as a memory implies that the output of a LSTM network depends on the entire history of the training data, not only on the current input sample. Moreover, using memory blocks allows to relate the current data that is being processed with the data processed long before, solving the problem experienced by common RNNs. For this reason, LSTM networks have had a positive impact on sequence prediction tasks. As stated for RNN, a bidirectional layer using two hidden LSTMs can be leveraged to process data both forward and backward.

4.4 Convolutional Neural Network (CNN)

Convolutional neural networks (CNNs) typically perform filtering operations on the input nodes of a layer, abstracting and selecting only meaningful input features. When CNNs are trained, the weights of links between nodes acting as filter are defined. Such networks have been historically applied in the computer vision field and, hence, are not directly applicable on texts as they are. To overcome this limitation, a text must be converted into a vector representation, and the convolutional filters are applied on this representation. A convolutional filter is composed by a kernel that slides on the vector representation and repeats the same function on each element until all the vectors have been covered. In Text Mining, CNNs can be useful to detect the words characterizing a classes.

4.5   Normalization Layer (NL)

During training, the output of a given layer is affected by parameters and processes used in previous layers. Small changes in the parameters set for a layer can hence be propagated as the network becomes deeper, resulting in large changes in the final output, i.e., the output of the last layer. Considering that the parameters are continuously changed to better fit the prediction task and that the data distribution changes across levels, the variations within the parameters can negatively influence the training, making it computationally expensive. To shape input data with a standard distribution and improve training performance, some normalization layers (NLs) can be introduced. One of the most common normalization layers is represented by *Batch Normalization*. This layer makes possible to reduce the dependence of the optimization parameters from the input values, avoiding over-fitting and making the training process more stable.

4.6   Attention Layer (AL)

Attention layers (ALs) are often adopted before the last fully connected layers of a model. Attention mechanisms in neural networks serve to orient perception as well as memory access. Attention layers filter the perceptions that can be stored in memory and filter them again on a second pass when they need to be retrieved from memory. Neural networks can allocate attention, and they can learn how to do so, by adjusting the weights they assign to various inputs. This makes possible to solve traditional limits in various NLP tasks. For instance, traditional word vectors presume that a words meaning is relatively stable across sentences. However, there could be massive differences in meaning for a single word, e.g., lit (an adjective that describes something burning) and lit (an abbreviation for literature); or get (a verb for obtaining) and get (an animals offspring). ALs can capture the shades of meaning for a given word that only emerge due to its situation in a passage and its inter-relations with other words. Moreover, ALs learn how to relate an input sequence to the output of the model in order to pay selective attention on more relevant input features. For example, in order to reflect the relation between inputs and outputs, an attention layer may compute an arithmetic mean of results of various layers according to a certain relevance.

4.7   Other Layers

There are also other layers that can be leveraged in order to fine-tune the performance of a model. The most representative ones are described below.

Embedding layer. An embedding layer turns positive integers (indexes) into dense vectors of fixed size chosen from a pre-initialized matrix. For an integer-encoded text, the dense vector corresponding to those integers is selected.

Noise layer. A noise layer is usually employed to avoid model over-fitting. It consists in modifying a fraction of input of layers, adding and subtracting some values following a predefined distribution (e.g., Gaussian).

Dropout layer. A dropout layer may be seen as a particular type of a noise layer. It assigns the value 0 to a randomly chosen fraction of its input data. The name *dropout* comes from the action to dropping some units of the input.

Dense layer. A dense layer is a densely connected layer that is used to map large unit inputs in a few unit results. For example, it may be used to define the number of classes that a model returns, mapping hundred and thousand nodes in a few number of classes.

5 Our Sentiment Predictor for E-Learning Reviews

This section serves as guide on designing sentiment prediction models for educational reviews and presents practical information on how to implement such systems. The main components of the proposed solution (Fig. 2) and the benefits of the designing choices for each of these components will be described. This helps readers have a broader view of difficulties and solutions behind sentiment prediction models and make appropriate decisions during their design.

5.1 Review Splitting

The *review splitting* step serves to define the various input dataset splits while developing the sentiment prediction model. Firstly, we consider the input dataset as a set D of N reviews organized as follows:

$$D = \{(text_1, score_1), \ldots, (text_N, score_N)\} \tag{1}$$

where $text_i$ is a textual review and $score_i$ is an integer rating belonging to the set $C = \{score_1, \ldots, score_M\}$.

During this step, we thus need to split input data D in three subsets, each for a specific phase of the development:

1. $D_{creation}$: the sample of data used to create Word Embeddings.
2. D_{train}: the sample of data used to fit the model (i.e., weights and biases).
3. D_{test}: the sample of data used as the gold standard to evaluate the model.

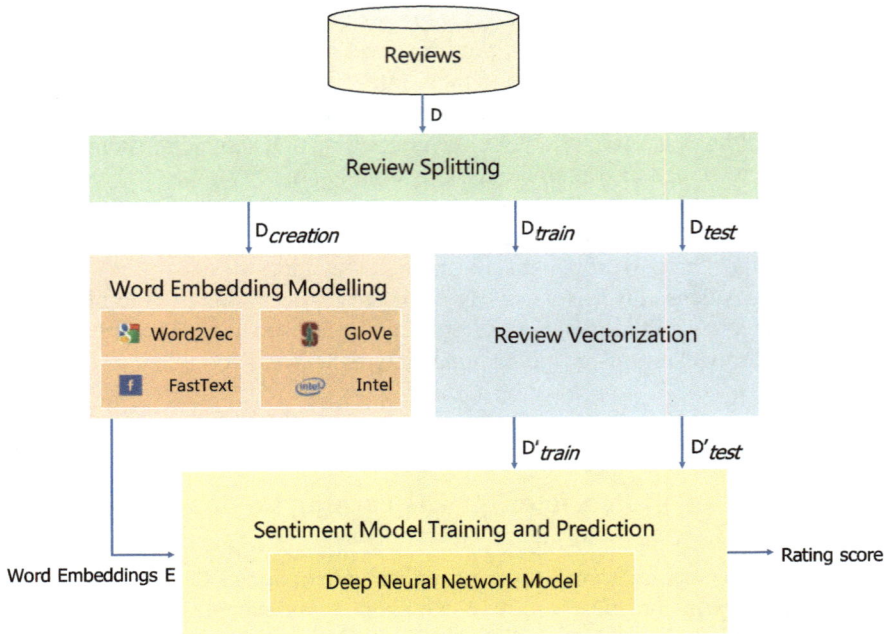

Fig. 2 Base components of our sentiment prediction model

In order to do this, we set up two split ratios and we assign the text-score pairs in D to the different subsets $D_{creation}$, D_{train}, D_{test} according to them:

1. $s_{creation} \in [0, 1]$: the percentage of reviews for each class $c \in C$ that are randomly chosen from the set D to create Word Embeddings, yielding $D_{creation}$.
2. $s_{training} \in [0, 1]$: the percentage of reviews for each class $c \in C$ that are randomly chosen from $D \setminus D_{creation}$ to train the model, yielding D_{train}.

The remaining reviews represent D_{test}. The overall procedure ensures that the subsets are disjoint and their union covers the entire dataset D.

5.2 Word Embedding Modeling

The state-of-the-art method to model a word with a vector is using Word Embeddings; it is common to see Word Embeddings that are 256-dimensional, 512-dimensional, or 1024-dimensional when dealing with very large vocabularies. There are two ways to generate and leverage Word Embeddings:

1. Learn Word Embeddings jointly with the same context we are interested in by starting with random word vectors and, then, learning word vectors along the process, iteratively.
2. Load into the sentiment prediction model the Word Embeddings pre-computed using a different ML task than the one we are interested in. If the amount of training data in D_{train} is small, this is the common solution.

To the best of our knowledge, no Word Embedding database specifically targets the e-learning context. Therefore, this step goes through the first most general solution of learning Word Embeddings from scratch, while we also use Word Embedding pre-computed on other contexts for comparison along the chapter.

In order to generate Word Embeddings from scratch, the subset of pre-processed reviews $D_{creation}$ was employed. We concatenated them into a large corpus, and this corpus was fed into a given Word Embedding generation algorithm selected among the following ones: *Word2Vec*, *GloVe*, *FastText*, or *Intel*. Each of them outputs a set of feature vectors E for words in that corpus. The feature values are non-negative real numbers. For each distinct word w in the vocabulary in $D_{creation}$, there exists a corresponding feature vector $e \in E$ which represents the Word Embedding for that word. All the feature vectors share the same size. The size of the resulting Word Embeddings and of the window where Word Embeddings generator algorithms look at contextual words can be arbitrarily selected.

5.3 Review Vectorization

The *review vectorization* is the process of transforming each review into a numeric sequence. This can be done in multiple ways (e.g., segment text into words and transform each word into a vector, segment text into characters and transform each character into a vector, extract n-grams of words or characters, and transform each n-gram into a vector). The different units into which the text is broken (words, characters, or n-grams) are called *tokens*, and breaking text into such tokens is called *tokenization*. The process consists of applying some tokenization schemes and then associating numeric vectors with the generated tokens. These vectors, packed into sequences, are needed for manipulating text during sentiment model training and inference.

In order to be treated by machines, we need to turn the datasets D_{train} and D_{test} into a set of integer-encoded pre-processed reviews defines as follows:

$$D'_{train} = \{(text'_1, score_1), \ldots, (text'_K, score_K)\} \forall (text_i, score_i) \in D_{train} \quad (2)$$

$$D'_{test} = \{(text'_1, score_1), \ldots, (text'_J, score_J)\} \forall (text_i, score_i) \in D_{test} \quad (3)$$

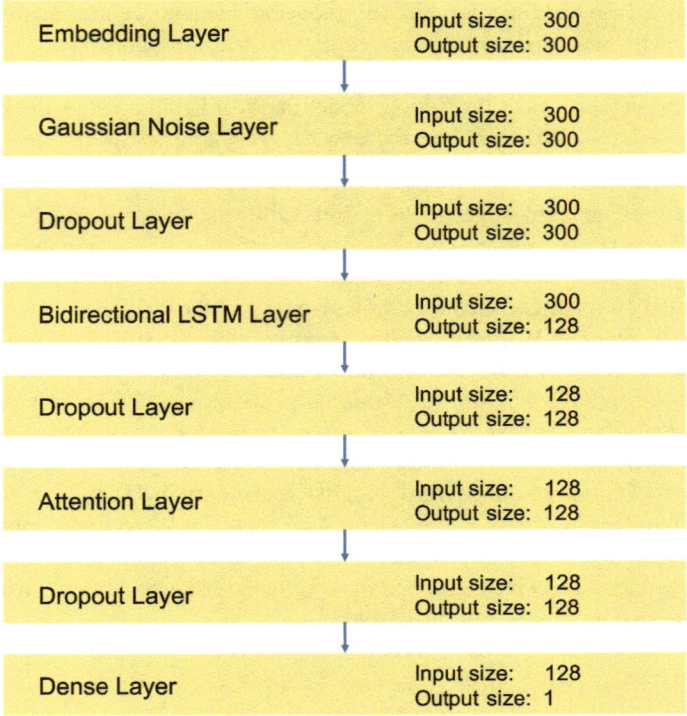

Fig. 3 Proposed deep learning model for sentiment score regression designed to leverage 300-dimensional input text sequences

where each pair $(text_i', score_i)$ includes an integer encoding of the text comment $text_i$ and the original rating $score_i$ from D_{train} and D_{test}, respectively.

The process for generating D_{train}' and D_{test}' works as follows. Each word has a unique associated integer value chosen from a range going from 0 to $|V| - 1$, where V is the vocabulary of words in D. For each input review $(text_i, score_i)$, we build an integer-encoded vector $text_i'$ from $text_i$, where an integer value at position j in $text_i'$ represents the mapped value for word w for that position in $text_i$. The sets D_{train}' and D_{test}' are thus vectorized.

5.4   Sentiment Model Definition

This step is necessary for defining the architecture of the deep neural network which takes pairs of integer-encoded texts and sentiment scores, maps such texts into Word Embeddings, and tries to predict the sentiment score from them.

The proposed architecture tailored for sentiment score prediction is shown in Fig. 3. Given that our training process requires running the network on a rather large corpus, our design choices are mainly driven by the computational efficiency of the network. Hence, differently from [2], which presents an architecture with two bidirectional LSTM layers, we adopt a single bidirectional LSTM layer architecture. Moreover, we configure the last layer to return a single continuous value, i.e., the predicted sentiment score. Therefore, our network is composed by an embedding layer followed by a bidirectional LSTM layer, a neural attention mechanism, and a dense layer. Each layer works as follows:

1. **Embedding Layer** takes a two-dimensional tensor of shape (N, M) as input, where N represents the number of integer-encoded text comment samples, while M the maximum sequence length of such samples. Each entry is a sequence of integers passed by the input layer. The output of the embedding layer is a two-dimensional vector with one embedding for each word w in the input sequence of words of each text comment t. Before receiving data, the embedding layer loads the pre-trained Word Embeddings computed during the previous step as weights. Such weights are frozen, so that the pre-trained parts are not updated during training and testing to avoid forgetting what they already know.

2. **Bidirectional LSTM Layer** is an extension of the traditional LSTM that generally improves model performance on sequence classification problems. It trains two LSTM instead of just one: The first is trained on the input sequence as it is and the second on a reversed copy of the input sequence. The forward and backward outputs are then concatenated before being passed on to the next layer, and this is the method often used in studies of bidirectional LSTM. Through this layer, the model is able to analyze a reviews as a whole, binding first and last words coming up with a more precise score. Moreover, exploiting the bidirectional version of a LSTM, the model is able to get patterns that depend on the learners' writing style.

3. **Attention Layer** enables the network referring back to the input sequence, instead of forcing it to encode all the information forward into one fixed-length vector. It takes n arguments $y_1, ..., y_n$ and a context c. It returns a vector z which is supposed to be the summary of the y_i, focusing on information linked to the context c. More specifically, in our model it returns a weighted arithmetic mean of the y_i, and the weights are chosen according to the relevance of each y_i given the context c. This step can improve performance, detecting which words more influence the sentiment assignments.

4. **Dense layer** is a regular densely connected layer implementing a function $output = activation(dot(input, kernel) + bias)$ where $activation$ is the element-wise activation function, while $kernel$ and $bias$ are a weights matrix and a bias vector created by the layer, respectively. The layer uses a linear activation $a(x) = x$ and provides a single output unit representing the sentiment score.

To mitigate the over-fitting, the network augments the cost function within layers with l_2-norm regularization terms for the parameters of the network. It also uses Gaussian noise and dropout layers to prevent feature co-adaptation.

5.5   Sentiment Model Training and Prediction

The fresh instance of the sentiment model takes a set of neural Word Embeddings E together with a set of pre-processed reviews D'_{train}, as input. With these embeddings and reviews, the component trains the deep neural network. As an objective, the network measures the mean squared error (MSE) of the predicted sentiment score against the gold standard value for each input sequence. Parameters are optimized using root mean square propagation (RMSProp) [37] with $learning\_rate = 0.001$. The network was configured for training on batches of size 128 along 20 epochs, shuffling batches between consecutive epochs. The trained deep neural network takes a set of unseen reviews D'_{test} and returns the sentiment score $score'$ predicted for that text comment $text'$, as output.

6   Experimental Evaluation

6.1   Dataset

The dataset used for our experiments is COCO [14], which includes data collected from one of the most popular online course platforms. It contains more than 43 K courses distributed in 35 languages, involving over 16 K instructors and 2.5M learners who provided 4.5M reviews about online courses.

   In our experiments, we considered only reviews with non-empty English text comments. They are 1,396,312 in COCO. Each review includes a rating ranging from 0.5 to 5 with step of 0.5. Considering that our approach supports only integer ratings, we mapped COCO ratings on a scale from 0 to 9 with steps of 1. The dataset D included 1,396,312 reviews, and the split ratios were $s_{creation} = s_{train} = 0.90$. Those values were selected since we wanted to keep both training and testing sets with balanced rating distributions. Moreover, we performed tenfold stratified cross validation. Hence, $1,396,312 - 6,500 * 10$ reviews were put in $D_{creation}$ to create embeddings, while $5,850 * 10$ were put in D_{train} for training the model and $650 * 10$ were put in D_{test} for testing it during each fold.

6.2   Baselines

We experimented and compared our deep learning approach with the following common ML algorithms:

- *Support Vector Machine.* Support vector machine (SVM) algorithm works by defining boundaries through hyperplanes in order to separate a class from the others. The aim of this algorithm is building hyperplanes among data samples in such a way that the separation between classes is as large as possible. The algorithm takes labeled pairs (x_i, y_i) where x_i is a vector representation of input data, and y_i is a numerical label. The algorithm then applies an optimization function in order to separate classes. When it is used for textual input, it is common to transform the text input into vectors of numbers representing features. To name an example, authors in [12] used vectors of numbers for assigning words with syntactical and semantic features to apply a SVM classifier. In the regression variant of SVM, generally named support vector regressor (SVR), the algorithm tries to find hyperplanes that can predict the distribution of information.

- *Random Forests.* Random forests (RF) is based on an ensemble of decision trees, where each tree is independently trained and votes for a class for the data presented as an input [5]. We use a random forest with 10 trees with depth 20. Essentially, each decision tree splits data into smaller data groups based on the features of the data until there are small enough sets of data that only have data points with the same label. These splits are chosen according to a purity measure and, at each node, the algorithm tries to maximize the gain on it. For our regression problem, we consider MSE.
- *Feed-forward Neural Network.* We used a common feed-forward neural network (FF) with ten hidden layers, as described in Sect. 4.1.

We exploited the regression algorithm implementations available within the scikit-learn library.[8] To feed data into these baseline models, we compute the average of Word Embeddings for each review. More specifically, given a review r with terms $\{t_0, \ldots, t_{n-1}\}$, we took the associated Word Embeddings $\{w_0, \ldots, w_{n-1}\}$ and computed their average w, which is used to represent the review text.

6.3 Metrics

In order to evaluate the performance of our model, we measured the MSE and the mean absolute error (MAE) scores. More precisely, MAE and MSE are defined as follows:

$$\text{MAE}(y, \hat{y}) = \frac{1}{n} \cdot \sum_{i=0}^{n-1} |y_i - \hat{y}_i| \tag{4}$$

$$\text{MSE}(y, \hat{y}) = \frac{1}{n} \cdot \sum_{i=0}^{n-1} (y_i - \hat{y}_i)^2 \tag{5}$$

where y_i is a true target value, $\hat{y}_i$ is a predicted target value, and n is the number of samples for both (4) and (5). It follows that given two tests t_1 and t_2, t_1 is better than

---

[8]https://scikit-learn.org/stable/index.html.

t_2 if its related MSE and MAE are lower. During the experiments, we maintained the proportion of the reviews for each of the ten classes of the original dataset for both training and test sets [41].

6.4 Deep Neural Network Model Regressor Performance

Figure 4 reports the MAE of regressors used in our experiments. First of all, our results confirm that Neural Networks, both using a single feed-forward layer and using our model, perform better than common ML algorithms, showing a lower error. Comparing the feed-forward baseline with our deep neural network model, there is a little error difference. It is possible to note that the combination *FF + FastText* obtains similar performances of both *DNNR + GloVe* and *DNNR + FastText*. The best performance was obtained by *DNNR + Word2Vec*. Similar considerations also apply when analyzing the MSE (Fig. 5). In fact, the *DNNR* model gets best performance as well. In contrast with the *MAE*, no baseline obtains performances similar to our model.

6.5 Contextual Word Embeddings Performance

This further experiment aims to show how the context-trained Word Embeddings we generated have advantage over reference generic-trained Word Embeddings, when they are fed into our deep neural network as frozen weights of the embedding layer. In

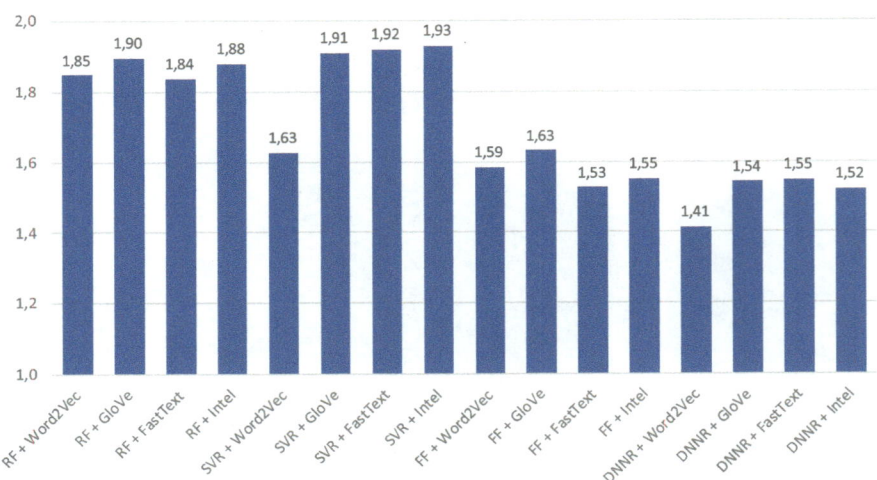

Fig. 4 MAE of experimented regressors

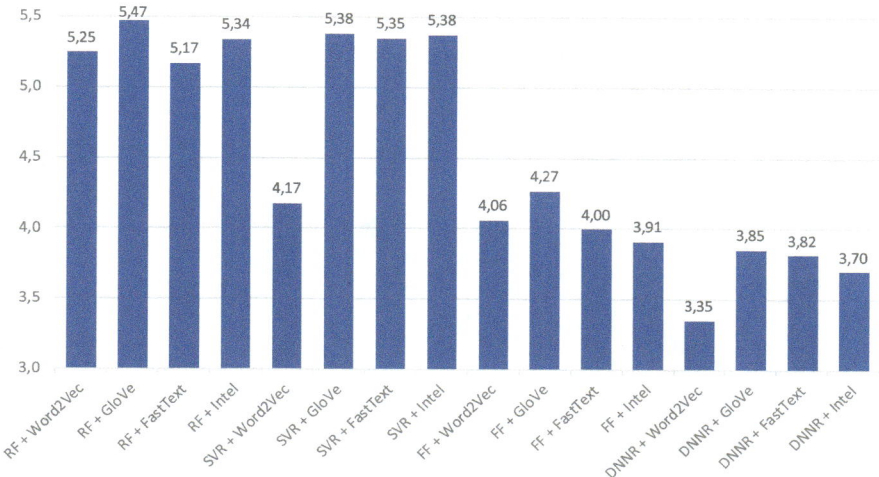

Fig. 5 MSE of experimented regressors

order to evaluate the effectiveness of our approach, we performed experiments using embeddings of size 300 trained on COCO's online course reviews. We compared them against the following reference generic-trained Word Embeddings of size 300 commonly adopted in literature:

- The *Word2Vec*[9] Word Embeddings trained on a part of the Google News dataset including 100 billion words with a vocabulary of 3 million words.
- The *GloVe*[10] Word Embeddings trained on a Wikipedia dataset including one billion words with a vocabulary of 400 thousand words.
- The *FastText*[11] Word Embeddings trained on a Wikipedia dataset including four billion words with a vocabulary of 1 million thousand words.

Context-trained *Intel* Word Embeddings are compared with generic *Word2Vec* Word Embeddings because (i) there are not public generic *Intel* Word Embeddings, and (ii) the *Intel* algorithm is an evolution of *Word2Vec* algorithm.

Figure 6 shows that there is not a relevant difference between context-trained word and generic-trained embeddings when the MAE is used for the comparison. Nevertheless, it is worth underling how the type of embeddings enables to obtain better results in the e-learning domain. Context-trained *Word2Vec* embeddings show the lowest values of MAE compared to other embeddings types. In contrast, when the MSE is considered, context-trained embeddings perform better, as shown in Fig. 7. In this case, context-trained embeddings have low values of MSE in almost all cases except for the *GloVe* Word Embeddings. The best performance was obtained by

---

[9]https://code.google.com/archive/p/word2vec/.

[10]https://nlp.stanford.edu/projects/glove/.

[11]https://s3-us-west-1.amazonaws.com/fasttext-vectors/wiki.en.vec.

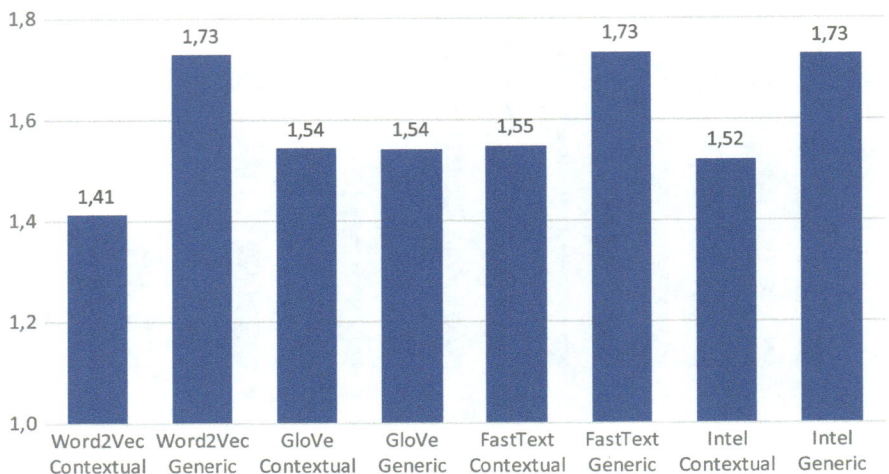

Fig. 6 Comparison between contextual Word Embeddings and generic Word Embeddings considering the MAE

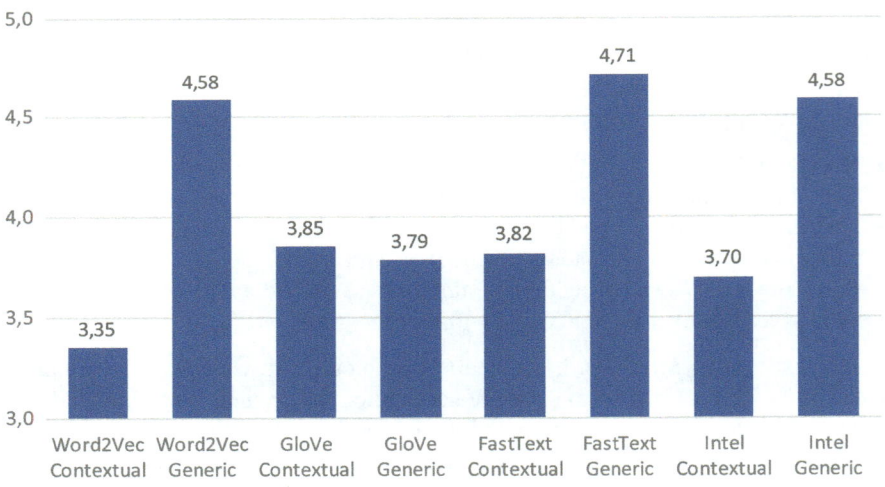

Fig. 7 Comparison between contextual Word Embeddings and generic Word Embeddings considering the MSE

context-trained *Word2Vec* embeddings, proving that (i) *Word2Vec* is the best algorithm to learn word representations from our dataset, and (ii) context-trained Word Embeddings are able to capture specific patterns of the e-learning domain. This makes possible to adapt our DL model on the e-learning domain and improve the results in sentiment score prediction.

7 Conclusions, Open Challenges, and Future Directions

This chapter was structured around a case study on SA within the e-learning context. By combining state-of-the-art DL methodologies and Word Embedding text representations, we introduced and described a deep neural network aimed to predict a sentiment score for text reviews left by learners after attending online courses, as a targeted educational context.

The proposed chapter guides the readers on how to address this task by providing an overview of the common building blocks of a SA system, from input text representations to neural network components, followed by a recipe which combines them into a model able to outperform common ML baselines. As most of the current approaches for SA are built on top of different Word Embedding representations, we showed how some types of Word Embeddings can better represent the semantics behind the e-learning context and help the model to well predict the sentiment score. Furthermore, considering that Word Embeddings tend to be sensitive to the context where they are trained in and that the current publicly available Word Embeddings were trained on general-purpose resources, we proved that the use of Word Embeddings generated from e-learning resources enables the model to capture more peculiarities from this target context.

Research on SA has produced a variety of solid methods, but still poses some interesting challenges that require further investigation:

1. **Public Datasets and Models**. Most SA studies in education have still used rather small datasets which were not made public available and make difficult to train a neural network. As education is a very heterogeneous research field, differentiated into informal, non-formal, and formal learning, a larger collection with more diverse datasets is needed. Furthermore, sharing code and pre-trained models has not been a common practice. Only few authors made their code available, while, for others, people need to re-implement it from scratch. More datasets and models should be shared.

2. **Text Representation Modeling**. Current approaches, such as *Word2Vec*, exhibit limits which can help us understand future trends. For instance, there is only one Word Embedding per word, i.e., Word Embeddings can only represent one vector for each word. Therefore, the term *"learned"* only had one meaning for *"I have learned that information last week"* and *"The instructor was a very learned individual"*. Moreover, Word Embeddings are difficult to train on large datasets, and, to tailor them to another context, they should be trained from scratch. This requires large datasets on the target context and high storage and computational resources. Finally, Word Embeddings have been generally trained on a neural network with only few hidden layers, and this has limited the semantic power of the corresponding representations.

3. **Sentiment Prediction Model Design**. SA systems have traditionally used an FNN as the underlying architecture. However, its densely connected layers have access only to the current input and have no memory of any other input that was already processed. Recent research showed that RNNs can provide state-of-the-

art embeddings that address most of the shortcomings of previous approaches. Emerging systems were trained on a multi-layer RNN and learned Word Embeddings from context, enabling it to store more than one vector per word based on the context it was used in. More advanced architectures need further investigation.

4. **Transfer Learning across Contexts**. Existing models tend to be sensitive to the context targeted by the underlying training data. This has favored the creation of semantic models that, after being trained with data from a given context, do not generalize well in other contexts. With the new availability of public datasets and pre-trained models, it will become easier to plug them into a task different from the one they were originally thought. Previously, performing SA required to train a model or use an API to get the sentiment predictions. By sharing more pre-trained models, cooperation between researchers in SA for education and researchers from other application areas can be promoted. Hence, people could build a new service on top of pre-trained models and quickly train them with small amounts of context-specific data.

5. **Model Explainability and Interpretability**. Most ML and DL algorithms built into automation and artificial intelligence systems lack transparency, and may contain an imprint of the unconscious biases of the data and algorithms underlying them. Hence, it becomes important to understand how w e can predict what is going to be predicted, given a change in input or algorithmic parameters. Moreover, it requires attention how the internal mechanics of the ML or DL system can be explained in human terms. As a context like education looks to deploy artificial intelligence and DL systems, understanding how an algorithm is actually working can help to better align the activities of data scientists and analysts with the key questions and needs of the involved stakeholders.

6. **Data and Algorithmic Bias Impact**. With the advent of ML and DL, addressing bias within education analytics will be a core priority due to several reasons. For instance, some biases can be introduced through the use of training data which is not an accurate sample of the target population or is influenced by socio-cultural stereotypes. Moreover, the methods used to collect or measure data and the algorithms leveraged for predicting sentiments can propagate biases. Future research should control these biases in the developed models, promoting fair, transparent, and accountable systems.

7. **Multi-aspect Sentiment Modeling**. SA in education has focused on determining either the overall polarity (i.e., positive or negative) or the sentiment rating (e.g., one-to-five stars) of a review. However, only considering overall ratings does not allow to represent the multiple potential aspects on which an educational element can be reviewed (e.g., the course content, the instructor, and the platform). To get insightful knowledge on how people perceive each of them, more research taking into account these various, potentially related aspects discussed within a single review is needed.

We expect that the case study on SA within the e-learning context covered in this chapter will help researchers, developers, and other interested people to get a clear view and shape the future research on this field.

Acknowledgements Danilo Dessì and Mirko Marras acknowledge Sardinia Regional Government for the financial support of their Ph.D. scholarship (P.O.R. Sardegna F.S.E. Operational Programme of the Autonomous Region of Sardinia, European Social Fund 2014–2020, Axis III "Education and Training," Specific Goal 10.5).

The research leading to these results has received funding from the EU's Marie Curie training network PhilHumans—Personal Health Interfaces Leveraging HUman-MAchine Natural interactionS under grant agreement 812882.

Furthermore, we gratefully acknowledge the support of NVIDIA Corporation with the donation of the Titan X GPU used for this research.

References

1. Araque, O., I. Corcuera-Platas, J.F. Sanchez-Rada, and C.A. Iglesias. 2017. Enhancing deep learning sentiment analysis with ensemble techniques in social applications. *Expert Systems with Applications* 77: 236–246.
2. Atzeni, M., and Reforgiato, D. 2018. Deep learning and sentiment analysis for human-robot interaction. In *European Semantic Web Conference*, 14–18. Springer
3. Bojanowski, P., E. Grave, A. Joulin, and T. Mikolov. 2017. Enriching word vectors with subword information. *Transactions of the Association for Computational Linguistics* 5, 135–146. URL https://transacl.org/ojs/index.php/tacl/article/view/999
4. Boratto, L., S. Carta, G. Fenu, and R. Saia. 2016. Using neural word embeddings to model user behavior and detect user segments. *Knowledge-Based Systems* 108: 5–14. https://doi.org/10.1016/j.knosys.2016.05.002..
5. Breiman, L. 2001. Random forests. *Machine Learning* 45 (1): 5–32. https://doi.org/10.1023/A:1010933404324.
6. Buscaldi, D., A. Gangemi, and D. Reforgiato Recupero. 2018. Semantic web challenges. In *Fifth SemWebEval Challenge at ESWC 2018*, Heraklion, Crete, Greece, June 3–June 7, 2018, Revised Selected Papers, 3rd ed. Springer Publishing Company, Incorporated
7. Cela, K.L., M.Á. Sicilia, and S. Sánchez. 2015. Social network analysis in e-learning environments. *Educational Psychology Review* 27 (1): 219–246.
8. Chauhan, G.S., P. Agrawal, and Y.K. Meena. 2019. Aspect-based sentiment analysis of students feedback to improve teaching-learning process. In *Information and Communication Technology for Intelligent Systems*, 259–266. Berlin: Springer.
9. Clarizia, F., F. Colace, M. De Santo, M. Lombardi, F. Pascale, and A. Pietrosanto. 2018. E-learning and sentiment analysis: a case study. In *Proceedings of the 6th International Conference on Information and Education Technology*, 111–118. ACM
10. Deng, L. 2014. A tutorial survey of architectures, algorithms, and applications for deep learning. *APSIPA Transactions on Signal and Information Processing*, 3
11. Dessì, D., M. Dragoni, G. Fenu, M. Marras, and D. Reforgiato Recupero. 2019. Evaluating neural word embeddings created from online course reviews for sentiment analysis. In *The 34th ACM/SIGAPP Symposium on Applied Computing*, 2124–2127. SAC
12. Dessì, D., G. Fenu, M. Marras, and D.R. Recupero. 2017. Leveraging cognitive computing for multi-class classification of e-learning videos. In *European Semantic Web Conference*, 21–25. Springer
13. Dessì, D., G. Fenu, M. Marras, and D.R. Recupero. 2019. Bridging learning analytics and cognitive computing for big data classification in micro-learning video collections. *Computers in Human Behavior* 92: 468–477.
14. Dessì, D., G. Fenu, M. Marras, and D. Reforgiato Recupero. 2018. Coco: Semantic-enriched collection of online courses at scale with experimental use cases. In *Trends and Advances in Information Systems and Technologies*, 1386–1396. Berlin: Springer.

15. Dos Santos, C., and M. Gatti. 2014. Deep convolutional neural networks for sentiment analysis of short texts. In *Proceedings of COLING 2014, the 25th International Conference on Computational Linguistics: Technical Papers*, 69–78
16. Dragoni, M., and D. Reforgiato Recupero. 2016. Challenge on fine-grained sentiment analysis within eswc2016. In *Semantic Web Challenges*, ed. H. Sack, S. Dietze, A. Tordai, and C. Lange, 79–94. Cham: Springer International Publishing.
17. Dragoni, G., and M. Petrucci. 2017. A neural word embeddings approach for multi-domain sentiment analysis. *IEEE Transactions on Affective Computing* 8 (4): 457–470.
18. Dridi, A., and D. Reforgiato. 2017. Leveraging semantics for sentiment polarity detection in social media. *International Journal of Machine Learning and Cybernetics* 10 (8): 2045–2055.
19. Esparza, G., A. de Luna, A.O. Zezzatti, A. Hernandez, J. Ponce, M. Álvarez, E. Cossio, and J. de Jesus Nava. 2017. A sentiment analysis model to analyze students reviews of teacher performance using support vector machines. In *International Symposium on Distributed Computing and Artificial Intelligence*, 157–164. Springer
20. Giatsoglou, M., M.G. Vozalis, K. Diamantaras, A. Vakali, G. Sarigiannidis, and K. Chatzisavvas. 2017. Sentiment analysis leveraging emotions and word embeddings. *Expert Systems with Applications* 69: 214–224.
21. Ji, S., N. Satish, S. Li, and P. Dubey. 2016. Parallelizing word2vec in multi-core and many-core architectures. arXiv preprint arXiv:1611.06172
22. Joulin, A., E. Grave, P. Bojanowski, M. Douze, H. Jégou, and T. Mikolov. 2016. Fasttext. zip: Compressing text classification models. arXiv:1612.03651
23. Le, Q., and T. Mikolov. 2014. Distributed representations of sentences and documents. In *International Conference on Machine Learning*, 1188–1196
24. Li, Y., Q. Pan, T. Yang, S. Wang, J. Tang, and E. Cambria. 2017. Learning word representations for sentiment analysis. *Cognitive Computation* 9 (6): 843–851.
25. Lin, C.C., W. Ammar, C. Dyer, and L. Levin. 2015. Unsupervised pos induction with word embeddings. arXiv preprint arXiv:1503.06760
26. Maas, A.L., R.E. Daly, P.T. Pham, D. Huang, A.Y. Ng, and C. Potts. 2011. Learning word vectors for sentiment analysis. In *Proceedings of the Annual Meeting of the Association for Computational Linguistics: Human Language Technologies*, vol. 1, 142–150
27. Mikolov, T., K. Chen, G. Corrado, and J. Dean. 2013. Efficient estimation of word representations in vector space. arXiv preprint arXiv:1301.3781
28. Mikolov, T., I. Sutskever, K. Chen, G.S. Corrado, and J. Dean. 2013. Distributed representations of words and phrases and their compositionality. In *Advances in Neural Information Processing Systems*, 3111–3119
29. Pang, B., and L. Lee. 2005. Seeing stars: Exploiting class relationships for sentiment categorization with respect to rating scales. CoRR **abs/cs/0506075**. URL http://arxiv.org/abs/cs/0506075
30. Pennington, J., R. Socher, and C. Manning. 2014. Glove: Global vectors for word representation. In *Proceedings of the 2014 Conference on Empirical Methods in Natural Language Processing (EMNLP)*, 1532–1543
31. Poria, S., E. Cambria, and A. Gelbukh. 2015. Deep convolutional neural network textual features and multiple kernel learning for utterance-level multimodal sentiment analysis. In *Proceedings of the 2015 Conference on Empirical Methods in Natural Language Processing*, 2539–2544
32. Reforgiato Recupero, D., Cambria, E.: Eswc'14 challenge on concept-level sentiment analysis. In: Presutti, V., Stankovic, M., Cambria, E., Cantador, I., Di Iorio, A., Di Noia, T., Lange, C., Reforgiato Recupero, D., Tordai, A. (eds.) Semantic Web Evaluation Challenge, pp. 3–20. Springer International Publishing, Cham (2014)
33. Reforgiato Recupero, D., E. Cambria, and E. Di Rosa. 2017. Semantic sentiment analysis challenge eswc2017. *Semantic Web Challenges*, 109–123. Berlin: Springer.
34. Reforgiato Recupero, D., M. Dragoni, and V. Presutti. 2015. Eswc 15 challenge on concept-level sentiment analysis. In *Semantic Web Evaluation Challenges*, ed. F. Gandon, E. Cabrio, M. Stankovic, and A. Zimmermann, 211–222. Cham: Springer International Publishing.

35. Rodrigues, M.W., L.E. Zárate, and S. Isotani. 2018. Educational data mining: A review of evaluation process in the e-learning. *Telematics and Informatics* 35 (6): 1701–1717.
36. Rodriguez, P., A. Ortigosa, and R.M. Carro. 2012. Extracting emotions from texts in e-learning environments. In: *2012 Sixth International Conference on Complex, Intelligent, and Software Intensive Systems*, 887–892. IEEE
37. Ruder, S. 2016. An overview of gradient descent optimization algorithms. arXiv preprint arXiv:1609.04747
38. Rudkowsky, E., M. Haselmayer, M. Wastian, M. Jenny, S. Emrich, and M. Sedlmair. 2018. More than bags of words: Sentiment analysis with word embeddings. *Communication Methods and Measures* 12 (2–3): 140–157.
39. Saif, H., Y. He, A. Fernandez, and H. Alani. 2014. Semantic patterns for sentiment analysis of twitter. In: *International Semantic Web Conference*, 324–340. Springer
40. Socher, R., J. Pennington, E.H. Huang, A.Y. Ng, and C.D. Manning. 2011. Semi-supervised recursive autoencoders for predicting sentiment distributions. In *Proceedings of the Conference on Empirical Methods in Natural Language Processing*, 151–161. Association for Computational Linguistics
41. Sokolova, M., and G. Lapalme. 2009. A systematic analysis of performance measures for classification tasks. *Information Processing & Management* 45 (4): 427–437.
42. Tang, D., B. Qin, and T. Liu. 2015. Document modeling with gated recurrent neural network for sentiment classification. In: *Proceedings of the 2015 Conference on Empirical Methods in Natural Language Processing*, 1422–1432
43. Tang, D., F. Wei, B. Qin, N. Yang, T. Liu, and M. Zhou. 2016. Sentiment embeddings with applications to sentiment analysis. *IEEE Transactions on Knowledge and Data Engineering* 28 (2): 496–509.
44. Tang, D., F. Wei, N. Yang, M. Zhou, T. Liu, and B. Qin. 2014. Learning sentiment-specific word embedding for twitter sentiment classification. In: *Proceeding of the Annual Meeting of the Association for Computational Linguistics*, 1555–1565
45. Tripathy, A., A. Agrawal, and S.K. Rath. 2016. Classification of sentiment reviews using n-gram machine learning approach. *Expert Systems with Applications* 57: 117–126.
46. Turney, P.D., and P. Pantel. 2010. From frequency to meaning: Vector space models of semantics. *Journal of artificial intelligence research* 37: 141–188.
47. Vu, T., D.Q. Nguyen, X. Vu, D.Q. Nguyen, M. Catt, and M. Trenell. 2018. NIHRIO at semeval-2018 task 3: A simple and accurate neural network model for irony detection in twitter. CoRR **abs/1804.00520**. URL http://arxiv.org/abs/1804.00520
48. Zhang, L., S. Wang, and B. Liu. 2018. Deep learning for sentiment analysis: A survey. *Wiley Interdisciplinary Reviews: Data Mining and Knowledge Discovery* 8 (4): e1253.
49. Zhang, Z., and M. Lan. 2015. Learning sentiment-inherent word embedding for word-level and sentence-level sentiment analysis. In: *2015 International Conference on Asian Language Processing (IALP)*, 94–97

Toxic Comment Detection in Online Discussions

Julian Risch and Ralf Krestel

Abstract Comment sections of online news platforms are an essential space to express opinions and discuss political topics. In contrast to other online posts, news discussions are related to particular news articles, comments refer to each other, and individual conversations emerge. However, the misuse by spammers, haters, and trolls makes costly content moderation necessary. Sentiment analysis can not only support moderation but also help to understand the dynamics of online discussions. A subtask of content moderation is the identification of toxic comments. To this end, we describe the concept of toxicity and characterize its subclasses. Further, we present various deep learning approaches, including datasets and architectures, tailored to sentiment analysis in online discussions. One way to make these approaches more comprehensible and trustworthy is fine-grained instead of binary comment classification. On the downside, more classes require more training data. Therefore, we propose to augment training data by using transfer learning. We discuss real-world applications, such as semi-automated comment moderation and troll detection. Finally, we outline future challenges and current limitations in light of most recent research publications.

Keywords Deep learning · Natural language processing · User-generated content · Toxic comment classification · Hate speech detection

1 Online Discussions and Toxic Comments

Posting comments in online discussions has become an important way to exercise one's right to freedom of expression in the web. This essential right is however under attack: Malicious users hinder otherwise respectful discussions with their toxic

J. Risch (✉) · R. Krestel
Hasso Plattner Institute, University of Potsdam,
Prof.-Dr.-Helmert-Str. 2–3, 14482 Potsdam, Germany
e-mail: julian.risch@hpi.de

R. Krestel
e-mail: ralf.krestel@hpi.de

© Springer Nature Singapore Pte Ltd. 2020
B. Agarwal et al. (eds.), *Deep Learning-Based Approaches for Sentiment Analysis*, Algorithms for Intelligent Systems,
https://doi.org/10.1007/978-981-15-1216-2_4

comments. A *toxic* comment is defined as a rude, disrespectful, or unreasonable comment that is likely to make other users leave a discussion. A subtask of sentiment analysis is a toxic comment classification. In the following, we introduce a fine-grained classification scheme for toxic comments and motivate the task of detecting toxic comments in online discussions.

1.1 News Platforms and Other Online Discussions Forums

Social media, blogs, and online news platforms nowadays allow any web user to share his or her opinion on arbitrary content with a broad audience. The media business and journalists adapted to this development by introducing comment sections on their news platforms. With more and more political campaigning or even agitation being distributed over the Internet, serious and safe platforms to discuss political topics and news in general are increasingly important. Readers' and writers' motivations for the usage of news comments have been subject to research [15]. Writers' motivations are very heterogeneous and range from expressing an opinion, asking questions, and correcting factual errors, to misinformation with the intent to see the reaction of the community. According to a survey among US American news commenters [51], the majority (56%) wants to express an emotion or opinion. This reason is followed by wanting to add information (38%), to correct inaccuracies or misinformation (35%) or to take part in the debate (31%).[1]

Toxic comments are a problem for these platforms. First, they lower the number of users who engage in discussions, and consequently, the number of visitors to their platform. As a result, an exchange of diverse opinions becomes impossible. With subscription models and ads as a way to earn money, a lower number of visitors means losing money. Second, legal reasons might require the platforms to deploy countermeasures against hate speech and to delete such content or not publish it at all. For example, in Germany, platform providers are obliged by the Network Enforcement Act[2] to delete "obviously illegal" content within 24 h of being notified. Some comments might be legal but still prohibited by the terms of use or discussion guidelines by the platform. To exemplify the reasons for comment deletion, we summarize nine rules that comprise the discussion guidelines by a German news platform.[3] A team of moderators enforces these rules. Most rules are not platform-specific but are rather part of the "Netiquette" — the etiquette on the Internet.

1. **Insults** are not allowed. Criticize the content of the article and not its author!
2. **Discrimination and defamation** are not allowed.
3. **Non-verifiable allegations and suspicions** that are not supported by any credible arguments or sources will be removed.

---

[1] Multiple reasons could be selected.
[2] https://germanlawarchive.iuscomp.org/?p=1245.
[3] https://www.zeit.de/administratives/2010-03/netiquette/seite-2.

4. **Advertising and other commercial content** should not be part of comments.
5. **Personal data** of others may not be published.
6. **Copyright** must be respected. Never post more than short excerpts when quoting third-party content.
7. **Quotations** must be labeled as such and must reference its source.
8. **Links** may be posted but may be removed if the linked content violates our rules.

These rules and our interest in understanding what makes a particular comment toxic motivate the creation of a classification scheme for toxic comments. Also, it helps to distinguish what toxic comment detection focuses on (e.g., insults, discrimination, defamation) and what it does not (e.g., advertising, personal data, copyright). Such a scheme is defined in the next section.

1.2 Classes of Toxicity

Toxicity comes in many different forms and shapes. For this reason, a classification scheme for toxic comments has evolved, which is inspired by annotations provided in different datasets as described in Sect. 2.1. Research on a classification scheme for toxic comments is a connection between computer science on the one hand and media and communication studies on the other hand. Waseem et al. proposed a two-dimensional scheme of abusive language with two dimensions "generalized/directed" and "explicit/implicit [55]." "Directed" means a comment addresses an individual, while "generalized" means it addresses a group. "Explicit" means, for example, outspoken name-calling, while "implicit" means, for example, sarcasm or other ways of obfuscation. Other terms for this dimension are overtly and covertly abusive.

Still, researchers have not reached a consensus about what constitutes harassment online and the lack of a precise definition complicates annotation [21]. Waseem et al. compare the annotations by laymen (users of a crowdsource platform) and by experts ("theoretical and applied knowledge of hate speech") [54]. They find that models trained on expert annotations significantly outperform models trained on laymen annotations.

In the following, we discuss one such classification scheme consisting of five different toxicity classes. We show examples for the different classes of toxic comments for illustration.[4]

---

[4]Warning: The remainder of this chapter contains comment examples that may be considered profane, vulgar, or offensive. These comments do not reflect the views of the authors and exclusively serve to explain linguistic patterns. The following examples stem from a dataset of annotated Wikipedia article page comments and user page comments [58], which is publicly available under Wikipedia's CC-SA-3.0 (https://creativecommons.org/licenses/by-sa/3.0/).

1.2.1 Obscene Language/Profanity

Example: "That guideline is bullshit and should be ignored." The first class considers swear or curse words. In the example, the single word "bullshit" comprises the toxicity of this comment. Typical for this class, there is no need to take into account the full comment if at least one profane word has been found. For this reason, simple blacklists of profane words can be used for detection. To counter blacklists, malicious users often use variations or misspellings of such words.

1.2.2 Insults

Example: "Do you know you come across as a giant prick?" While the previous class of comments does not include statements about individuals or groups, the class "insults" does. "Insults" contain rude or offensive statements that concern an individual or a group. In the example, the comment directly addresses another user, which is common but not necessary.

1.2.3 Threats

Example: "I will arrange to have your life terminated." In online discussions, a common threat is to have another user's account closed. Severely toxic comments are the threats against the life of another user or the user's family. Statements that announce or advocate for inflicting punishment, pain, injury, or damage on oneself or others fall into this class.

1.2.4 Hate Speech/Identity Hate

Example: "Mate, sound like you are jewish. Gayness is in the air." In contrast to insults, identity hate aims exclusively at groups defined by religion, sexual orientation, ethnicity, gender, or other social identifiers. Negative attributes are ascribed to the group as if these attributes were universally valid. For example, racist, homophobic, and misogynistic comments fall into the category of identity hate.

1.2.5 Otherwise Toxic

Example: "Bye! Don't look, come or think of coming back!" Comments that do not fall into one of the previous four classes but are likely to make other users leave a discussion are considered "toxic" without further specification. Trolling, for example, by posting off-topic comments to disturb the discussion falls into this class. Similarly, an online discussion filled with spam messages would quickly become abandoned by users. Therefore, spam falls into this class, although spam detection is not the focus of toxic comment detection.

The listed classes are not mutually exclusive. Comment classification problems are sometimes modeled as multi-class classification and sometimes as multi-label classification. Multi-class means that different labels are mutually exclusive, e.g., a comment can be an insult or a threat but not both at the same time. In contrast, multi-label means that a comment can have multiple labels at the same time. Multi-label classification better mirrors real-world applications, because a comment can, for example, be both an insult and a threat at the same time. In research, this problem is often slightly simplified by assuming analyzed classes are mutually exclusive. We will discuss research datasets later, e.g., Table 3 gives an overview of datasets used in related work.

2 Deep Learning for Toxic Comment Classification

Deep learning for sentiment analysis and, in particular, toxic comment classification is mainly based on two pillars: large datasets and complex neural networks. This section summarizes available datasets and explains neural network architectures used for learning from this data.

2.1 Comment Datasets for Supervised Learning

Online comments are publicly available and every day the number of data samples increases. For example, in 2018, 500 million tweets have been posted on Twitter per day.[5] However, without labeling this data, it can only be used for unsupervised learning, such as clustering or dimensionality reduction. Semi-supervised and supervised learning approaches require labeled data. Examples of labels are the before-mentioned classes of toxicity. In a rather costly process, human annotators check for each and every comment whether it fits into one of the pre-defined classes. Because of the inherent ambiguity of natural language, annotators might not always agree on the label. Further, a comment might be perceived abusive in one context but not abusive in a different context. Different annotation guidelines, low annotator agreement, and overall low quality of annotations are one of the current research challenges in toxic comment classification [1].

Another issue is repeatability. Comments are publicly available, but typically, researchers are not allowed to distribute datasets that they annotated. This is because both — the original author of an online comment and the platform provider — hold the rights of the data. Alternatively, researchers can distribute their annotations alongside the web scrapers that they used to collect online comments. However, it is impossible to rebuild the exact same dataset from scratch by scraping the original web pages again. In the meantime, comments are added, edited, or deleted entirely. It

---

[5]https://www.omnicoreagency.com/twitter-statistics/.

has been proposed to address this issue by measuring the extent of data changes with fingerprinting techniques [44]. The idea of *partial* data repeatability is to use finger-prints to identify unchanged subsets of the data and repeat experiments only on these subsets. This novel idea has not (yet) prevailed, and therefore, today's research on toxic comment classification focuses on a small set of publicly available datasets: the "Yahoo News Annotated Comments Corpus" (522k unlabeled and 10k labeled comments) [33], the "One Million Posts Corpus" (1M unlabeled and 12k labeled comments) [49], and a collection of Wikipedia discussion pages (100k human-labeled and 63M machine-labeled comments) [58]. Wulczyn et al. also publish their annotation guidelines. Thereby, other researchers can understand and potentially reproduce the annotation process. Further, publishing annotation guidelines and annotated data is necessary to allow other researchers to verify/falsify the findings.

The annotation process is crucial for unbiased training datasets and a necessity for training unbiased models. Collecting a large number of labeled, toxic comments is complicated for several reasons. First, moderators edit or delete toxic comments. Moderation might happen shortly after publication so that the comment is shown to the public only for a short time frame. Only in this short time frame, the comment can be collected by a web scraper. Alternatively, moderation takes place before publication, when web scrapers cannot obtain the comment.

Nevertheless, web scrapers use pre-defined lists of abusive language to find large numbers of toxic comments. This approach introduces a bias: Toxic comments that do not match with the pre-defined list will not be included in the dataset. Although this bias is unintended, datasets with such bias are still valuable for research, simply because there is a lack of alternatives. One such dataset comprises 25k labeled tweets that have been collected by searching the Twitter API for tweets that contain words and phrases from a hate speech lexicon [12]. Overall, most related work analyzes datasets extracted from Twitter and makes the tweet IDs publicly available to support the re-creation of the dataset for repeatability [12, 18, 36, 54, 56].

Another challenge is the inherent class imbalance of available datasets. Table 1 lists statistics for two of these datasets. The class distribution of the dataset by Wulczyn et al. [58] is strongly imbalanced with a bias to "clean" comments; whereas,

Table 1 Statistics of the datasets by Wulczyn et al. (left) [58] and Davidson et al. (right) [12] show that both datasets are highly imbalanced

Class	# of occurrences
Clean	201,081
Toxic	21,384
Obscene	12,140
Insult	11,304
Identity hate	2117
Severe toxic	1962
Threat	689

Class	# of occurrences
Offensive	19,190
Clean	4163
Hate	1430

the dataset by Davidson et al. [12] is strongly imbalanced with a bias to "offensive" comments. These class distributions are not representative of the underlying data in general. In fact, most comment platforms contain only a tiny percentage of toxic comments. Since these datasets are collected with a focus on toxic comments, they are biased in a significant way. This needs to be taken into account when deploying deep neural models trained on these datasets in real-world scenarios.

2.2   Neural Network Architectures

Large datasets of toxic comments allow training complex neural networks with millions of parameters. Word embeddings are the basis of neural networks when working with text data in general and also in the specific context of toxic comment classification. They translate each word to a vector of typically 50–300 floating-point numbers, and thus, serve as the input layer. As opposed to sparse, one-hot encoded vectors, these dense vectors can capture and represent word similarity by cosine similarity of the vectors. Beyond simple distance measurements, arithmetics with words can be performed as presented with the Word2Vec model [30]. The similar approaches GloVe [39] and FastText [9] provide alternative ways to calculate word embeddings. FastText is particularly suited for toxic comments because it uses subword embeddings. The advantage of subword embeddings is that they overcome the out-of-vocabulary problem. Toxic comments often use obfuscation, for example "Son of a B****," "****k them!!!!" but also misspelled words, which are common in online discussions. Fast-paced interaction, small virtual keyboards on smartphones, and the lack of editing/correction tools reinforce this problem. Word2Vec and GloVe fail to find a good representation of these words at test time because these words never occurred at training time. These words are out-of-vocabulary. In contrast, FastText uses known subwords of the unknown word to come up with a useful representation. The ability to cope with unknown words is the reason why previous findings [34] on the inferiority of word embeddings in comparison with word n-grams have become outdated.

Similar to other text classification tasks, neural networks for toxic comment classification use recurrent neural network (RNN) layers, such as long short-term memory (LSTM) [23] or gated recurrent unit (GRU) [11] layers. Standard neuronal networks suffer from the vanishing gradient problem. Back-propagation through time might cause the gradients used for the weight updates to become vanishingly small with the increasing number of time steps. With gradients close to zero, no updates are made to the weights of the neural network, and thus, there is no training process. LSTM and GRU layers overcome the vanishing gradient problem with the help of gates. Each cell's state is conveyed to the next cell and gates control changes to these states. Long-range dependencies can be conveyed for an arbitrary number of time steps if the gates block changes to the states for the respective cells. An extension to standard LSTM and GRU layers is bidirectional LSTM or GRU layers, which process the sequence of words in correct and reverse order.

All recurrent layers, regardless whether it is a simple RNN, LSTM, or GRU layer, can either return the last output in the output sequence or the full sequence. If the last output in the sequence is returned, it serves as a representation of the full input comment. However, the outputs of each step in the sequence can be used as an alternative. So-called pooling layers can combine this sequence of outputs. Pooling in neural networks is typically used to reduce an input with many values to an output of fewer values. In neural networks for computer vision, pooling is widespread because it makes the output translation invariant. Pooling on the word level can also make neural networks in natural language processing translation invariant so that the exact position in a sequence of words is irrelevant. For toxic comment classification, both average-pooling and max-pooling are common with a focus on the latter. An intuitive explanation for the use of max-pooling over average-pooling is the following. If a small part of a comment is toxic, max-pooling will focus on the most toxic part and finally result in classifying the comment as toxic. In contrast, with average-pooling, the larger non-toxic part overrules the small toxic part of the comment, and thus, the comment is finally classified as non-toxic. The definition of toxicity classes typically assumes that there is no way to make up a toxic part of a comment by appeasing with other statements. Therefore, max-pooling is more suited than average-pooling for toxic comment classification. As an extension to max-pooling, k-max-pooling outputs not only the largest activation but also the second largest (up to k-largest). It has been shown to further improve classification accuracy in some scenarios [45].

An alternative to pooling after the recurrent layer is an attention layer. Graves has originally introduced the attention mechanism for neural networks in 2013 with an application to handwriting synthesis [20]. It was quickly followed by an application to image classification [31] and neural machine translation to align words in translations [11]. It has been successfully applied also to toxic comment classification [37]. The attention mechanism is basically a weighted combination of all outputs from the preceding recurrent layer. The model can thereby put more emphasis on selected words (or outputs of the recurrent layer) that are decisive for the classification. In semi-automated moderation scenarios, attention can be imagined as a spotlight that highlights abusive or otherwise suspicious words. The final dense layer handles the classification output. For multi-label classification, the dense layer uses a sigmoid activation, and for multi-class classification problems, it uses a softmax activation.

Due to relatively small amounts of training data, overfitting can be an issue. Dropout is a countermeasure against this issue. It does only alter the training process and has no influence on validation or testing. The different kinds of dropouts used in neural networks for toxic comment classification are not task-specific:

1. **Standard dropout** randomly selects neurons and blocks their incoming and outgoing connections. The neuron is therefore ignored during forward and backward propagation.
2. **Spatial dropout** aims to block not only the connections of single neurons but of correlated groups of neurons. For example, if a single value of a 300-dimensional word embedding is dropped, it can be estimated based on the other 299 values. To prevent this, the full embedding vector with its 300 values is dropped at once.

Table 2 Overview on neural network architectures used in related work

Study	Model	Embeddings	Metric
[16]	–	Paragraph2vec	roc-auc
[7]	CNN/LSTM/FastText	GloVe, FastText	p,r,f1
[49]	LSTM	Word2Vec	p,r,f1
[38]	GRU	Word2Vec	roc-auc
[37]	CNN/GRU/RNN+Att	Word2Vec	roc-auc,spearman
[58]	Multi-layer perceptron	–	roc-auc,spearman
[18]	CNN	Word2Vec	p,r,f1
[45]	GRU	FastText	f1
[43]	LSTM	FastText	f1
[60]	CNN+GRU	Word2Vec	f1
[46]	–	Word2Vec	p,r,f1
[41]	LSTM	–	p,r,f1
[1]	CNN/LSTM/GRU/RNN+Att	GloVe, FastText	p,r,f1,roc-auc

3. **Recurrent dropout** is a special kind of dropout that is used in recurrent neural networks. It affects the updates of recurrent cell states.

Table 2 lists published approaches for toxic comment detection with deep learning. It provides an overview of used model architectures, embeddings, and evaluation metrics. For example, for the particular task of hate speech classification (three classes: sexist, racist, or neither), Badjatiya et al. identify a combination of LSTM and gradient boosted decision trees as the best model [7]. Their neural network approaches outperform their various baseline methods (tf-idf or BOW and SVM classifier; char n-gram and logistic regression). Comparing convolutional neural networks (CNNs) and recurrent neural networks (RNNs), there is no clear favorite in Table 2. Both network architectures are of comparable popularity because they achieve comparable performance. However, the training of CNNs is, in general, faster than the training of RNNs because it can be better parallelized. Djuric et al. [16] use comment embeddings based on paragraph2vec [30] and refrain from using both CNNs and RNNs.

Table 2 also shows that several different metrics are used for evaluation. Because the datasets are imbalanced, accuracy is not used but precision, recall, and (weighted) macro- (or micro-) f1-score. Weighted f1-score focuses on the classification of the minority class by emphasizing the respective penalty for misclassification. Further roc-auc and Spearman correlation are used, which we explain in more detail in the following. Spearman's rank correlation coefficient is used to compare ground truth annotations with the model predictions. To this end, the correlation between the fraction of annotators voting in favor of toxic for a particular comment and the probability for the class toxic as predicted by the model is calculated. The receiver operating characteristics area under the curve (roc-auc) is used to measure how good a model is at distinguishing between two classes, e.g., toxic and non-toxic comments. For that purpose, the majority class label in the set of annotations is considered the ground truth and is compared to the predicted probability.

3 From Binary to Fine-Grained Classification

In real-world applications, toxic comment classification is used to support a decision-making process: Does a particular comment need moderation or can it be published right away? This problem is a binary classification problem, which oversimplifies the different nuances in language and abstracts from the classification scheme that we described earlier. A more fine-grained classification, on the other hand, gives insights into why a comment is not suitable for publication. This can help the moderators in making a final decision but also the benevolent offender to avoid infringement of comment rules in the future. Therefore, different classes of toxicity, such as insult, threat, obscene language, profane words, and hate speech have to be distinguished. With this fine-grained classification, it is also possible to distinguish between merely bad comments and criminal offenses. The following explains why fine-grained comment classification is a much harder task than binary comment classification. Further, we discuss two related topics: *transfer learning* to deal with limited training data, and *explanations* to help moderators to understand and trust neural network predictions.

3.1 Why Is It a Hard Problem?

Binary classification is already difficult. Nobata et al. list several reasons why abusive language detection is a difficult task [34]. For example, simple detection approaches can be fooled by users who obfuscate and conceal the true meaning of their comments intentionally. Another difficulty is the use of stylistic devices in online discussions such as *irony* to express sarcasm or quoting possible problematic content. Further, language is not static: New words are introduced, other words change their meaning, and there is an ever-shifting fine line of what is barely considered legitimate to state and what not. This flexible and ever *changing language* requires a detection approach to adapt over time, for example, to neologisms. It is also unclear what classification scheme to use and how to precisely distinguish classes from each other. As a consequence of this uncertainty, researchers have come up with various annotation guidelines and resulting datasets use different labels, as seen earlier (e.g., in Table 3).

If we now switch to more fine-grained labels, we face two additional problems:

1. Reduced available training data per class
2. Increased difficulty for annotation

Table 3 Overview on datasets used in related work

Study	# Annotated comments	Available	Classes
[16]	950k Yahoo finance	No	Hate-speech, other
[7]	16k Twitter	Yes	Sexist, racist, neither
[49]	12k news	Yes	8 classes[a]
[38]	1.5m news	Yes	Accepted, rejected
[37]	1.5m news, 115k Wikipedia	Yes	Reject, accept/personal attack, other
[58]	100k Wikipedia	Yes	Personal attack, other
[18]	6.7k Twitter	Yes	Racism, sexism[b]
[45]	30k Facebook	Yes	Overtly, covertly aggressive, neither
[43]	5k Twitter/Facebook	Yes	Profanity, insult, abuse, neither
[60]	2.5k Twitter	No	Hate, non-hate
[46]	3m news	No	Accepted, rejected
[41]	16k Twitter	Yes	Sexist, racist, neither
[1]	25k Twitter, 220k Wikipedia	Yes	Offense, hate, neither/7 classes[c]

[a]Negative sentiment, positive sentiment, off-topic, inappropriate, discriminating, feedback, personal stories, argumentative
[b]Multi-label
[c]Toxic, obscene, insult, identity hate, severe toxic, threat, neither (multi-label)

With a fine-grained classification, the number of available samples per class gets lower. It is a major challenge to collect enough samples per class without introducing a problematic bias to the sampling from a basic population of comments. The class imbalance complicates training neural networks, and therefore, countermeasures become necessary. Downsampling and upsampling alter the dataset so that there is an equal number of samples from every class. To achieve this, unnecessary samples of the majority class can be discarded or samples of the minority class can be sampled repeatedly. Another technique is to use a weighted loss function, which influences the training process: Penalties for errors in the minority class are made higher than the majority class. Another idea is the synthetic minority oversampling technique (SMOTE) [10], which has already been used to augment a dataset of aggressive comments [43]. For both SMOTE and class weights, similar gains in increased f1-score have been reported [43].

It is essential to keep the number of trainable parameters, and thus, the model's capacity as small as possible if training data is limited. While GRU units have only two gates, LSTM units have three gates. GRU units are preferable because of their smaller number of parameters. The aim to keep the number of parameters small also explains the popularity of pooling layers, because they do not contain any trainable parameters. The alternative of using dense layers to combine the outputs of recurrent layers increases the number of parameters. Depending on the network architecture, multiple layers can also share their weights and thereby reduce the number of parameters. Last but not least, weight regularization can be used to limit the value range of parameters.

The second problem relates to the increased effort to annotate the training data. The inter-annotator agreement is already relatively low for binary labels when looking at all but the most obvious examples. Moreover, it gets even lower with more fine-grained classes. The boundaries between those classes are often fuzzy, and the meaning of sentences depends on context, cultural background, and many more influencing factors. An insult for one person could be regarded as a legitimate utterance by another. The inherent vagueness of language makes the annotation process even for domain experts, such as forum moderators, extremely difficult. This means that the focus on training data generation lies on quality, not on quantity. The flip side of this is that there is not much high-quality annotated data available. One way to cope with the limited availability of annotated data besides adapting the network architecture as mentioned earlier is to make the most of the available data, e.g., by using transfer learning.

3.2 Transfer Learning

For English language texts, large amounts of training data are available. However, for less common languages, training data is sparse and sometimes no labeled data is available at all. One way to cope with this problem is to machine-translate an English language dataset to another language. If the machine translation is of good quality, the annotations of the English language comments also apply to the translated comments. For offensive language detection on German language comments, 150,000 labeled, English comments were machine-translated to German and then used as training data [43].

In a similar way, datasets for the English language can also be augmented. The idea is to make use of slight variations in language introduced by translating a comment from, for example, English to German and then back to English. The following comments exemplify this idea (example by Risch et al. [45]):

- Original comment: "Happy Diwali.!!let's wish the next one year health, wealth n growth to our Indian economy."
- Comment translated to German and then back to English: "Happy Diwali, let us wish the next year health, prosperity and growth of our Indian economy."

The word *wealth* is substituted by *prosperity*, the short form *let's* is substituted by *let us*, and *n* is correctly extended to *and*. The augmentation by machine translation increases the variety of words and phrases in the dataset, and it also normalizes colloquial expressions. A dataset that has been augmented with this approach is available online.[6]

---

[6]https://hpi.de/naumann/projects/repeatability/text-mining.html.

Another idea to overcome the problems of small amounts of training data is to pre-train a neural network on different data or for a different task first. Afterward, only the last layer or several last layers of the network are fine-tuned on the actual, potentially much smaller dataset. During the fine-tuning, parameters on all other layers are fixed, because these layers are assumed to have learned a generic representation of comments on the larger dataset. Only task-specific parameters are trained during fine-tuning. For example, this approach has been successfully used to first pre-train on 150,000 comments with coarse-grained labels and to afterward fine-tune on 5000 comments with fine-grained labels [43].

In the paper titled "Attention Is All You Need," Vaswani et al. propose a novel attention mechanism called transformer [52]. This attention mechanism has laid the groundwork for the following progress in pre-training deep neural networks on large text corpora and transferring these models easily to a variety of tasks. With ELMo, a technique to learn contextualized word embeddings has been proposed [40]. The key idea is that a word can be represented with different embeddings depending on its surrounding words in a particular sentence. Technically, the approach is to train bidirectional LSTMs to solve a language modeling task. With ULMFiT, a fine-tuning method called "discriminative fine-tuning" has been introduced, which allows to transfer and apply pre-trained models to a variety of tasks [24]. BERT overcomes the limitation of all previous models that input needs to be processed sequentially left-to-right or right-to-left [13].

With fine-grained classification for toxic comment detection, we can not only distinguish comments that are allowed to be published online from comments that should be deleted by moderators. The fine-grained classes can also provide a first explanation of why a comment is deleted. For example, it could be deleted because it contains an insult or a threat to the news article author. Similarly, a hate speech comment could be fine-grained classified by the target group of the attack, e.g., a particular religious or ethnic group. Such explanations for classification results increase trust in the machine learned model. The following section goes into more detail and shines a light on the explanations of neural networks for toxic comment classification.

3.3   *Explanations*

Explanations play an essential role in real-world recommender and classification systems. Users trust recommendations and algorithmic decisions much more if they provide an explanation as well. One example is the "other customers also bought" recommendations in e-commerce applications. By explaining why a particular product was recommended, the recommendations are considered better and more trustworthy.

In the context of user comment classification, explanations are also very much needed to establish trust in the (semi-)automatic moderation process. If no reason is provided why a user's comment was deleted or not published in the first place, this user might get the feeling of being censored or her opinion otherwise oppressed. Therefore, a fine-grained classification is inevitable. Even if results for binary classification ("delete or not delete") are slightly better compared to fine-grained classification results ("deleted because of x"), the latter is preferred. Explaining to the users why their comment was deleted does not only help to dispel worries about censorship but also to keep the users engaged on the platform. In addition, they get educated about the way the comment sections are supposed to be used in this particular community ("Netiquette").

A fine-grained classification of deleted content helps to broadly categorize an offending comment but does not explain why a comment was classified into a particular class. To this end, explanations of the machine learning algorithm are needed. There is a large volume of research concerned with explaining deep learning results. For text classification, it is necessary to point toward the phrases or words that make a comment off-topic, toxic, or insulting. These kinds of explanations are beneficial to monitor the algorithm and identify problems early on. If a comment was classified as insulting because of a very common, neutral word, it can mean that the algorithm needed to be recalibrated or retrained to make comprehensible decisions.

Naive Bayes can serve as a baseline approach for explanations because it is simple and yet gives some insights. For each word in the vocabulary, we calculate the probability that a comment containing this word is classified as toxic. The Naive assumption of word independence is inherent to this approach, which means word correlations are not taken into account. As a consequence, the same word is assigned the same probabilities across all comments.

Another approach, *Layer-wise relevance propagation* (LRP), has been first proposed to explain image classifications by neural networks [6]. More recently, LRP has been successfully applied to natural language processing and to sentiment analysis in particular [4, 5]. Figure 1 shows heatmaps for two example comments based on Naive Bayes probabilities and LRP relevance scores for an LSTM-based neural network.[7] For Naive Bayes, red boxes indicate a high conditional probability that given the occurrence of the word the comment is toxic. For LRP, red boxes indicate the relevance score in favor of the class "toxic."

The Naive Bayes approach highlights only a small number of words as decisive for the classification. This problem is known as *over-localization* and has been reported as a problem also for other explanation approaches [53]. The LRP visualization reveals that the LSTM correctly identifies word pairs that refer to each other, such as "article deletion" and "fuck u." In contrast, for the Naive Bayes approach, "fuck" and "u" are independent words and therefore "u" is not highlighted. Figure 2 shows heatmaps for an exemplary toxic comment based on four different techniques. The comparison includes a Naive Bayes approach, an LSTM-based network visualized

---

[7]The visualizations are based on a tool called "innvestigate" by Alber et al. [2]: https://github.com/albermax/innvestigate.

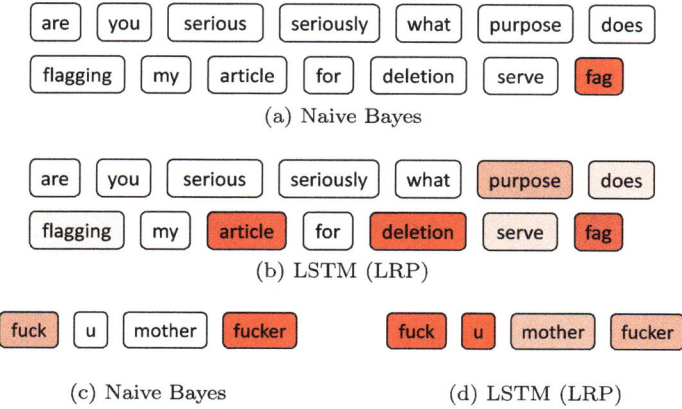

Fig. 1 Heatmaps highlight the most decisive words for the classification with a Naive Bayes approach and an LSTM-based network

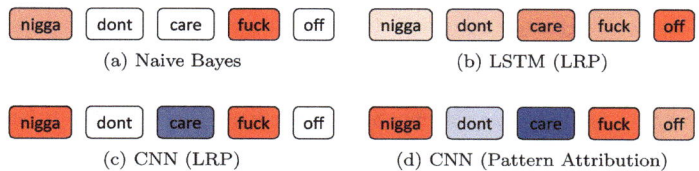

Fig. 2 Heatmaps highlight the most decisive words for the classification with a Naive Bayes approach, an LSTM, and two CNNs

with LRP, and a CNN visualized with LRP and pattern attribution [25]. Again, red boxes indicate the probability or relevance score in favor of the class "toxic," while blue boxes indicate the opposite class "not toxic."

4 Real-World Applications

Overwhelmed by the recent shift from a few written letters to the editor to online discussions with dozens of participants on a 24/7 basis, news platforms are drowning in vast numbers of comments. On the one hand, moderation is necessary to ensure respectful online discussions and to prevent misuse by spammers, haters, and trolls. On the other hand, moderation is also very expensive in terms of time, money, and working power. As a consequence, many online news platforms have discontinued their comment sections. Different lines of machine learning research aim to support online platforms in keeping their discussion sections open. This section covers a selection.

4.1 Semi-automated Comment Moderation

For example, semi-automated comment moderation can support human moderators but does not completely replace them [46]. A machine learning model is trained on a binary classification task: a set of presumably appropriate comments that can be published without further assessment and a set of presumably inappropriate comments that need to be presented to a human moderator for assessment. Today's industrial applications so far refrain from using deep learning models for comment moderation due to the lack of explainability. Such black box models do not fulfill the requirement of comprehensible classification results. Moderators and readers both want to understand the reasons behind a classification decision. Future improvements in explaining the decisions of deep neural networks are needed to apply them for comment moderation. Until then, the industry will fall back to less complex models, such as logistic regression models. These models can explain which features make a comment inappropriate in a specific context [46].

Ambroselli et al. propose a logistic regression model based on article metadata, linguistic, and topical features to predict the number of comments that an article will receive [3]. Based on these predictions, news directors can balance the distribution of highly controversial topics across a day. Thereby, readers are enabled to engage in more discussions, and the moderation workload is distributed evenly. Further, guiding the attention of moderators toward potentially disrespectful discussions facilitates efficient moderation. There are several studies of implemented systems that support the moderation of online discussions [3, 46, 48]. These discussions can also be mined to predict the popularity of news stories [47], to measure how controversial a comment is [19] or to rank comments by persuasiveness [57]. Figure 3 shows how the fraction of moderated comments varies over time. Interestingly, the peeks correlate with breaking news events.

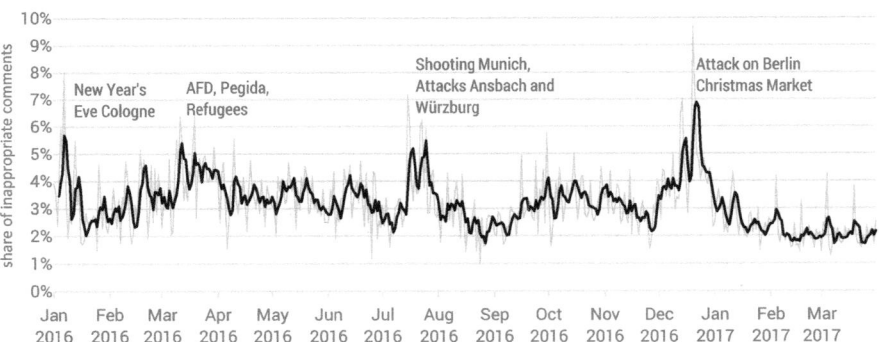

Fig. 3 Share of inappropriate comments (light gray) aggregated with a four-day centered moving average (black) stands out at the date of specific news events

4.2 Troll Detection

We consider malicious users of comment sections, as users who post comments with a motivation to disturb otherwise respectful discussions. In contrast to toxic comment classification, the focus is on users who attract negative attention with multiple misbehaviors. Research on malicious users in online discussions distinguishes trolls and sockpuppets [28]. Trolls characterizes that they try to disturb on-topic discussions with provoking or off-topic utterances. Hardaker defines trolls as users "…whose real intentions are to cause disruption and/or to trigger or exacerbate conflict for the purposes of their own amusement." [22].

Sockpuppets are multiple user accounts that are under the control of the same person. The latter can have multiple reasons and is not per se a problem for a discussion — although the platform's terms of use typically forbid it. For example, users who access a platform from multiple different devices might use multiple user accounts to protect their privacy and prevent tracking across their devices. Some users who forgot their account password create a new account. If they, later on, remember their old account's password, they sometimes continue to use both accounts. However, there are also malicious intents, such as to upvote own comments or argue in favor of own comments and create the impression of consensus if there is not. If there actually is a broad consensus, malicious users can use multiple accounts to create the impression of strong dissent and controversial discussions with divisive comments.

There is a publicly available dataset of 3 million tweets by almost 3000 Twitter troll accounts.[8] These accounts are considered trolls because of their connection to a Russian organization named Internet Research Agency (IRA). IRA is a defendant in an indictment filed by the U.S. Justice Department in February 2018. The organization is characterized as a "troll factory" and is accused of having interfered with the U.S. presidential election in 2016 in a way that is prohibited by US law. Fake profiles posing as US activists allegedly tried to influence the election systematically. Linvill and Warren defined five different classes of IRA-associated Twitter accounts[9]:

1. **Right Trolls** support Donald Trump and other Republicans, while attacking Democrats.
2. **Left Trolls** support Bernie Sanders and criticize, for example, Hillary Clinton with divisive tweets. They also discuss socially liberal topics.
3. **News Feeds** post local, regional, and US news and links to news platforms.
4. **Hashtag Gamers** post their tweets in context of a particular hashtag. By choosing popular hashtags, they maximize the visibility of their tweets.
5. **Fearmongers** spread fear and panic by posting hoaxes.

---

[8]https://about.twitter.com/en_us/values/elections-integrity.html#data.

[9]Their article was originally published on the Resource Center on Media Freedom in Europe according to the terms of Creative Commons: Attribution-NonCommercial 4.0 International (CC BY-NC 4.0). https://www.rcmediafreedom.eu/Publications/Academic-sources/Troll-Factories-The-Internet-Research-Agency-and-State-Sponsored-Agenda-Building, https://creativecommons.org/licenses/by-nc/4.0/.

Galán-García et al. trained a machine learning model to detect troll profiles in Twitter [17]. Their publication focuses on real-world applications, and they prove that current models are already good enough to be beneficial for selected tasks. However, the next section deals with an error analysis for state-of-the-art models and identifies their weaknesses. We outline different directions for further research based on this analysis.

5 Current Limitations and Future Trends

Common challenges for toxic comment classification among different datasets comprise out-of-vocabulary words, long-range dependencies, and multi-word phrases [1]. To cope with these challenges, subword embeddings, GRUs and LSTMs, and phrase mining techniques have been developed. A detailed error analysis by van Aken et al. for an ensemble of several state-of-the-art approaches [12, 34, 34, 42, 50, 59, 60] reveals open challenges [1]. We discuss this analysis and its implications in the following.

5.1 Misclassification of Comments

Based on the analysis by van Aken et al. we discuss six common causes for misclassification [1]. We distinguish causes for false positives (non-toxic comments that are misclassified as toxic) and false negatives (toxic comments that are misclassified as non-toxic). The following examples are Wikipedia talk page and user page comments [58]. This dataset was also used in the Kaggle Challenge on Toxic Comment Classification.[10]

5.1.1 Toxicity Without Swear Words

Toxicity can be conveyed without mentioning swear words. The toxic meaning is only revealed with the help of context knowledge and understanding the full sentence, as exemplified by the toxic comment: "she looks like a horse." The word "horse" is not insulting in general. To understand the toxicity of the comment, a model needs to understand that "she" refers to a person and that "looking like a horse" is generally considered insulting if directed to a person. However, this insult is not revealed by looking at the words of the sentence independently.

In contrast to these false negatives, there are false positives that contain toxic words, although they are overall non-toxic. If a user posts a self-referencing comment, human annotators rarely consider these comments toxic, for example: "Oh, I feel

---

[10]https://kaggle.com/c/jigsaw-toxic-comment-classification-challenge/.

like such an asshole now. Sorry, bud." However, the learned model focuses on the mentioned swear words, which triggers the misclassification. Taking into account a full sentence and getting its meaning still remain a challenge for deep learning approaches.

5.1.2 Quotations, References, Metaphors, and Comparisons

A problem is that state-of-the-art models are not able to take into account the context of a comment, which includes other comments in the discussion. On the one hand, examples of false positives are otherwise non-toxic comments that cite toxic comments. Because of the toxic citation, the overall comment can be misclassified as toxic. Example: "I deleted the Jews are dumb comment."

On the other hand, an example of false negatives is the comment: "Who are you a sockpuppet for?" The word sockpuppet is not toxic in itself. However, the accusation that another user is a sockpuppet attacks the user without addressing his or her comment itself. In Paul Graham's hierarchy of disagreement, which lists types of arguments in a disagreement, this is the second-lowest type of argument called "Ad Hominem."[11]

5.1.3 Sarcasm, Irony, and Rhetorical Questions

Sarcasm, irony, and rhetorical questions have in common that the meaning of the comment is different from its literal meaning. This disguise can cause false negatives in the classification. While they are not the focus of this book chapter, we at least give examples for this reported problem for toxic comment detection [34, 42]. Example comment: "hope you're proud of yourself. Another milestone in idiocy." If the first sentence in this example is taken literally, there is nothing toxic about the comment. However, the user who posted the comment actually means the opposite, which is revealed by the second sentence. Other examples are rhetorical questions, which do not ask for real answers. Example: "have you no brain?!?!" This comment is an insult because it alleges another user to act without thinking. Rhetorical questions in toxic comments often contain subtle accusations, which current approaches hardly detect.

5.1.4 Mislabeled Comments

The annotation of toxic comments is a challenging task for several reasons. Annotation guidelines cannot consider each and every edge case. For example, a comment that criticizes, and therefore, cites a toxic comment is not necessarily toxic itself. Example: "No matter how upset you may be there is never a reason to refer to another editor as 'an idiot' ." State-of-the-art approaches classify this comment as

---

[11] http://www.paulgraham.com/disagree.html.

not toxic, although it is labeled as toxic. We argue that this comment is actually not toxic. Thus, this false negative is not a misclassification by the current models but rather a mislabeling by the annotators.

Similar to false negatives, there are false positives caused by wrong annotations. Ill-prepared annotators, unclear task definition, and the inherent ambiguity of language may cause a minority of comments in training, validation, and test dataset to be annotated wrongly. Example: "IF YOU LOOK THIS UP UR A DUMB RUSSIAN."

5.1.5   Idiosyncratic and Rare Words

Intentionally, obfuscated words, typos, slang, abbreviations, and neologisms are a particular challenge in toxic comment datasets. If there are not enough samples with these words in the training data, the learned representations (e.g., word embeddings) may not account for the true meaning of a word. Thus, wrong representations may cause misclassification. Example: "fucc nicca yu pose to be pullin up." Similarly, the classification of the comment: "WTF man. Dan Whyte is Scottish" depends on the understanding of the term "WTF." The amount of slang used is platform-specific. For this reason, misclassification due to rare words is twice as high for tweets than for Wikipedia talk page comments [1].

5.2   Research Directions

What is the opposite of toxic comments? High-quality, engaging comments! Finding them automatically is a growing research field [14, 26, 27, 32, 35]. A possible application is to automatically choose editor picks, which are comments highlighted by the editors of a news platform. State-of-the-art work involves supervised machine learning approaches in order to classify comments. However, all these approaches require large annotated datasets (30k annotated comments [27]), which are costly to obtain. Lampe and Resnick study whether a similar task can be accomplished by a large team of human moderators [29]. On the Web site Slashdot, the moderators need to distinguish high- and low-quality comments in online conversations.

A different direction is to improve classification by taking into account the context of a comment. Instead of using a single comment as input, the full discussion and other context, such as the news article or user history, can be used. A motivation for this additional input is the way that humans read online comments. Because of the web page layout of social networks and news platforms and the chronological order of comments, early comments receive the most attention. To read later comments, users typically need to click through dozens of subpages. For this reason, research assumes that the first few comments play a special role in setting the tone of further discussion as respectful or disrespectful [3, 8].

Dealing with biased training data is another research challenge common to many supervised machine learning approaches. One reason why this problem occurs is that the sampling of the training data is biased. For example, an annotated comment training set might include only comments from discussions of the politics section, not including comments from other sections, such as sports. This distribution might not mirror the distribution in the test set. A second type of bias is due to prejudices and stereotypes inherent to the data. A representative sample would contain this bias, although we might want to prevent our model from learning it. The research question of how to reduce bias in trained models is also addressed by a data science challenge and the corresponding dataset[12] by Kaggle and Jigsaw.

Another challenge, especially for deployed systems, is the explainability of classification decisions. This is also true for other deep learning models and not unique to comment classification. For comment moderation, explanations are not just nice to have but play an essential part in the process. As discussed earlier, explaining the automatic deletion of a comment is crucial in the context of freedom of expression. Besides, no news outlet wants to be perceived as censoring undesired opinions. Finding good, convincing explanations is therefore essential for successful comment moderation.

Good explanations are essential in semi-automated comment moderation tools to help the moderators to make the right decision. For fully automated systems, explanations are even more critical. Moreover, with the growing number of comments on platforms without moderation, such as Facebook or Twitter, more automatic systems are needed. Finding a balance between censorship and protecting individuals and groups on the web will be challenging. However, this challenge is not only a technical but also a societal and political one, with not less than democracy on the line.

6 Conclusions

In this chapter, we discussed sentiment analysis for toxic comment detection. One motivation for this task is the overwhelming number of comments posted online, which needs moderation to remain engaging, respectful, and informative. Real-world applications, such as semi-automated comment moderation, can benefit from research on toxic comment detection. We defined and discussed fine-grained classification schemes for toxicity to support further progress in this field, and we gave an overview of publicly available datasets and state-of-the-art neural network architectures. Toxic comment detection was also set into context with the most recent research on transfer learning and on explaining neural networks. Finally, we outlined the current challenges and future directions for research in this field.

Acknowledgements We thank Betty van Aken, Carl Ambroselli, Samuele Garda, Lasse Kohlmeyer, Niklas Köhnecke, Eva Krebs, Alexander Riese, Andreas Loos, Alexander Löser, and Felix Naumann for valuable input, either through implementation or discussion of the topic.

---

[12]https://www.kaggle.com/c/jigsaw-unintended-bias-in-toxicity-classification.

References

1. van Aken, B., J. Risch, R. Krestel, and A. Löser. 2018. Challenges for toxic comment classification: An in-depth error analysis. In *Proceedings of the Workshop on Abusive Language Online (ALW@EMNLP)*, 33–42
2. Alber, M., S. Lapuschkin, P. Seegerer, M. Hägele, K.T. Schütt, G. Montavon, W. Samek, K.R. Müller, S. Dähne, and P.J. Kindermans. 2018. Innvestigate neural networks! arXiv preprint arXiv:1808.04260
3. Ambroselli, C., J. Risch, R. Krestel, and A. Loos. 2018. Prediction for the newsroom: Which articles will get the most comments? In *Proceedings of the Conference of the North American Chapter of the Association for Computational Linguistics (NAACL)*, 193–199. ACL
4. Arras, L., F. Horn, G. Montavon, K.R. Müller, and W. Samek. 2016. Explaining predictions of non-linear classifiers in nlp. In *Proceedings of the Workshop on Representation Learning for NLP*, 1–7. Association for Computational Linguistics
5. Arras, L., G. Montavon, K.R. Müller, and W. Samek. 2017. Explaining recurrent neural network predictions in sentiment analysis. In *Proceedings of the Workshop on Computational Approaches to Subjectivity, Sentiment and Social Media Analysis*, 159–168. Association for Computational Linguistics, Copenhagen, Denmark
6. Bach, S., A. Binder, G. Montavon, F. Klauschen, K.R. Müller, and W. Samek. 2015. On pixelwise explanations for non-linear classifier decisions by layer-wise relevance propagation. *PloS one* 10 (7)
7. Badjatiya, P., S. Gupta, M. Gupta, and V. Varma. 2017. Deep learning for hate speech detection in tweets. In *Proceedings of the International Conference on World Wide Web (WWW)*, 759–760. International World Wide Web Conferences Steering Committee
8. Berry, G., and S.J. Taylor. 2017. Discussion quality diffuses in the digital public square. In *Proceedings of the International Conference on World Wide Web (WWW)*, 1371–1380. International World Wide Web Conferences Steering Committee, Republic and Canton of Geneva, Switzerland
9. Bojanowski, P., E. Grave, A. Joulin, T. Mikolov, and T. 2017. Enriching word vectors with subword information. *Transactions of the Association for Computational Linguistics (TACL)* 5 (1): 135–146
10. Chawla, N.V., K.W. Bowyer, L.O. Hall, and W.P. Kegelmeyer. 2002. Smote: synthetic minority over-sampling technique. *Journal of Artificial Intelligence Research* 16 (1): 321–357
11. Cho, K., B. van Merrienboer, C. Gulcehre, D. Bahdanau, F. Bougares, H. Schwenk, and Y. Bengio. 2014. Learning phrase representations using rnn encoder–decoder for statistical machine translation. In *Proceedings of the Conference on Empirical Methods in Natural Language Processing (EMNLP)*, 1724–1734. Association for Computational Linguistics
12. Davidson, T., D. Warmsley, M. Macy, and I. Weber. 2017. Automated hate speech detection and the problem of offensive language. *Proceedings of the International Conference on Web and Social Media (ICWSM)*, 512–515
13. Devlin, J., M.W. Chang, K. Lee, and K. Toutanova. 2018. Bert: Pre-training of deep bidirectional transformers for language understanding. arXiv preprint arXiv:1810.04805
14. Diakopoulos, N.: Picking the nyt picks: Editorial criteria and automation in the curation of online news comments. International Symposium on Online Journalism (ISOJ) **6**(1), 147–166 (2015)
15. Diakopoulos, N., and M. Naaman. 2011. Towards quality discourse in online news comments. In *Proceedings of the Conference on Computer Supported Cooperative Work (CSCW)*, 133–142. ACM
16. Djuric, N., J. Zhou, R. Morris, M. Grbovic, V. Radosavljevic, and N. Bhamidipati. 2015. Hate speech detection with comment embeddings. In *Proceedings of the International Conference on World Wide Web (WWW)*, 29–30. International World Wide Web Conferences Steering Committee

17. Galán-García, P., and J.G.d.l. Puerta, C.L. Gómez, I. Santos, and P.G. Bringas. 2016. Supervised machine learning for the detection of troll profiles in twitter social network: Application to a real case of cyberbullying. *Logic Journal of the IGPL* 24 (1): 42–53
18. Gambäck, B., and U.K. Sikdar. 2017. Using convolutional neural networks to classify hatespeech. In *Proceedings of the Workshop on Abusive Language Online (ALW@ACL)*, 85–90
19. Gómez, V., Kaltenbrunner, A., and V. López. 2008. Statistical analysis of the social network and discussion threads in slashdot. In *Proceedings of the International Conference on World Wide Web (WWW)*, 645–654. ACM
20. Graves, A. 2013. Generating sequences with recurrent neural networks. arXiv preprint arXiv:1308.0850
21. Guberman, J., C. Schmitz, and L. Hemphill. 2016. Quantifying toxicity and verbal violence on twitter. In *Proceedings of the Conference on Computer Supported Cooperative Work (CSCW)*, 277–280. ACM, New York, NY, USA
22. Hardaker, C.: Trolling in asynchronous computer-mediated communication: From user discussions to academic definitions. Journal of Politeness Research. Language, Behaviour, Culture **6**, 215–242 (2010)
23. Hochreiter, S., Schmidhuber, J.: Long short-term memory. Neural Computation **9**(8), 1735–1780 (1997)
24. Howard, J., and S. Ruder. 2018. Universal language model fine-tuning for text classification. In *Proceedings of the Annual Meeting of the Association for Computational Linguistics (ACL)*, 328–339. Association for Computational Linguistics, Melbourne, Australia
25. Kindermans, P.J., K.T. Schütt, M. Alber, K.R. Müller, D. Erhan, B. Kim, and S. Dähne. 2017. Learning how to explain neural networks: PatternNet and PatternAttribution. arXiv preprint arXiv:1705.05598
26. Kolhatkar, V., and M. Taboada. 2017. Constructive language in news comments. In *Proceedings of the Workshop on Abusive Language Online (ALW@ACL)*, 11–17
27. Kolhatkar, V., and M. Taboada. 2017. Using new york times picks to identify constructive comments. In *Proceedings of the Workshop: Natural Language Processing meets Journalism@EMNLP*, 100–105
28. Kumar, S., and N. Shah. 2018. False information on web and social media: A survey. arXiv preprint arXiv:1804.08559
29. Lampe, C., and P. Resnick. 2004. Slash (dot) and burn: Distributed moderation in a large online conversation space. In *Proceedings of the Conference on Human Factors in Computing Systems (CHI)*, 543–550. ACM
30. Mikolov, T., I. Sutskever, K. Chen, G.S. Corrado, and J. Dean. 2013. Distributed representations of words and phrases and their compositionality. *Advances in Neural Information Processing Systems (NIPS)*, 3111–3119
31. Mnih, V., N. Heess, A. Graves, and K. Kavukcuoglu. 2014. Recurrent models of visual attention. *Advances in Neural Information Processing Systems (NIPS)*, 2204–2212
32. Napoles, C., A. Pappu, and J.R. Tetreault. 2017. Automatically identifying good conversations online (yes, they do exist!). *Proceedings of the International Conference on Web and Social Media (ICWSM)*, 628–631
33. Napoles, C., J. Tetreault, A. Pappu, E. Rosato, and B. Provenzale. 2017. Finding good conversations online: The yahoo news annotated comments corpus. *Proceedings of the Linguistic Annotation Workshop (LAW)*, 13–23
34. Nobata, C., J. Tetreault, A. Thomas, Y. Mehdad, and Y. Chang. 2016. Abusive language detection in online user content. In *Proceedings of the International Conference on World Wide Web (WWW)*, 145–153. International World Wide Web Conferences Steering Committee
35. Park, D., S. Sachar, N. Diakopoulos, and N. Elmqvist. 2016. Supporting comment moderators in identifying high quality online news comments. In *Proceedings of the Conference on Human Factors in Computing Systems (CHI)*, 1114–1125. ACM
36. Park, J.H., and P. Fung. 2017. One-step and two-step classification for abusive language detection on twitter. In *Proceedings of the Workshop on Abusive Language Online (ALW@ACL)*, 41–45. Association for Computational Linguistics, Vancouver, BC, Canada

37. Pavlopoulos, J., P. Malakasiotis, and I. Androutsopoulos, I. 2017. Deeper attention to abusive user content moderation. In *Proceedings of the Conference on Empirical Methods in Natural Language Processing (EMNLP)*, 1125–1135. Association for Computational Linguistics, Copenhagen, Denmark

38. Pavlopoulos, J., P. Malakasiotis, J. Bakagianni, and I. Androutsopoulos. 2017. Improved abusive comment moderation with user embeddings. In *Proceedings of the Workshop on Natural Language Processing meets Journalism (co-located with EMNLP)*, 51–55. Association for Computational Linguistics, Copenhagen, Denmark

39. Pennington, J., R. Socher, and C. Manning. 2014. Glove: Global vectors for word representation. *Proceedings of the Conference on Empirical Methods in Natural Language Processing (EMNLP)*, 1532–1543

40. Peters, M., M. Neumann, M. Iyyer, M. Gardner, C. Clark, K. Lee, and L. Zettlemoyer. 2018. Deep contextualized word representations. In *Proceedings of the Conference of the North American Chapter of the Association for Computational Linguistics (NAACL)*, 2227–2237. Association for Computational Linguistics, New Orleans, Louisiana

41. Pitsilis, G.K., Ramampiaro, H., Langseth, H.: Effective hate-speech detection in twitter data using recurrent neural networks. Applied Intelligence **48**(12), 4730–4742 (2018)

42. Qian, J., M. ElSherief, E.M. Belding-Royer, and W.Y. Wang. 2018. Leveraging intra-user and inter-user representation learning for automated hate speech detection. *Proceedings of the Conference of the North American Chapter of the Association for Computational Linguistics (NAACL)*, 118–123

43. Risch, J., E. Krebs, A. Löser, A. Riese, and R. Krestel. Fine-grained classification of offensive language. In *Proceedings of GermEval (co-located with KONVENS)*, 38–44

44. Risch, J., and R. Krestel. Measuring and facilitating data repeatability in web science. *Datenbank-Spektrum* 19(2): 117–126.

45. Risch, J., and R. Krestel. 2018. Aggression identification using deep learning and data augmentation. In *Proceedings of the Workshop on Trolling, Aggression and Cyberbullying (TRAC@COLING)*, 150–158

46. Risch, J., and R. Krestel. 2018. Delete or not delete? Semi-automatic comment moderation for the newsroom. In *Proceedings of the Workshop on Trolling, Aggression and Cyberbullying (TRAC@COLING)*, 166–176

47. Rizos, G., S. Papadopoulos, and Y. Kompatsiaris. 2016. Predicting news popularity by mining online discussions. In *Proceedings of the International Conference on World Wide Web Companion (WWW)*, 737–742. International World Wide Web Conferences Steering Committee

48. Schabus, D., and M. Skowron. 2018. Academic-industrial perspective on the development and deployment of a moderation system for a newspaper website. *Proceedings of the Language Resources and Evaluation Conference (LREC)*, 1602–1605

49. Schabus, D., M. Skowron, and M. Trapp. 2017. One million posts: A data set of german online discussions. *Proceedings of the International Conference on Research and Development in Information Retrieval (SIGIR)*, 1241–1244

50. Schmidt, A., and M. Wiegand. 2017. A survey on hate speech detection using natural language processing. *Proceedings of the International Workshop on Natural Language Processing for Social Media*, 1–10

51. Stroud, N.J., E. Van Duyn, and C. Peacock. 2016. News commenters and news comment readers. *Engaging News Project*, 1–21

52. Vaswani, A., N. Shazeer, N. Parmar, J. Uszkoreit, L. Jones, A.N. Gomez, Ł. Kaiser, and I. Polosukhin. 2017. Attention is all you need. *Advances in Neural Information Processing Systems (NIPS)*, 5998–6008

53. Wang, C. 2018. Interpreting neural network hate speech classifiers. In *Proceedings of the Workshop on Abusive Language Online (ALW@EMNLP)*, 86–92. Association for Computational Linguistics, Brussels, Belgium

54. Waseem, Z. 2016. Are you a racist or am i seeing things? Annotator influence on hate speech detection on twitter. In *Proceedings of the Workshop on NLP and Computational Social Science*, 138–142. Association for Computational Linguistics, Austin, Texas

55. Waseem, Z., T. Davidson, D. Warmsley, and I. Weber. 2017. Understanding abuse: A typology of abusive language detection subtasks. In *Proceedings of the Workshop on Abusive Language Online (ALW@ACL)*, 78–84. Association for Computational Linguistics, Vancouver, BC, Canada

56. Waseem, Z., and D. Hovy. 2016. Hateful symbols or hateful people? predictive features for hate speech detection on twitter. In *Proceedings of the Student Research Workshop@NAACL*, 88–93. Association for Computational Linguistics, San Diego, California

57. Wei, Z., Liu, Y., Li, Y.: Is this post persuasive? Ranking argumentative comments in online forum. Proceedings of the Annual Meeting of the Association for Computational Linguistics (ACL) **2**, 195–200 (2016)

58. Wulczyn, E., N. Thain, and L. Dixon. 2017. Ex machina: Personal attacks seen at scale. In *Proceedings of the International Conference on World Wide Web (WWW)*, 1391–1399. International World Wide Web Conferences Steering Committee

59. Zhang, Z., and L. Luo. Hate speech detection: A solved problem? the challenging case of long tail on twitter. *Semantic Web Journal*, 1–21

60. Zhang, Z., D. Robinson, and J. Tepper. 2018. Detecting hate speech on twitter using a convolution-gru based deep neural network. In *European Semantic Web Conference*, 745–760. Berlin: Springer

Aspect-Based Sentiment Analysis of Financial Headlines and Microblogs

Hitkul, Simra Shahid, Shivangi Singhal, Debanjan Mahata, Ponnurangam Kumaraguru and Rajiv Ratn Shah

Abstract To improve the performance in e-commerce markets, big giants like Amazon, Myntra and Flipkart are providing consumers with a platform to review their services and also give them an opportunity to provide a useful insight of the service to the future buyers. On the other hand, companies use such reviews to make a significant upgradation in their products (or services) to survive in the competition from others in the market. This shows the importance of studying user views or opinions on a particular product (or service) consumed by users. In Natural Language Processing (NLP), the process of studying such user opinion is termed as opinion mining. It is a task of finding out overall sentiment present in a review. Past research in this area has assumed that a sentence cannot have multiple sentiments associated with it. However, this is not true. For example, "This car looks beautiful, but does not handle very well." comprises a positive sentiment towards the looks of the car but a negative sentiment towards its handling. To address such issues, aspect-based sentiment analysis (ABSA) was introduced. ABSA aims to detect an aspect (i.e. features) in a given text and then perform sentiment analysis of the text with respect to that aspect. The chapter aims to discuss the concept of ABSA for the problem introduced as a FiQA 2018 challenge subtask 1 (https://sites.google.com/view/fiqa) in WWW 2018 shared task. It highlights all the state-of-the-art models in the domain and discusses some new approaches. We propose neural network models combined with hand-engineered features and attention mechanism, to perform ABSA on financial headlines and microblogs. Our proposed model outperformed the existing state-of-the-art results in sentiment part by 50% and in the aspect part by 20%.

Hitkul · S. Singhal · D. Mahata · P. Kumaraguru · R. R. Shah (✉)
Indraprastha Institute of Information Technology, Delhi, India
e-mail: rajivratn@iiitd.ac.in

Hitkul
e-mail: hitkuli@iiitd.ac.in

S. Shahid
Delhi Technological University, Delhi, India
e-mail: simrashahid_bt2k16@dtu.ac.in

© Springer Nature Singapore Pte Ltd. 2020
B. Agarwal et al. (eds.), *Deep Learning-Based Approaches for Sentiment Analysis*, Algorithms for Intelligent Systems, https://doi.org/10.1007/978-981-15-1216-2_5

Keywords Aspect based sentiment analysis · Finance news · Classification · Sentiment score · Prediction · Stocks

1 Introduction

Social media has become an integral part of our lives. Nowadays, people have not only restricted themselves to use such platform for communication with others but are also active in raising their opinions on any event they found worthy talking about. This shows the need for analysing such vast amount of multimedia data [35]. One such scenario where people have started raising their opinions is the e-commerce markets. Customers provide valuable feedback on the items purchased to help other future buyers in making the correct decision. This also provides companies with an opportunity to use social media, not only as an interaction platform but also to get useful insights on their products via consumer feedback.

Consumer feedback is an intrinsic part of e-commerce society. It is proved that around 90% of consumers believe that online reviews impact the buying decisions. There are myriad of options available out today to chose from, so consumers often turn to reviews of the services they wish to use before making a final decision. On the other side, consumer feedback helps companies to improve the quality of their products in hand to retain the customers from diverging to their competitors or leaving their services.

In order to study useful information from such reviews, it is extremely important to design an automated system that can extract correct sentiments from them. This is so because the sheer amount of reviews that are encountered each day makes it impossible to be studied by manual intervention. In NLP, a tool that is used to do such a large-scale study is termed as sentiment analysis.

It is a process of determining whether a piece of a text is positive, negative or neutral; i.e. it aims is to identify the right attitude towards the topic in conversation. Many organizations around the world use some form of sentiment analysis to generate valuable feedback. For instance, Wall Street uses sentiment analysis in trading algorithms and provides up-to-date sentiment tracking of financial news and headlines. In addition to it, IBM uses sentiment summaries of the surveys to help their companies understand consumer attitudes towards the services they provide. But the traditional approach followed by sentiment analysis method suffers from an inherent disadvantage. It assumes that only one sentiment is associated with every sentence. But this is false. For example, the phone has a great display, but battery life is poor. It is extremely hard to judge whether the overall sentiment of this sentence shows a positive or negative attitude. In real scenario, the above sentence shows a positive sentiment towards the aspect "display" of the phone and negative attitude towards the aspect "battery" of the phone. Though many systems will declare this as a neutral sentence, it will be incorrect. To solve such issues, concept of ABSA was introduced.

ABSA [18, 44] is a problem of identifying all the different aspects in a given sentence and finds sentiment with respect to each aspect. There are several subtasks involved in ABSA such as correctly identifying aspects present in a sample, among which identifying relevant entities and aspects and determining the corresponding sentiment/polarity are the most studied. Generally, there are three types of sentences that ABSA deals with.

- **Sentence having single aspect**: These are those sentences that have only one aspect associated with it. For example, Sensex prices will dip. Here, the only aspect that is present is the "price".
- **Sentence having multiple aspects**: These are those sentences that have multiple aspects associated with them. There are two subdivisions associated with such sentences.
  - **Aspects having same polarity**: In such case, all the aspects associated with a sentence have either positive, negative or neutral polarity. For example, the restaurant has really good food and ambience. Here, both the aspects "food" and "ambience" have positive polarity associated with it, i.e. good.
  - **Aspects having dissimilar polarity**: In such case, all the aspects associated with a sentence have different polarities. For example, the food is very good, but the ambience is not good. Here, the aspects are "food" and "ambience" but it carries positive attitude towards food and negative attitude towards ambience.

The emphasis on studying financial domain is to provide customers with better financial analysis and decision-making. Similar to reviews received by e-commerce companies on their products, financial review received by business analysts is worthy of study. Such reviews are also characterized with positive and negative opinions on specific aspects of a certain investment opportunity. The few benefits of incorporating ABSA for financial domain are: (i) financial statements are generally written as free-form essays that can either have positive, negative or neutral attitude towards the aspect present in a sentence. To correctly identify the overall sentiment of a sentence using sentiments of each aspect can be beneficial to provide structured information to users from an unstructured set of write-ups. (ii) Another use case could be to employ the aspect-based sentiments as features to classify future performance or volatility of investment ideas.

With reference to [12, 36], previous experiments in Financial ABSA makes use of : (i) multichannel convolutional neural network with different word embeddings at different channels, (ii) use of hand-engineered features and financial lexicons to pay attention to clues in a given sentence, (iii) use of dependency tree models to give weighted attention to words with respect to the position of the target word.

The chapter gives a brief overview of ABSA for financial domain. This problem was introduced in WWW 2018 shared task under FiQA 2018 challenge subtask 1. The organizers are provided with a data set that consists of financial tweets and headlines. Each tweet/headline contains multiple relevant snippets. Each snippet has an associated aspect and sentiment score with respect to that aspect. The aim of the challenge is to build a model that can predict the aspects present in a given snippet and its sentiment score. A tool like this can be used for monitoring public reaction on stocks and companies. These insights can be useful for making business decisions, investments and stock predictions.

The rest of the chapter is organized as follows. Section 2 deals with the detailed overview of the past research done in this domain. This is followed by Sect. 3 that describes different SOTA models. Next comes the detailed description of the proposed model designed by our team in Sects. 4–6. Finally, experimental results are presented in Sect. 7, concluding with remarks and highlights of the future work in Sect. 7.1.

2 Related Work

Opinion mining is a process of extracting overall sentiment of the sentence. But it has been observed that a sentence can have multiple sentiments associated with it. This poses problem as available machine learning and deep learning models [2, 3, 11, 14, 19, 20, 22, 26, 31–34, 48] are not sufficient enough to solve such issues. They often result in getting neutral attitudes towards such sentences. Moreover, an early study by Vo. et al. [42] also showed that 40% of classification errors are caused by approaches that are independent of the aspects. To solve this issue, ABSA was introduced. The different techniques used to solve ABSA task are as follows:

- Rule-based approaches: Poria et al. [28] proposed a rule-based approach that uses domain knowledge and dependency trees to detect both explicit and implicit aspects. Chikersal et al. [1] used a rule-based classifier with supervised learning which helped in refining the support vector machine's predictions for sentiment analysis.
- Deep learning-based approaches: In recent years, deep learning techniques have shown a great success for sentiment analysis. Kar et al. [13] combine hand-engineered lexical, sentiment and metadata features with the representations learned from convolutional neural networks (CNNs) and bidirectional gated recurrent unit (Bi-GRU) having attention model applied on top. Poriya et al. [27] used a seven-layer deep CNN to mark aspect/non-aspect word in opinion mining sentences. But CNNs were not sufficient in itself. CNNs are able to capture both local features of phrases and global sentence semantics. Moreover, it has the ability to learn local responses from temporal or spatial data but such kind of structures lacks the ability to learning sequential correlations. To overcome this drawback, recurrent neural networks were designed. They specialize for sequential modelling but were unable to extract features in a parallel way. So, the concept of CLSTM was introduced by [50]. It combines the best of both CNN and long short-term memory (LSTM) to model sentences. Recently, Hazarika et al. [8] highlighted that neural network-based models can detect aspects and its related information from a sentence but fail to learn some inter-aspect dependencies. Moreover, Piao and Breslin [25] introduced Deep-FASP, an ensemble-based approach for sentiment

predictions and aspect classification. This approach uses CNN and RNN with a ridge regression and a voting strategy.

– Adding contextual information: It has been observed that leveraging contextual information has shown its benefits in the sentiment analysis task. Ma et al. [17] proposed that both the targets and contexts can learn from each other via interactive learning. It interactively learns attentions in the contexts and targets, and generates the representations for the targets and contexts separately. Tang et al. [39] introduced a deep memory network for ABSA that is able to capture the importance of each contextual word. However, this model requires the aspects to appear in the training data many times, which is not the case of the provided FiQA data set.

Recently, a challenge was organized by WWW 2018 shared task for ABSA for the financial domain. There were a total of six submissions for this challenge. The top two models from the leader board are:

– **ALA model**: The model [36] is similar to [44] attention-based LSTM. The ALA model comprised of stacked LSTMs and aspect embeddings that was used for attention to capture the most important part of a sentence for a target aspect. No pre-trained embeddings were used. Instead, own embeddings were trained from the training corpus. Jangid et al. [12] used a bidirectional long short-term memory (BiLSTM) units to extract the aspect from a given headline and microblog and used targets to get enhanced vectors using dependency tree. It was then fed to a multichannel CNN with different word embeddings at different channels (Godin and Google News) for sentiment analysis. The dependency tree was then used to give weighted attention to the words with respect to position of the target word.

– **IIIT Delhi model**: The system has two major parts, an aspect model and a sentiment model. It uses a multichannel convolutional neural network for sentiment analysis. Each channel deals with a different embedding type including GloVe, Google News Word2Vec and Godin.

The next section gives an in-depth detail of the above two models proposed as a solution for the FiQA task.

3 State-of-the-Art Models

This section gives an insight about the two different state-of-the-art models proposed in the FiQA challenge. These two models were the top two ranked solutions for this task.

3.1 ALA Model

The model [36] was the winning model in the FiQA challenge. ALA model makes use of the attention mechanism and aspect embeddings that is randomly initialized

Fig. 1 Overall framework of
ALA model

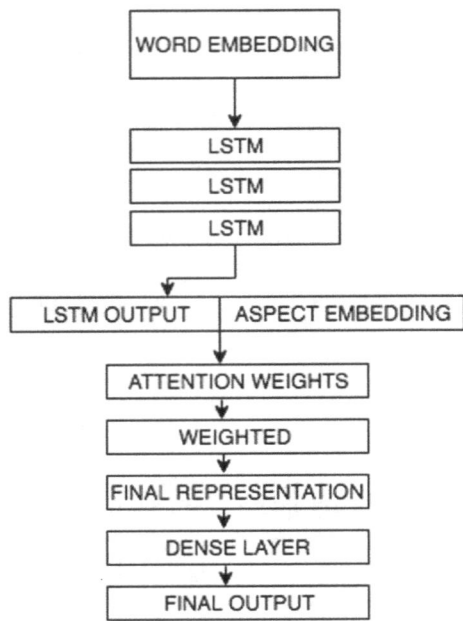

by the sampling from a uniform distribution and optimized during training phase. It is helpful in determining the attention weights. Benefiting from the attention mechanism, the ALA model incorporates the aspect information effectively and is able to learn the alignment between microblogs and aspect. This enables the model to focus on different parts of a text instance when different aspects are considered. Figure 1 shows the overall framework of the model proposed by authors.

The model makes use of aspect information to assist the prediction of the sentiment score. It assumes that all the samples in the test set have the same aspect. The model does not use any aspect information and still manages to perform better than any other models . Though this is not a valid assumption for real-world scenario, neural networks have the capability to find a way to the problem without any explanation.

3.2 IIIT Delhi Model

The model [12] secured the second position in the challenge. Figure 2 represents the proposed framework designed by our team. The system has two major parts, an aspect model and a sentiment model. It uses a multichannel convolutional neural network for sentiment analysis. Each channel deals with a different embedding type including GloVe [24], Google News Word2Vec [21] and Godin. Choice of word embeddings is a hyperparameter. A BiLSTM extracts aspect from a given headline or

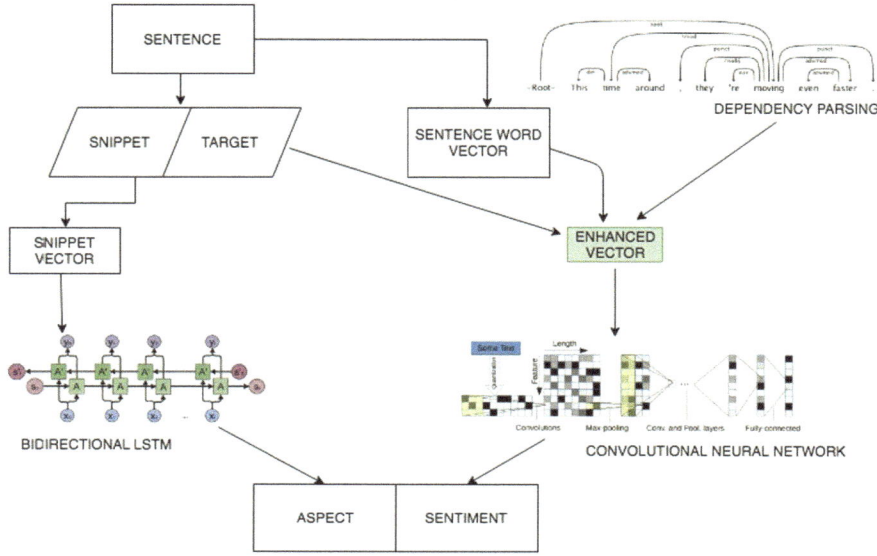

Fig. 2 A schematic diagram of our proposed model (IIIT Delhi)

microblog snippet. Each target present in the sentence is used to calculate an enhanced word vector using dependency parsing technique. A dependency tree was chosen over constituency tree as the former were designed to represent the relationship between words in a sentence. The final enhanced word vector is calculated by using the given target, original word vectors and the dependency tree made using spaCy[1] and NetworkX.[2] These enhanced word vectors link the sentiment score with targets. Finally, enhanced word vector is passed into the sentiment model to generate the sentiment intensity score of a sentence towards the target.

The key takeaway from the proposed method was the way fine-tuning was performed using Bayesian optimization [37]. This resulted in finding the best combination of the hyperparameters. Our team trained over 250 models and found the best set of hyperparameters, hence emphasizing on the importance of fine-tuning a model.

4   Our Methodology

In this section, we aim to discuss in detail our upgraded version of the proposed model for the FiQA task. The model designed for the challenge by our team titled as "IIIT Delhi" secured second position in the challenge. After looking at the shortcomings

---

[1] https://spacy.io/.

[2] http://networkx.github.io/.

of the previously designed model, our team improved their architecture to beat all the baseline performances done for this task so far.

The rest of this section discusses the details of the steps used for construction of this upgraded model. Section 4 discusses the features used followed by in-depth view of the aspect model in Sect. 5. This is followed by detailed discussion of the sentiment model in Sect. 6. This gets concluded in Sect. 7 by giving a brief overview of the evaluation metric along with a list of hyperparameters used to achieve this task.

4.1 Features

In the FiQA data set (see Sect. 7.1), important snippets and targets present in tweet/headlines are given. Task 1 i s t odentify aspect present in each snippet, and Task 2 is to find sentiment polarity towards each target. To kick-start with the problem, features from the data are extracted. The following features are used to extract important information from the data.

Word Embeddings Word embeddings are vector representation of the words that are used to capture context of a word in a document followed by its semantic and syntactic similarity [38, 40, 45]. For our proposed neural network architecture, word embedding is given as the input. We used pre-trained Stanford GloVe [24] and Google News Word2Vec [21] embeddings. In word vector representation, each sentence is represented as a matrix $R^{n \times d}$, where n is the vocab size. For the word that does not appear in the provided training set, we simply use a zero vector to represent them. The pre-trained word vectors are given the initial weights of the embedding layer, and the word embedding can be updated during the training process.

Lexicons Sentiment analysis on short text like microblogs is more challenging as compared to longer text like headlines. Problems faced in microblog sentiment analysis are: (i) short-length messages, (ii) informal words like ⟨lol, laugh out loud⟩, (iii) abbreviations like ⟨cmd, command⟩, (iv) spelling variation like ⟨frie, free⟩ and (v) emoticons. Also, sentiment analysis of the financial text largely depends on the quality of sentiment lexicons. Thus, the use of good lexicons is very critical. Especially in financial corpus, certain words like "dip" have a different impact on an aspect than it does in usual opinion mining of the reviews. For example, "Apple is facing a major dip in prices". Here, "dip" will give a negative sentiment to the aspect "price" for the company Apple. On the other hand, for review mining, the sentence "The food has really tasty dips" will give a positive sentiment to the aspect "food" for the word dips. This shows the importance of lexicons in financial corpus.

For our experimentation, we made use of the following lexicons:

- Opinion Lexicon [4, 43]: A list of English positive and negative opinion words or sentiment words (around 6800 words).
- Loughran and McDonald Sentiment Word Lists [43]: The dictionary provides a means of determining which tokens (collections of characters) are actual words,

which is important for consistency in word counts. Within the dictionary spread-sheet, flags are also provided for the sentiment dictionaries.

– MPQA Lexicon (Wilson et al. 2005) [46]: The multi-perspective question answering (MPQA) subjectivity lexicon is maintained by Theresa Wilson, Janyce Wiebe and Paul Hoffmann.

Hand-Engineered Features Feature engineering plays a key role in determining the outcome of a model. There are various factors such as (i) kind of data available, (ii) features prepared using feature engineering, (iii) model chosen. All these together determine the results achieved by the proposed architecture. This is represented in Eq. (1).

$$\text{Data} \rightarrow \text{Features} \rightarrow \text{Model} \rightarrow \text{Results} \tag{1}$$

Kar et al. [13] showed benefits of using hand-engineered features. We use the following hand-engineered features in conjunction with word vectors as an input to our models.

1. **Word and char N-grams**: tf–idf weights are calculated for each word, and character n-grams was extracted, where $n = 1, 2, 3$ and $2, 3, 4$, respectively.
2. **Agreement Score**: It is the agreement value of the positive and negative words in the data instance. This was calculated for the sentences as in [15].

$$\text{Agreement Score} = 1 - \sqrt{1 - \left| \frac{\text{Mpos} - \text{Mneg}}{\text{Mpos} + \text{Mneg}} \right|} \tag{2}$$

where Mpos and Mneg mean a number of positive and negative words in a span of text, respectively.

3. **Pointwise Mutual Information (PMI)**: We also calculated PMI value for each word in the training corpus and made a sentence vector of the score values of all the words.

$$\text{score}(w) = \text{PMI}(w, \text{pos}) - \text{PMI}(w, \text{neg}) \tag{3}$$

$$\text{PMI}(w, \text{pos}) = \log_2 \frac{\text{freq}(w, \text{pos}) * N}{\text{freq}} (w) * \text{freq}(\text{pos}) \tag{4}$$

where pos and neg mean collection of all positive and negative reviews, respectively. freq(w, pos) means the number of times token w appeared in collection pos, and N means total the number of tokens in pos or neg. The steps followed to get the final score value are: (i) generate and stemmed lemmatized word lists and frequencies from the corpus, (ii) count the positive and negative words according to the financial lexicons, (iii) calculate relative frequencies, PMI score and agreement score for each word, (iv) compute the total score of each feature for the entire sentence.

Sentence Vectors: We make use of sentence vectors as features. For every word, in a sentence, a sentence vector was computed by taking an average across all word vectors of a sentence. We made use of pre-trained Google News Word2Vec [6] embeddings. The experiment was also performed with Glove [24] to make the optimal choice for generating the sentence vectors.

5 Aspect Classification Models

This section gives an in-depth detail of the different classification models that were worked upon for this challenge. The aim of the models was to extract the aspects from a sentence and then classify them into one of the 27 classes of the aspect level 2.

5.1 Models

Bidirectional LSTM with Attention In this section, we discuss simple bidirectional long-term short memory (BiLSTM) architecture for aspect classification tasks.

Let the output of BiLSTM, $H \in R^{d \times N}$, be the hidden vector matrix , where d is the size of hidden layers and N is the length of the given sentence. H consists of output vectors $[h_1, h_2, \ldots, h_N]$. To perform attention mechanism, we compute α as follows:

$$z = \tanh(W_1 * H) \tag{5}$$

$$\alpha = \text{softmax}(w^\mathrm{T} z) \tag{6}$$

$$r = H\alpha^\mathrm{T} \tag{7}$$

where $W_1 \in R^{d \times d}, w \in R^d, z \in R^{d \times N} \alpha \in R^N$. tanh activation is performed on the weighted representation of the sentence $r \in R^d$.

Final hidden representation, h_{final}, is computed as follows:

$$h_{\text{final}} = \tanh(W_2 * r + b_2) \tag{8}$$

where $W_2 \in R^{d \times d}, b_2 \in R^d$.

The outputs of the final sentence representation are passed to a softmax classifier to predict label $\hat{y}$ from a discrete set of 27 classes for a sentence S.

$$\hat{y} = \text{softmax}(W_3 * h_{\text{final}} + b_3) \tag{9}$$

where $W_3 \in R^{1 \times d}$, $b_3 \in R^{\text{classes}}$. The weights and bias parameters W_1, W_2, W_3, b_1, b_2 are projection parameters learned during training.

Vector Averaging For every word w in a sentence s, a sentence vector $\vec{s}$ was computed by taking an average across all word vectors of a sentence. The sentence vector is represented as:

$$\overrightarrow{\text{sentence}} = \text{Average}(w_1, w_2, ...w_N) \tag{10}$$

where N is the length of the $\overrightarrow{\text{sentence}}$ with dimensions $N * d$ where d is the dimension of the word vector. As shown in Fig. 3, the sentence vectors are input to MLP having different combinations of neurons, respectively. Each neuron of a hidden layer calculates a vector $\vec{v}$:

$$\vec{v} = \text{ReLU}(W * \vec{x} + b) \tag{11}$$

where W is the weight matrix and b is the bias vector that are learned during the training. In each layer, we use ReLU activation. The output of the last layer is passed to a softmax classifier to predict the label $\hat{y}$ from a discrete set of 27 classes for a sentence S.

Word Vectors with tf–idf Features In this model, we pass the tf–idf features of the word and char n-gram to a multilayer perceptron to classify the aspects. We also tried combining the features from both tf–idf and sentence vectors to fed as an input to the MLP, but no improvement in results was encountered. Figure 4 shows architecture of this model.

Fig. 3 Vector averaging model

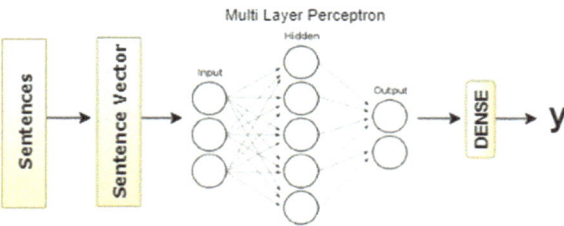

Fig. 4 tf–idf model

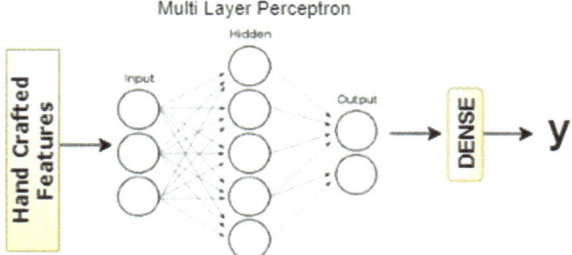

5.2 Classification Model Training

The classification models are trained by back-propagation to optimize the cross-entropy loss function. If y is the actual aspect class, and $\hat{y}$ is the predicted aspect class, we aim to minimize the cross-entropy error between y and $\hat{y}$.

L_2 regularization along with cross-entropy is defined as:

$$L = -\sum g_i \log(y_i) + \lambda_1 \left(\sum \theta^2 \right)$$

where $y_i \in R^{\text{classes}}$ denotes the actual, $\hat{y} \in R^{\text{classes}}$ is the predicted probability for each class and λ_1 is the coefficient for L_2 regularization.

We use the back-propagation method to compute the gradients and update all the parameters by:

$$\Theta = \Theta - \lambda_2 \left(\frac{\delta L(\Theta)}{\delta \Theta} \right)$$

where λ_2 is the learning rate. We trained our models with *Adam* optimizer. In order to avoid over-fitting, we use dropout to randomly omit half of the features on each training case.

6 Sentiment Models

6.1 Models

Sentiment models predict a score between $[-1, 1]$ for the extracted aspect. We use various combinations of models to achieve SOTA results.

Interactive Attention Model The model was proposed by [17]. Figure 5 shows an overall framework for this architecture. The paper proposes that the target and context words play an important role in learning each other representation. The model comprises two interdependent networks for the target and context words. Here, the target words are the snippets present in the data set and context are all words other than the snippet. The architecture uses an interdependent or interactive approach that uses targets to get important information from the context and uses context to supervise the target which is helpful to sentiment prediction. Finally, the two representations, target and context, are concatenated and fed to softmax layer for final prediction.

Interactive Attention Model with Features This uses a similar framework as interactive attention model. The architecture is depicted in Fig. 6. The only difference from the above approach is the use of handcrafted features like agreement score, PMI score and others that were discussed in Sect. 4.1. This improves attention scores of the context in addition to the target as shown in interactive attention model [17].

Fig. 5 A high-level diagram of interactive attention model

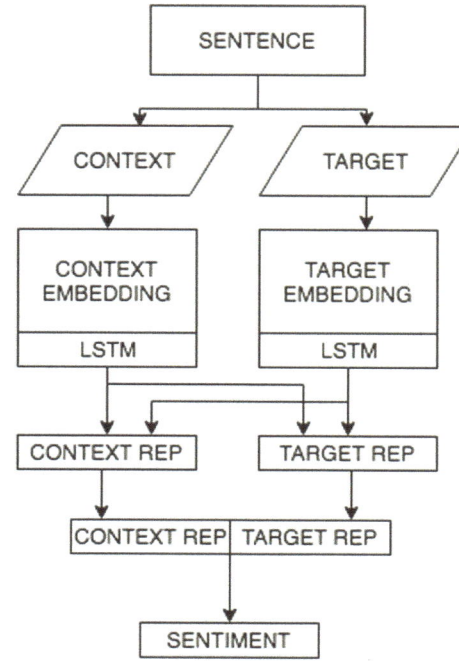

Fig. 6 Interactive attention model with features

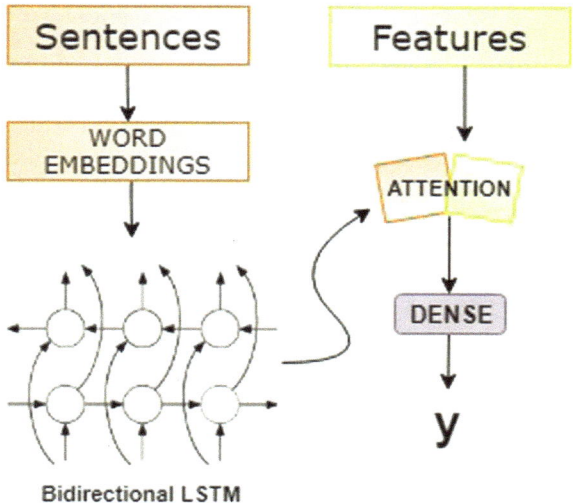

Features play an important role in assigning attention to the sentence for sentiment prediction and forming the final representation. The final representations are then passed through fully connected layer with *sigmoid* [49] activation. Different combinations of features were tried before making the final optimal choice.

Attention In text classification and neural machine translation tasks, certain relation exists between the input text and the output text, where each word is highly related to a certain part of the input text. This intuition inspires the attention mechanism [16, 18, 23, 30, 39, 41, 47, 51].

Past research shows that the use of LSTM [5, 9] helps in learning only from the previous time steps. BiLSTM [7] summarizes the context from both forward and backward passes of a sequence. The forward LSTM $\overrightarrow{h}$ reads a sequence and calculates a sequence of forward hidden states, and the backward LSTM $\overleftarrow{h}$ reads the same sequence in the opposite direction to calculate a sequence of backward hidden. For each word w_j, we get the representation h_j by concatenation of the forward hidden state $\overrightarrow{h}$ with the backward hidden state $\overleftarrow{h}$.

$$h = [\overrightarrow{h}, \overleftarrow{h}] \tag{12}$$

This shows that a standard LSTM or BiLSTM fails at successfully highlighting the important part for aspect-level sentiment classification. In order to solve this issue, we designed an attention architecture that can highlight the key part of a sentence in response to a given aspect.

The two architectures designed for the purpose are as follows:

1. **Attention with features** Let the output of BiLSTM, $H \in R^{d \times N}$, be the hidden vector matrix , where d is the size of hidden layers and N is the length of the given sentence. A set of features discussed in Sect. 4.1 are passed to four hidden dense layers having different combinations of neurons, respectively. The choice of features and the combination of neurons used is a hyperparameter.

   The output of MLP is concatenated with the H matrix to produce the feature matrix F. A high-level diagram of the proposed framework is depicted in Fig. 7.

$$F = \text{concatenate}(H, \text{Features}) \tag{13}$$

$$\alpha = \text{ReLU}(w, F) \tag{14}$$

A representation score r is calculated using weighted sum of attention vector α and hidden vector matrix $\mathbf{H}$. α is computed using the feature matrix $\mathbf{F}$.

$$r = \sum H * \alpha \tag{15}$$

The final sentence representation is given by:

$$h_{\text{final}} = \text{ReLU}(W_1 * r) \tag{16}$$

Fig. 7 Attention with features

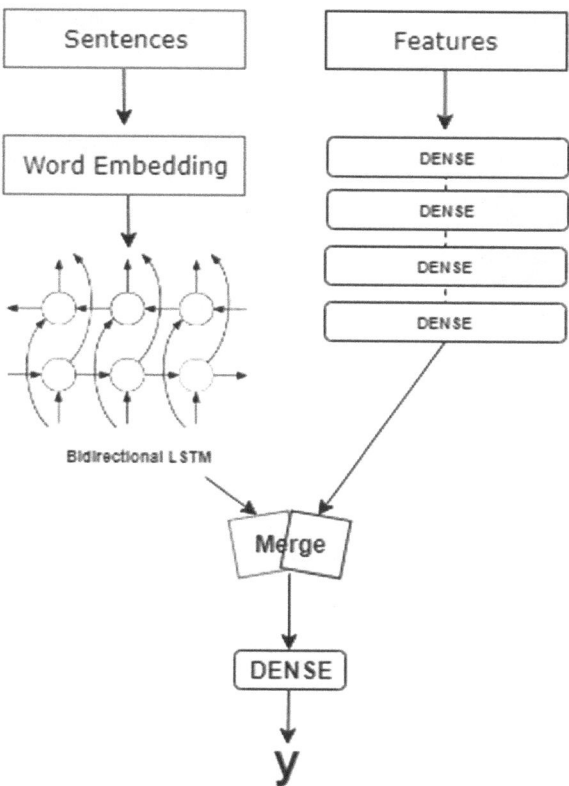

2. **Attention with target information** The model is proposed by Wang et al. [44]. Here, the aspect embeddings are the target embeddings obtained from the snippets. We can incorporate these embeddings as suggested in the paper in three ways:

   (a) **Concatenate with the context embeddings**: Target embeddings are concatenated with context embeddings and then passed to the LSTM.
   (b) **Concatenate with the context representation of LSTM**: Word embeddings are passed to LSTM, and their representation is concatenated with the target embeddings. A high-level representation of the model is shown in Fig. 8.
   (c) **Concatenate with the context embeddings and the context representations of LSTM**: Both (a) and (b) can be combined such that we have target embeddings before passing it to LSTM and after LSTM too. This is depicted in Fig. 9.

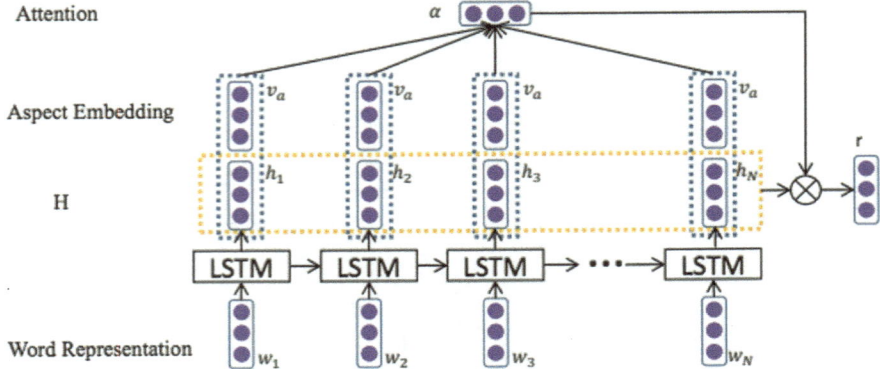

Fig. 8 Diagram represents attention with targets when concatenation is performed with the context representation of LSTM

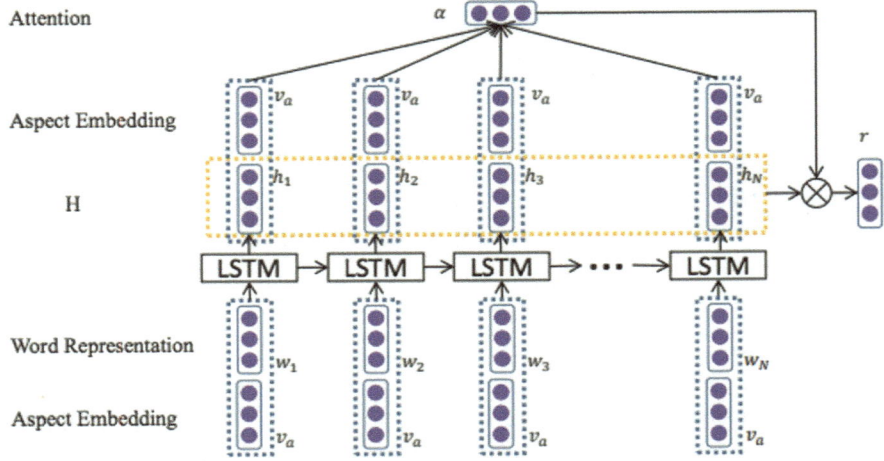

Fig. 9 Diagram represents attention with targets when concatenation is performed with the context embeddings and the context representations of LSTM

CLSTM with Aspect In this architecture, we use aspect embeddings to represent aspect information instead of using handcrafted features. A high-level representation of the model is shown in Fig. 10. This framework is similar to the one depicted in the ALA model [36]. Here, CNN [14] and LSTM [5] are used rather than stacked LSTM as proposed in the ALA model. CNNs are used to capture broader and prominent features of the sentence, while LSTM is used to capture long-term dependencies. This architecture takes sentence and aspect as input and gets sentence embedding and aspect embedding, respectively.

We apply convolution operation on window vectors with a filter q to generate a feature map $m \in R^{Lk+1}$.

Fig. 10 CLSTM with aspect embeddings

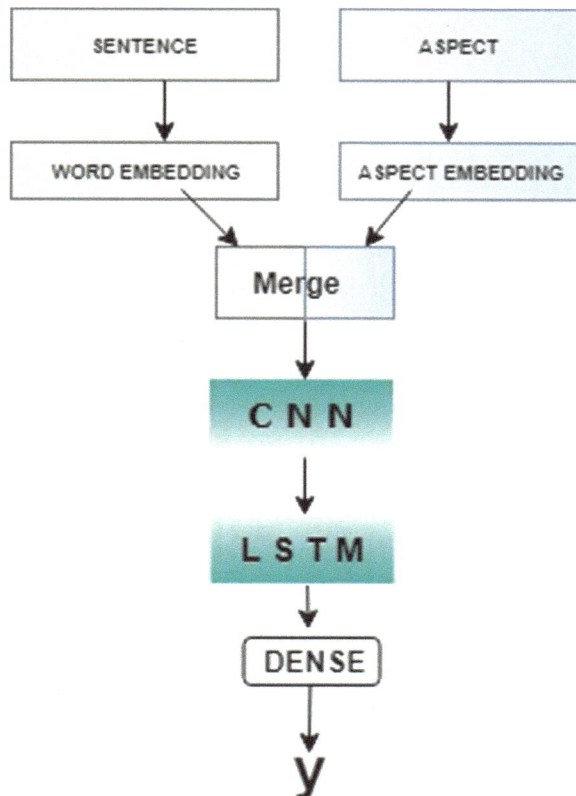

Each element m_j of the feature map for each window vector w_j is produced as follows:

$$m_j = f(w_j * q + b) \tag{17}$$

where * is element-wise multiplication, $b \in R$ is a bias term and f is a nonlinear transformation function tanh or ReLU. The CLSTM model uses multiple filters to generate multiple feature maps. Different feature maps are then concatenated together and sent as an input to the LSTM network.

6.2 Sentiment Model Training

Mean squared error (MSE) is used as a loss function for evaluating the sentiment models. If y is the actual sentiment, and $\hat{y}$ is the predicted sentiment, we aim to minimize the mean squared error between y and $\hat{y}$.

Mean squared error is defined as:-

$$\text{MSE} = \frac{1}{n} \sum (y - \hat{y})^2 \tag{18}$$

where y denotes the ground truth sentiment values and $\hat{y}$ is the predicted sentiment values. We use ReLU [29] as an activation function in the intermediate layers and *sigmoid* [49] in the last layer.

7 Evaluation

This section gives a detailed overview of the data set, introduced in the FiQA challenge, followed by the necessary steps taken to pre-process the data. Since data was highly imbalanced that could make results screwed, data augmentation was performed. This section concludes with details of hyperparameters used, metric chosen for evaluation and results obtained.

7.1 Data Set

The data set used for training is available at FiQA 2018 website. Organizers of the task provided 435 annotated financial headlines and 675 annotated financial tweets as the training data set. For each tweet/headline, targets and snippet relevant to the target were provided. Each target has sentiment score and an aspect. Examples of data block for a headline and a tweet are shown below.

```
"55": {
    "sentence": "Tesco Abandons Video-Streaming
     Ambitions in Blinkbox Sale",
    "info": [
      {
        "snippets": "['Video-Streaming Ambitions']",
        "target": "Blinkbox",
        "sentiment_score": "-0.195",
        "aspects": "['Corporate/Strategy']"
      },
      {
        "snippets": "['Tesco Abandons Video-Streaming
         Ambitions']",
        "target": "Tesco",
        "sentiment_score": "-0.335",
        "aspects": "['Corporate/Strategy']"
      }
    ]
  }
```

```
"14864": {
"sentence": "$TRX http://stks.co/1KkK Long setup. MACD cross.",
"info": [
  {
    "snippets": "['Long setup. MACD cross.']",
    "sentiment_score": "0.438",
    "target": "TRX",
    "aspects": "['Stock/Technical Analysis']"
  }
]
}
```

The aspect of each snippet can have up to six levels, but for this challenge, all the participants were required to report aspects up to level 2 only. Level 1 has 4 unique aspects, and level 2 has 27 unique aspects. A major problem that existed in the training data set was class imbalance problem. Table 1 clearly shows that the frequency distribution for aspect "corporate" ranges from 463 to "stock" with 649 and is insignificant for "economy" and "market". Similarly, for aspect level 2, there is a lot of imbalance with "price action" being the most frequent while others are negligible in its comparison. To solve this issue, concept of data augmentation is used, details of which are in the next subsection.

7.2   Data Augmentation

As shown in Table 1, the training data set was highly imbalanced that led to biased outcomes for more frequent class. To solve this problem, data augmentation is applied. It is a process of adding more data to training set to produce more robust models. For the current task in hand, we did data augmentation by Web scraping various financial news websites like Reuter using Beautiful Soup and Selenium tools. We upsampled only those classes which had very insignificant frequency like IPO, trade, insider activity, currency and buyside. We upsampled each of these aspects to about 50 more samples, getting a total data set of 1343 financial tweets. This is shown in Fig. 11. We too performed downsampling of the aspect "price action", followed by looking at samples that were oversampled and decided to take different sampling ratios. This kind of different experimentations with data using data augmentation techniques showed an improvement in our proposed model.

7.3   Data Pre-processing

The data set was collected using Twitter as a social media platform. So to perform text pre-processing on Twitter posts, standard steps were applied. This included removal of: (i) Twitter usernames, (ii) stop words, (iii) punctuation, (iv) website

Table 1 Frequency distribution of aspect in training data

Aspect level 1	Aspect level 2	Frequency
Corporate	Reputation	1
	Company communication	8
	Appointment	37
	Financial	26
	Regulatory	18
	Sales	92
	M&A	76
	Legal	28
	Dividend policy	26
	Risks	57
	Rumour	33
	Strategy	49
Stock	Options	12
	IPO	8
	Signal	26
	Coverage	45
	Fundamentals	13
	Insider activity	5
	Price action	437
	Buyside	5
	Technical analysis	98
Economy	Trade	2
	Central banks	5
Market	Currency	2
	Conditions	3
	Market	24
	Volatility	11

links and (iv) hashtags from all the sentences. All the occurrences of the aspect word "company" are replaced with "COMPANY", and all other general occurrences of company are replaced by "OCOMPANY". Each sentence is padded so that they can match the length of longest sentence. After applying all the relevant pre-processing steps, final length of sentence was fixed to 16. The sentiment score was scaled down to the range of [0, 1] from [−1, 1].

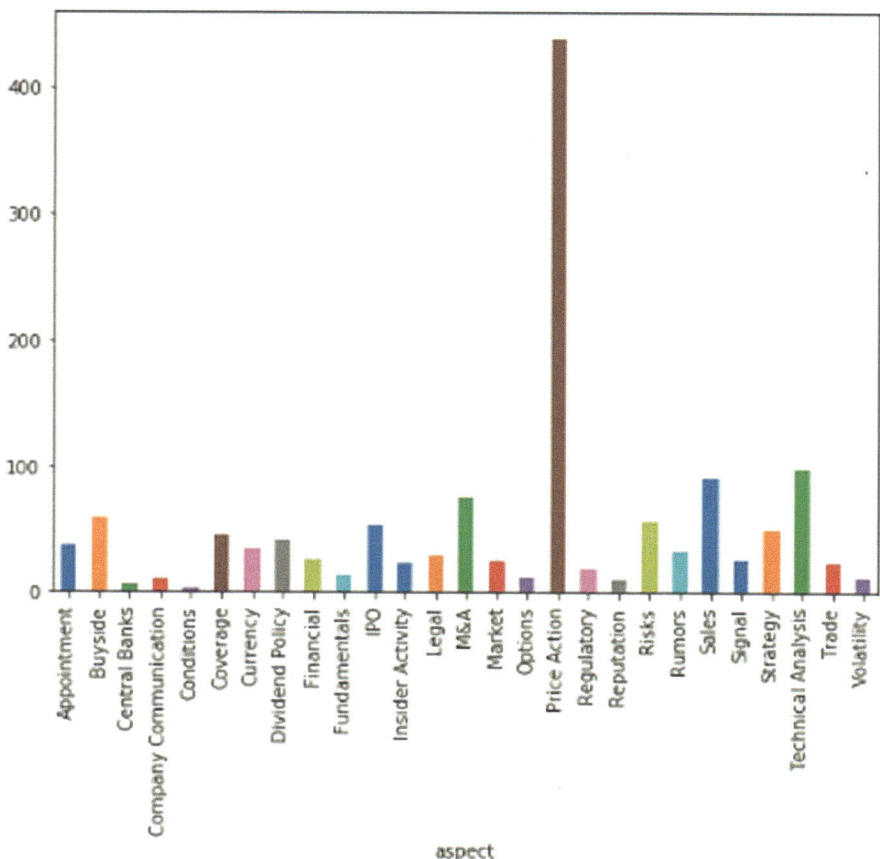

Fig. 11 Upsampling of less frequent classes after performing data augmentation

7.4 Metrics

The evaluation metric chosen by the FIQA challenge to evaluate the performance of the aspect classification task was macro-F1. To deal with class imbalance problem in training data, we also calculated weighted average F1 score as it keeps an account for the class imbalances while calculating the score. F1 score is based on precision and recall. The formula is as follows:

$$F_1 = 2 * \frac{\text{Precision} \times \text{Recall}}{\text{Precision} + \text{Recall}} \tag{19}$$

This is a better measure than accuracy when we need to seek a balance between precision and recall and when there is an uneven class distribution, which is the case here.

For measuring the performance of the sentiment model, mean squared error (MSE), $r^2\,score$ and cosine similarity were asked in the challenge. $r^2\,score$ is a statistical measure of how close the data is to the fitted regression line. It is also known as the coefficient of determination.[3] Generally, higher the score, better the model is. But $r^2\,score$ is not a good metric for evaluation when the data has a lot of variation. It is possible to have a low $r^2\,score$ for a well-fitted model and a higher one for a model that does not fit well with the data.[4] $r^2\,score$ should be computed with residual plots to get the entire picture of the regression model. We have computed the $r^2\,score$ as the challenge providers had asked for it.

MSE measures average squared error of our predictions. For each point, it calculates square difference between the predictions and the target and then averages those values. Higher the MSE score, worse the model is.

$$\text{MSE} = \frac{1}{n} \sum_{i=1}^{n} \left(Y_i - \hat{Y} \right)^2 \tag{20}$$

7.5 Results

We outperformed the existing state-of-the-art methods in the sentiment model's MSE score by 50% and for the aspect part by 20%.

In reference to Tables 2 and 3, we have a significant improvement in MSE and cosine similarity scores. However, a low $r^2\,score$ is not indicative of a bad fitted model as the data has a lot of variation. $r^2\,score$ is not a good metric for higly skewed data. It can be high for bad fitted model and low for a well-fitted one.

Tables 2, 3, 4 and 5 show the comparison of existing models in this domain with our proposed model in the decreasing order of their performances. The details of both sentiment and aspect are discussed below:

- **For Sentiment Prediction Task**: It is clearly visible from Tables 2 and 3 that our proposed models performed better than the existing models in sentiment prediction track for both headlines and microblogs. Though the use of lexicons and agreement score as features have improved the performance, PMI score did not made any contribution. After observing the results, we can understand the strength of the handcrafted features. We also experimented with support vector regression (SVR)

---

[3]https://blog.minitab.com/blog/adventures-in-statistics-2/regression-analysis-how-do-i-interpret-r-squared-and-assess-the-goodness-of-fit.

[4]https://blog.minitab.com/blog/adventures-in-statistics-2/five-reasons-why-your-r-squared-can-be-too-high.

Table 2 Sentiment model comparison for headlines

Model	Headlines		
	MSE	R^2	cosine sim
Attention with features	0.0661	0.0662	0.8945
BiLSTM with features	0.0666	0.0712	0.0712
Tongji-CUKG [36]	0.1345	0.4579	0.6768
CLSTM with aspect	0.1599	−1.2753	0.7849
IIIT Delhi [12]	0.2039	0.1779	0.4401
Inf-UFG	0.2067	0.1665	0.4153

Table 3 Sentiment model comparison for microblogs

Model	Microblogs		
	MSE	R^2	cosine sim
Attention with features	0.03042	−0.06134	0.9509
BiLSTM with features	0.0352	−0.0982	0.7021
CLSTM with aspect	0.08723	−1.8681	0.8882
Inf-UFG	0.0958	0.1642	0.5333
Tongji-CUKG [36]	0.1040	0.0923	0.6063
IIIT Delhi [12]	0.1049	0.0849	0.3422
NLP301	0.3058	−1.667	−0.0685

models with $C = [1, 10, 100]$. The best performing model was BiLSTM with attention.

– **For Aspect Classification Task**: In this task, for headlines, refer Table 4, and for microblogs refer Table 5. In the aspect classification track for headlines, vector averaging model and BiLSTM with attention outperformed other models. We achieved an accuracy of 35.48% in headlines and 77.7% in the post. We also experimented with AdaBoost, gradient boosting, logistic regression, k-nearest neighbours, random forests and decision trees classifiers. However, all of them failed in performing better than the existing models.

8   Conclusion and Future Work

In this chapter, we discussed ABSA for financial domain using microblogs and news headlines. We made use of features mentioned in Sect. 4.1 to identify important features from the texts. With the proposed framework, we clearly state the importance of hand-engineered features.

Table 4 Aspect model comparison for headlines

Model	Headlines		
	Accuracy	F1 score	
		Macro	Weighted
Sentence vectors	0.3548	0.2639	0.3633
BiLSTM with attention	0.3010	0.1525	0.2835
Tongji-CUKG [36]	0.2688	0.1399	–
tf–idf features	0.2473	0.1346	0.2330
IIIT Delhi [12]	0.0537	0.0149	0.327

Table 5 Aspect model comparison for microblogs

Model	Microblogs		
	Accuracy	F1 score	
		Macro	Weighted
Tongji-CUKG [36]	0.8484	0.4619	–
tf–idf features	0.7777	0.3320	0.7693
Naive Bayes count vectors	0.7171	0.2854	0.6443
BiLSTM with attention	0.6666	0.2127	0.6998
NLP301	0.7575	0.2832	–
IIIT Delhi [12]	0.2424	0.0250	0.8980

For future work, we aim to train our proposed models on a larger data set to capture the information about the aspect classes that are not studied due to unavailability of enough training samples in the data set proposed by challenge organizers. Other goals include: (i) evaluating the performance of recent transfer learning-based NLP technique (i.e. ULMFIT [10]) for the ABSA task on financial headlines and (ii) incorporating knowledge graphs for finance knowledge and enhancing the features and embeddings by using them to get better performance.

References

1. Chikersal, Prerna, Soujanya Poria, and Erik Cambria. 2015. SeNTU: sentiment analysis of tweets by combining a rule-based classifier with supervised learning. In *Proceedings of the 9th International Workshop on Semantic Evaluation (SemEval 2015)*, 647–651
2. Chowdhury, Arijit Ghosh, Ramit Sawhney, Puneet Mathur, Debanjan Mahata, and Rajiv Ratn Shah. 2019. Speak up, fight back! Detection of social media disclosures of sexual harassment. In *Proceedings of the 2019 Conference of the North American Chapter of the Association for Computational Linguistics: Student Research Workshop*, 136–146

3. Dashtipour, Kia, Soujanya Poria, Amir Hussain, Erik Cambria, Ahmad Y.A. Hawalah, Alexander Gelbukh, and Qiang Zhou. 2016. Multilingual sentiment analysis: state of the art and independent comparison of techniques. *Cognitive Computation* 8 (4), 757–771
4. Ding, Xiaowen, Bing Liu, and Philip S. Yu. 2008. A holistic lexicon-based approach to opinion mining. In *Proceedings of the 2008 International Conference on Web Search and Data Mining*, 231–240
5. Gers, Felix A, Nicol N Schraudolph, and Jürgen Schmidhuber. 2002 Aug. Learning precise timing with LSTM recurrent networks. *Journal of Machine Learning Research* 3, 115–143
6. Goldberg, Yoav, and Omer Levy. 2014. word2vec explained: deriving Mikolov et al.'s negative-sampling word-embedding method. arXiv preprint arXiv:1402.3722
7. Graves, Alex, and Jürgen Schmidhuber. 2005. Framewise phoneme classification with bidirectional LSTM and other neural network architectures. *Neural Networks* 18 (5–6), 602–610
8. Hazarika, Devamanyu, Soujanya Poria, Prateek Vij, Gangeshwar Krishnamurthy, Erik Cambria, and Roger Zimmermann. 2018. Modeling inter-aspect dependencies for aspect-based sentiment analysis. In *Proceedings of the 2018 Conference of the North American Chapter of the Association for Computational Linguistics: Human Language Technologies*, volume 2 (short papers), vol. 2, 266–270
9. Hochreiter, Sepp, and Jürgen Schmidhuber. 1997. Long short-term memory. *Neural Computation* 9 (8), 1735–1780
10. Howard, Jeremy, and Sebastian Ruder. 2018. Fine-tuned language models for text classification. CoRR **abs/1801.06146**. URL http://arxiv.org/abs/1801.06146
11. Jain, Roopal, Sawhney Ramit, and Mathur Puneet. 2018. Feature selection for cryotherapy and immunotherapy treatment methods based on gravitational search algorithm. In *International Conference on Current Trends Towards Converging Technologies (ICCTCT)*, 1–7. New York: IEEE
12. Jangid, Hitkul, Shivangi Singhal, Rajiv Ratn Shah, and Roger Zimmermann. 2018. Aspect-based financial sentiment analysis using deep learning. In *Companion of the the Web Conference 2018 on the Web Conference 2018, 1961–1966*. International World Wide Web Conferences Steering Committee. International World Wide Web Conferences Steering Committee
13. Kar, Sudipta, Suraj Maharjan, and Thamar Solorio. 2017. RiTUAL-UH at SemEval-2017 task 5: sentiment analysis on financial data using neural networks. In *Proceedings of the 11th International Workshop on Semantic Evaluation (SemEval-2017)*, 877–882
14. Kim, Yoon. 2014. Convolutional neural networks for sentence classification. CoRR **abs/1408.5882**. URL http://arxiv.org/abs/1408.5882
15. Kumar, Abhishek, Abhishek Sethi, Md Shad Akhtar, Asif Ekbal, Chris Biemann, and Pushpak Bhattacharyya. 2017. IITPB at SemEval-2017 task 5: sentiment prediction in financial text. In *Proceedings of the 11th International Workshop on Semantic Evaluation (SemEval-2017)*, 894–898
16. Luong, Minh-Thang, Hieu Pham, and Christopher D Manning. 2015. Effective approaches to attention-based neural machine translation. arXiv preprint arXiv:1508.04025
17. Ma, Dehong, Sujian Li, Xiaodong Zhang, and Houfeng Wang. 2017. Interactive attention networks for aspect-level sentiment classification. arXiv preprint arXiv:1709.00893
18. Ma, Yukun, Haiyun Peng, and Erik Cambria. 2018. Targeted aspect-based sentiment analysis via embedding commonsense knowledge into an attentive LSTM. In *Proceedings of AAAI*
19. Mathur, Puneet, Rajiv Shah, Ramit Sawhney, Debanjan Mahata. 2018a. Detecting offensive tweets in Hindi-English code-switched language. In *Proceedings of the Sixth International Workshop on Natural Language Processing for Social Media*, 18–26
20. Mathur, Puneet, Ramit Sawhney, Meghna Ayyar, and Rajiv Shah. 2018b. Did you offend me? classification of offensive tweets in Hinglish language. In *Proceedings of the 2nd Workshop on Abusive Language Online (ALW2)*, 138–148
21. Mikolov, Tomas, Ilya Sutskever, Kai Chen, Greg S. Corrado, and Jeff Dean. 2013. Distributed representations of words and phrases and their compositionality. In *Advances in Neural Information Processing Systems*, 3111–3119

22. Mishra, Rohan, Pradyumn Prakhar Sinha, Ramit Sawhney, Debanjan Mahata, Puneet Mathur, and Rajiv Ratn Shah. 2019. SNAP-BATNET: cascading author profiling and social network graphs for suicide ideation detection on social media. In *Proceedings of the 2019 Conference of the North American Chapter of the Association for Computational Linguistics: Student Research Workshop*, 147–156

23. Mnih, Volodymyr, Nicolas Heess, Alex Graves. 2014. Recurrent models of visual attention. Advances in neural information processing systems, 2204–2212

24. Pennington, Jeffrey, Richard Socher, and Christopher Manning. 2014. Glove: global vectors for word representation. In *Proceedings of the 2014 Conference on Empirical Methods in Natural Language Processing (EMNLP)*, 1532–1543

25. Piao, Guangyuan, and John G. Breslin. 2018. Financial aspect and sentiment predictions with deep neural networks: an ensemble approach. In *Companion of the the Web Conference 2018 on the Web Conference 2018, 1973–1977*. International World Wide Web Conferences Steering Committee. International World Wide Web Conferences Steering Committee

26. Poria, Soujanya. 2017. Novel symbolic and machine-learning approaches for text-based and multimodal sentiment analysis

27. Poria, Soujanya, Erik Cambria, Alexander Gelbukh. 2016. Aspect extraction for opinion mining with a deep convolutional neural network. *Knowledge-Based Systems* 108, 42–49

28. Poria, Soujanya, Erik Cambria, Ku Lun-Wei, Chen Gui, and Alexander Gelbukh. 2014. A rule-based approach to aspect extraction from product reviews. In *Proceedings of the Second Workshop on Natural Language Processing for Social Media (SocialNLP)*, 28–37

29. Radford, Alec, Luke Metz, and Soumith Chintala. 2015. Unsupervised representation learning with deep convolutional generative adversarial networks. arXiv preprint arXiv:1511.06434

30. Rush, Alexander M, Sumit Chopra, and Jason Weston. 2015. A neural attention model for abstractive sentence summarization. arXiv preprint arXiv:1509.00685

31. Sawhney, Ramit, Puneet Mathur, and Ravi Shankar. 2018a. A firefly algorithm based wrapper-penalty feature selection method for cancer diagnosis. In *International Conference on Computational Science and Its Applications*, 438–449

32. Sawhney, Ramit, Ravi Shankar, and Roopal Jain. 2018b. A comparative study of transfer functions in binary evolutionary algorithms for single objective optimization. In *International symposium on distributed computing and artificial intelligence*, 27–35. Berlin: Springer

33. Sawhney, Ramit, Prachi Manchanda, Raj Singh, and Swati Aggarwal. 2018c. A computational approach to feature extraction for identification of suicidal ideation in tweets. In *Proceedings of ACL 2018, Student Research Workshop*, 91–98

34. Sawhney, Ramit, Prachi Manchanda, Puneet Mathur, Rajiv Shah, and Raj Singh. 2018d. Exploring and learning suicidal ideation connotations on social media with deep learning. In *Proceedings of the 9th Workshop on Computational Approaches to Subjectivity, Sentiment and Social Media Analysis*, 167–175

35. Shah, Rajiv, and Roger Zimmermann. 2017. Multimodal analysis of user-generated multimedia content

36. Shijia, E, Li Yang, Mohan Zhang, and Yang Xiang. 2018. Aspect-based financial sentiment analysis with deep neural networks. In *Companion of the the Web Conference 2018 on the Web Conference 2018*, 1951–1954. International World Wide Web Conferences Steering Committee. International World Wide Web Conferences Steering Committee

37. Snoek, Jasper, Hugo Larochelle, and Ryan P. Adams. 2012. Practical Bayesian optimization of machine learning algorithms. In *Advances in Neural Information Processing Systems*, 2951–2959

38. Socher, Richard, Cliff C. Lin, Chris Manning, and Andrew Y. Ng. 2011. Parsing natural scenes and natural language with recursive neural networks. In *Proceedings of the 28th International Conference on Machine Learning (ICML-11)*, 129–136

39. Tang, Duyu, Bing Qin, and Ting Liu. 2016. Aspect level sentiment classification with deep memory network. arXiv preprint arXiv:1605.08900

40. Turney, Peter D., and Pantel, Patrick. 2010. From frequency to meaning: vector space models of semantics. *Journal of Artificial Intelligence Research* 37, 141–188

41. Vinyals, Oriol, Łukasz Kaiser, Terry Koo, Slav Petrov, Ilya Sutskever, and Geoffrey Hinton. 2015. Grammar as a foreign language. In*Advances in Neural Information Processing Systems*, 2773–2781
42. Vo, Duy-Tin, Yue Zhang. 2015. Target-dependent twitter sentiment classification with rich automatic features. *IJCAI*, 1347–1353
43. Wang, C-J et al. 2013. Financial sentiment analysis for risk prediction. In: Proceedings of the Sixth International Joint Conference on Natural Language Processing
44. Wang, Yequan, Minlie Huang, Li Zhao, et al., 2016. Attention-based LSTM for aspect-level sentiment classification. In *Proceedings of the 2016 Conference on Empirical Methods in Natural Language Processing*, 606–615
45. Weston, Jason, Samy Bengio, and Nicolas Usunier. 2011. Wsabie: scaling up to large vocabulary image annotation. *IJCAI* 11, 2764–2770
46. Wilson, Theresa, Janyce Wiebe, and Paul Hoffmann. 2005. Recognizing contextual polarity in phrase-level sentiment analysis. In *Proceedings of the Conference on Human Language Technology and Empirical Methods in Natural Language Processing*, 347–354. Association for Computational Linguistics. Association for Computational Linguistics
47. Xu, Kelvin, Jimmy Ba, Ryan Kiros, Kyunghyun Cho, Aaron Courville, Ruslan Salakhudinov, Rich Zemel, and Yoshua Bengio. 2015. Show, attend and tell: Neural image caption generation with visual attention. In *International Conference on Machine Learning*, 2048–2057
48. Young, Tom, Devamanyu Hazarika, Soujanya Poria, and Erik Cambria. 2018. Recent trends in deep learning based natural language processing. *IEEE Computational Intelligence Magazine* 13 (3), 55–75
49. Zadeh, Mehdi Rezaeian, Seifollah Amin, Davar Khalili, Vijay P. Singh. 2010. Daily outflow prediction by multi layer perceptron with logistic sigmoid and tangent sigmoid activation functions. *Water Resources Management* 24 (11), 2673–2688
50. Zhou, Chunting, Chonglin Sun, Zhiyuan Liu, and Francis Lau. 2015. A C-LSTM neural network for text classification. arXiv preprint arXiv:1511.08630
51. Zhou, Peng, Wei Shi, Jun Tian, Zhenyu Qi, Bingchen Li, Hongwei Hao, and Bo Xu. 2016. Attention-based bidirectional long short-term memory networks for relation classification. In *Proceedings of the 54th Annual Meeting of the Association for Computational Linguistics* (volume 2: Short papers), vol. 2, 207–212

Deep Learning-Based Frameworks for Aspect-Based Sentiment Analysis

Ashish Kumar and Aditi Sharan

Abstract Opinions are key influencers of almost all human practices. One can easily find a number of opinions about any product or services in the form of product reviews. These product reviews are available in a tremendous amount. It is not feasible or even impossible to go through each review and make a concise decision about any product. Aspect-based sentiment analysis (ABSA) comes as a solution to this problem. It gives an approach to examine online reviews and provides a summary based on these reviews. Machine learning techniques have been broadly utilized for ABSA. Recently with the evolution of processing power of computers and digitization of the society, deep learning is taking off. Deep learning methods produced state-of-the-art results in various NLP tasks without intensive feature engineering. In this chapter, we present an introduction about ABSA following a comprehensive overview of various deep learning models used in the field of ABSA.

Keywords Aspect-based sentiment analysis · Recurrent neural network · Long short-term memory · Convolution neural network · Natural language processing

1 Introduction

Because of the simple openness of the Web, people are often using Web portals to frame an opinion about a certain product, topic, and service. These online reviews expressed online opinions. These online opinions are valuable resources for decision making. The availability of enormous reviews does not ease our task; in fact, the complications are increased as it is not possible to read each and every review. In

A. Kumar (✉) · A. Sharan
School of Computer and Systems Sciences,
Jawaharlal Nehru University, New Delhi 110067, India
e-mail: ashishkumar2912@gmail.com; ashish29_scs@jnu.ac.in

A. Sharan
e-mail: aditisharan@mail.jnu.ac.in

© Springer Nature Singapore Pte Ltd. 2020   139
B. Agarwal et al. (eds.), *Deep Learning-Based Approaches
for Sentiment Analysis*, Algorithms for Intelligent Systems,
https://doi.org/10.1007/978-981-15-1216-2_6

order to make an opinion about a product based on its reviews, it is critical to analyze the sentiment associated with reviews.

Sentiment analysis on opinions can be done at various levels, viz. document, sentence, and aspect. Document and sentence levels deal with the identification of overall opinion in the designated document and sentence, respectively. But these levels ignore the fact that a document and sentence may talk about different features (aspects) of an entity. There is a need for extracting these aspects and their corresponding sentiment polarity. This process is called aspect-based sentiment analysis (in short ABSA).

ABSA is a task that involves various subtasks. Following are some of these subtasks [6].

1. Identification of aspect-terms
2. Identification of opinion-terms
3. Extraction of aspect-categories
4. Sentiment (Polarity) identification
5. Sentiment intensity identification
6. Sentiment shifters identification
7. Opinion holder and time identification
8. Generation of opinion tuple
9. Opinion summarization

An opinion can be expressed or understood in different ways. However, according to [16], an objective definition of opinion is given below:

Definition 1 (*Opinion*) An opinion is a quintuple.

$$(e_i, a_{ij}, s_{ijkl}, h_k, t_l) \tag{1}$$

Table 1 Example reviews

Reviews	Entity	Aspect-term	Aspect-category	Opinion-word
1. This laptop is great!	Laptop	–	–	Great
2. It is very overpriced and not very tasty	Restaurant	–	Food, Price	Overpriced, tasty
3. The pizza was pretty good and huge	Restaurant	Pizza	Food	Good, huge
4. it is the best service you will find in even the largest of restaurants	Restaurant	Service	Service	Best
5. It is not worth for that bucks	–	–	Price	Worth

In the problem of sentiment analysis (opinion mining), the relation between the items of tuples is extracted, for example, identifying the different aspects a_{ij} of an entity e_i, determining sentiment s_{ijkl} of an aspect a_{ij}, and finding the opinion holder h_k, who has expressed the opinion at time t_l. Sentiment s_{ijkl} can be positive, negative, or neutral, or can take any discrete value on a certain scale to define sentiment intensity.

Before proceeding to ABSA, it is important to objectively define the notion of aspect-terms and aspect-categories and some other related terms. There are various terms used in ABSA, i.e., entity, aspect-term, aspect-category, opinion-word, etc. Let us try to understand these terms with the help of some examples.

It is clear from Table 1 that reviews are regarding some entities like Laptop, Restaurants, etc. However, they may point out the sentiment about the entity as a whole (in Review 1) or about some features of the entity (in Reviews 2, 3, 4, 5). Different aspect-terms can be used to render an opinion about an aspect-category. Aspect-terms like *pizza*, *bagels*, etc., can be used to put an opinion about aspect-category Food.

Aspect-category is a generic notion/concept/property of an entity. The opinion is generally expressed about an aspect-category. Aspect-categories may be different in different domain. For an instance, let us take an example of Restaurant domain; then, the possible category list includes Service, Food, Price, and Ambience. Different aspect-terms may belong to an aspect-category. Aspect-term and aspect-category are two different things. In some cases, aspect-category may be explicitly mentioned by an aspect-term, (in Reviews 3 and 4). However, such cases are very few. An aspect-category is quite often a hypernym of an aspect-term. For example, Food is hypernym of pizza, but that is not always the case. Aspect-category can be implied implicitly also (in Reviews 2 and 5).

ABSA may be performed either on aspect-terms or on aspect-category. As aspect-terms are explicit, it is easier to identify these terms and relate them with the sentiment expressed. However, when we are performing ABSA on an entity, we may be interested in analyzing opinions about some well-defined aspects of the entity, as discussed earlier. Each of these represents an aspect-category. Thus, ABSA on aspect-categories is more appealing, but at the same time more demanding. When we speak about review sentences, a sentence may talk about a single aspect-category, however, that is not always the case. Some sentences may depict opinion about multiple categories also (in Review 1). This adds further challenges to ABSA.

With the accessibility of an enormous volume of data and increase in computational power of computers (GPUs) in the last decade, deep learning has become the first choice of the research community. Unlike traditional machine learning techniques that require intensive feature engineering and a separate model for classification, deep learning performs both tasks. It does feature engineering and classification with the help of input data only. Apart from good performance in image encouraging tasks, deep learning shows promising outcomes in various NLP tasks like named-entity recognition, text summarization, machine translation, sentiment analysis, etc.

This chapter highlights the diverse deep learning strategies for aspect-terms extraction, aspect-category detection, and sentiment polarity detection methods. This chapter shows the contribution and challenges of deep learning in ABSA. Chapter association is done in an accompanying manner. Section 2 describes the problem formulation in ABSA. Section 3 lighten up some useful observation/assumption by researchers in the field of ABSA. Section 4 talks about input representation for deep learning. In Sect. 5, introduction about some basic methods of deep learning is given. Section 6 highlights the different deep learning architectures used in ABSA. Finally, Sect. 7 finishes up the chapter.

2 Problem Formulation

As there is a lot of subjectivity involve in opinion mining and ABSA, it is important to formulate the problem of aspect-term extraction and aspect-category detection. This section formulates these two problems.

2.1 Aspect-Term Extraction

This task aims to extract explicit aspect expression presented in an online review. In most of the cases, ATE task can be examined as a sequence labeling problem where each review token is labeled to represent whether it is a piece of aspect-term or not. For this tagging process, popular BIO tagging scheme [22] is used generally, where B represents the beginning of aspect-term, I represents inside of aspect-term, and O is used for others that are not part of aspect-term.

For a given review sentence $x = (w^{<1>}, w^{<2>}, \ldots, w^{<T>})$, the output is a sequence of labels $y = (y^{<1>}, y^{<2>}, \ldots, y^{<T>})$, where $w^{<i>}$ represents word position in review sentence and each individual label $y^{<i>} \in \Sigma$; $\Sigma = \{B, I, O\}$. The problem can be considered as a multi-class classification problem with $|\Sigma|^T$ different classes.

2.2 Aspect-Category Detection

For a given predefined aspect-category set $C = \{c_1, c_2, c_3, \ldots, c_k\}$, where C denotes the category label space with k possible categories and a review dataset $R = \{r_1, r_2, r_3, \ldots, r_n\}$ containing n review sentences. The task of aspect-category detection can be formulated as a learning function $h : R \to 2^C$ from a multi-category training set $D = (r_i, Y_i) \mid 1 \leq i \leq n, Y_i \subseteq C$ is a set of category labels associated with r_i. For each unseen review $r \in R$, the aspect-category prediction function $h(\cdot)$ predicts $h(r) \subseteq C$ as the set of proper category label for r.

3 Observation/Assumption in ABSA

By conducting a survey on various articles regarding ABSA, we came up with the following observations/assumptions that are made by researchers. By considering these assumptions in mind while designing a model for ABSA, these assumptions can help to tackle the problem of ABSA in a very efficient way.

1. In ABSA, the context of words plays a very significant role. Words' location and how the words are interacting with each other matter a lot. A basic bag-of-words model is never again adequate, since all context information lost in the bag-of-words model [29]. To represent negative sentiment, generally negation of positive sentiment words is used. If we use bag-of-word, then it will be difficult to capture sentiment orientation in that case.
2. To construe the sentiment of a specific aspect in the review sentence, only some subsets of context words are required. Rather than focusing on full context, it is always beneficial to give attention to that subset of context words [28]. In the given review *"The price is reasonable although the service is poor."* The context word *reasonable* is more important as compared to *poor* for aspect "price." Oppositely *poor* is more important as compared to *reasonable* for aspect "service."
3. Aspect-term should co-occur with some opinion-words that helps in determining the sentiment polarity on it [15]. For example, Given review *"I've eaten at many different Indian restaurants."*, contains no opinion-words. Hence, the word *restaurants* should not be extracted as aspect-term.
4. In addition to the word-context association, handling the connection between sentiment words and aspects can likewise be valuable for ABSA [34]. For example, many sentiment words are aspect-specific like 'delicious' and 'tasty' are used for aspect *food* only while 'cheap' and 'costly' are used only for *price*. Dependency parsing will be helpful in capturing the connection between aspect-terms and sentiment (opinion) words.
5. In sequential labeling, the predictions at the previous time-steps are useful clues for reducing the error space of the current prediction. For example, in the B-I-O tagging, if the previous prediction is O, then the current prediction cannot be I [15].
6. In a sentiment classification task, while using deep neural networks like LSTM, casting sentence portrayal alone does not perform well. Fusing target information into LSTM helps in improvement of sentiment classification accuracy.
7. For aspect-term extraction task, if we are using CNN then do not apply any max-pooling layer after convolution layers because a sequence labeling model needs good representations for every position and max-pooling operation mixes the representations of different positions, which is undesirable [32].

4 Input Representation

Inputs for the neural networks should be represented appropriately for the desired outputs. One should consider the good representation of the input, so that designated neural network learn good features. Word-embedding forms the basic building block for representing text as an input to a deep neural network in the field of NLP. Following are some of the ways to represent a sentence/words/aspect-terms.

1. Each word of review sentence is represented as an embedding vector. Word-embeddings [17, 19] are distributed representation of words in a vector space. Words and phrases are encapsulated to vectors of real numbers. Word-embedding represents word meaning from its surrounding context which is learned from large corpora.
2. To represent any sentence, one can take the bag-of-words approach by averaging the word vectors of the input sentence.
3. Each aspect-term can be represented using word-embeddings. For the aspect-term that consists only single word can be represented using the word-embedding of that word only. But for multi-word aspect-term, averaging of each word can be a way to represent multi-word aspect-term [28].
4. Some time it is better to represent words as a consolidation of word-embeddings and character-embeddings [10] to illustrate the effect of word morphology.
5. However, it has been observed that aspect-terms are generally nouns or noun phrases. So passing the POS information along with word information can be useful. If there are six pos-tags (noun, verb, adverb, adjective, preposition, and conjunction), these can be encoded as a six-dimensional vector. If the word-embedding dimension is 300, then word + POS features dimension will be 306 [20].
6. Since aspect plays a key role in ABSA, aspect information can be taken into account by concatenating aspect vector into sentence hidden representations and by additionally appending aspect vector into the input word vectors [30].

5   Concepts Related to Deep Learning

5.1   Word-Embeddings

Different ways of generating semantical associations are Linked Statistical Data (LSD), WordNet, Word-embeddings, etc. WordNet is an ontology representation of relationships of words which is constructed manually. WordNet is a symbolic representation, computing the similarity between words is limited to its hierarchical representation, whereas word-embeddings represent word meaning from its surrounding context words which are learned from large corpora. Word2vec (word-embeddings model) represents words in multi-dimensional vector space, this enables similarity calculation in terms of vector distance. Hence in terms of similarity calculation, word2vec is more effective than WordNet. In this technique, words and phrases are encapsulated to vectors of real numbers. Various strategies have been proposed to obtain word-embeddings [17–19]. In this architecture, neural network language model first learns word vectors, and then, n-grams neural network language model is trained on top of these distributed representations of words. Out of two models, skip-gram and continuous bag-of-words (CBOW) proposed by [18], Skip-gram model anticipates the context based on the current word. The following mathematical formulation needs to be maximized as an objective function for a given sequence of words $w_1, w_2, w_3, \ldots, w_T$.

$$\frac{1}{T} \sum_{t=1}^{T} \sum_{-c \leq j \leq c, j \neq 0} \log p(w_{t+j} \mid w_t) \tag{2}$$

where w_t is the focused word and c is context window size. A larger value of c provides more samples for training which leads to high accuracy at the cost of training time. softmax function is used to calculate the probability $p(w_{t+j} \mid w_t)$.

$$p(w_o \mid w_I) = \frac{\exp(v'_{w_o}{}^T v_{w_I})}{\sum_{w=1}^{W} \exp(v'_w{}^T v_{w_I})} \tag{3}$$

where v_w and v'_w are the "input" and "output" vector representations of w, and W is vocabulary size. The cost of computing $\nabla \log p(w_O \mid w_I)$ is in proportion to W, which is often large (10^5 to 10^7 terms). So, one of the following two approximations are used, hierarchical softmax and negative sampling, to solve it.

5.2 Long Short-Term Memory (LSTM)

LSTMs are an extraordinary sort of RNNs [12], which have been devised to capture long-term dependencies that are difficult to be handled by simple RNNs. LSTM resolves the problem of vanishing gradient by offering the concept of gates into their state dynamics. The fundamental architecture of RNN and LSTM is same, but hidden state activation computation function is different in LSTM. The memory of LSTM is called cell and is treated as a black box whose inputs are the past state $a^{<t-1>}$ and present input $x^{<t>}$.

LSTM has the ability to add new information, update, and remove previous information stored in the cell states. It uses the concept of gates to regulate the information flow at each time-step. A standard LSTM consists many gates like input (i), forget (f), and output (o) gates. The input gate decides what new information from the input need to be updated in the cell state. While the forget gate determines which information is not needed anymore. So that it can be erased from the cell state. Output gate chooses what to output conditioned on input and the content of the memory cell.

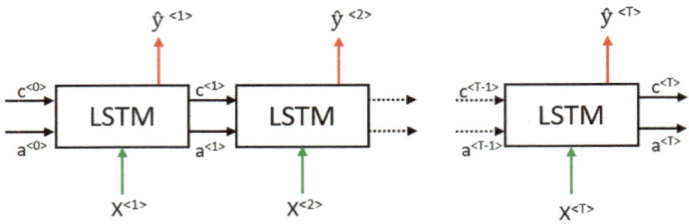

Fig. 1 LSTM Network

The LSTM cell at time-step t is formulated as below:

$$\hat{c}^{<t>} = \tanh(W_c[a^{<t-1>}, x^{<t>}] + b_c)$$
$$i^{<t>} = \sigma(W_i[a^{<t-1>}, x^{<t>}] + b_i)$$
$$f^{<t>} = \sigma(W_f[a^{<t-1>}, x^{<t>}] + b_f)$$
$$o^{<t>} = \sigma(W_o[a^{<t-1>}, x^{<t>}] + b_o)$$
$$c^{<t>} = i^{<t>} * \hat{c}^{<t>} + f^{<t>} * c^{<t-1>}$$
$$a^{<t>} = o^{<t>} * \tanh(c^{<t>})$$

where σ is the logistic sigmoid function, $*$ is element-wise multiplication function, $\hat{c}$, c are memory cell states, a is activation (hidden) state, W_c, W_i, W_f, W_o are weight matrices, and b_c, b_i, b_f, b_o are bias vectors of different gates for input x. Like RNN, LSTM cells' combination is used to represent LSTM architecture (illustrated in Fig. 1).

For sequence tagging problems like aspect-term extraction if the input sequence is $x^{<1>}, x^{<2>}, \ldots, x^{<T>}$ and output sequence is $y^{<1>}, y^{<2>}, \ldots, y^{<T>}$, then while making a prediction for $y^{<3>}$, LSTM uses the information not only from $x^{<3>}$ but from $x^{<2>}$ and $x^{<1>}$ also. However, one weakness of LSTM is that it only uses the information that is earlier in the sequence to make a prediction. In order to decide whether or not the word is part of an aspect-term, it would be really useful to know not just information from the preceding words but to know information from the later words in the sentence.

5.3   Bi-directional Long Short-Term Memory (Bi-LSTM)

The idea behind the concept of Bi-LSTMs is to use the information associated with previous and future elements while generating output for the current element [23]. To capture all available information, two different networks are used (one for each direction) and results from both networks are combined to predict the final output. In simple words, bidirectional LSTMs are just two LSTMs assembled side by side [8]. The Bi-LSTM architecture is depicted in Fig. 2.

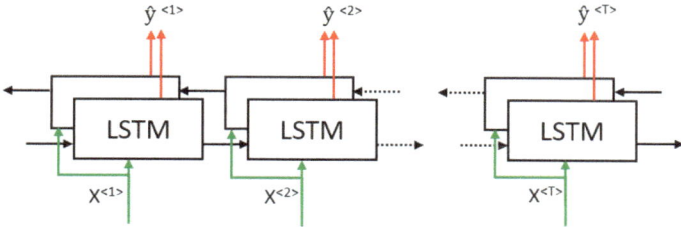

Fig. 2 Bi-LSTM network

5.4 RNN with Attention

Given a continuous input vector sequence $x^{<1>}, x^{<2>}, \ldots, x^{<T>}$, a standard RNN estimates the sequence of output of same length $y^{<1>}, y^{<2>}, \ldots, y^{<T>}$ using following equations [25].

$$h^{<t>} = \tanh(W_a[a^{<t-1>}, x^{<t>}] + b_a)$$
$$\hat{y}^{<t>} = \text{softmax}(W_y[a^{<t>}] + b_y)$$

But this structure fails, for the problems when the input sequence and output sequence have a different length (in machine translation). A simple way is to map the inputs to a vector of fixed length and adopt this vector in an output sequence generation (encoder–decoder architecture).

First RNN that computes this fixed-size context vector c is called encoder and second RNN that computes output from the encoded fixed-size vector is called decoder. This overall architecture is called encoder–decoder. It computes the following conditional probability.

$$p(y^{<1>}, \ldots, y^{<T_y>} | x^{<1>}, \ldots, x^{<T_x>}) = \prod_{t=1}^{T_y} p(y^{<t>} | c, y^{<1>}, \ldots, y^{<t-1>}) \qquad (4)$$

All the information of input sequence (sentence) is converted into a single vector. It must fully capture all the information (meaning) from the input sequence. This encoding process is senseless with a potentially very long input sequence. So the performance deteriorates when input sequence length increases.

To tackle this issue, attention mechanism was proposed by [2]. An attention mechanism was used to selective focus on sentence part while doing language translation. Using RNN, Eq. 4 can be modeled as

$$p(y^{<t>} | v, y^{<1>}, \ldots, y^{<t-1>}) = g(y^{<t-1>}, s^{<t>}, c^{<t>}) \qquad (5)$$

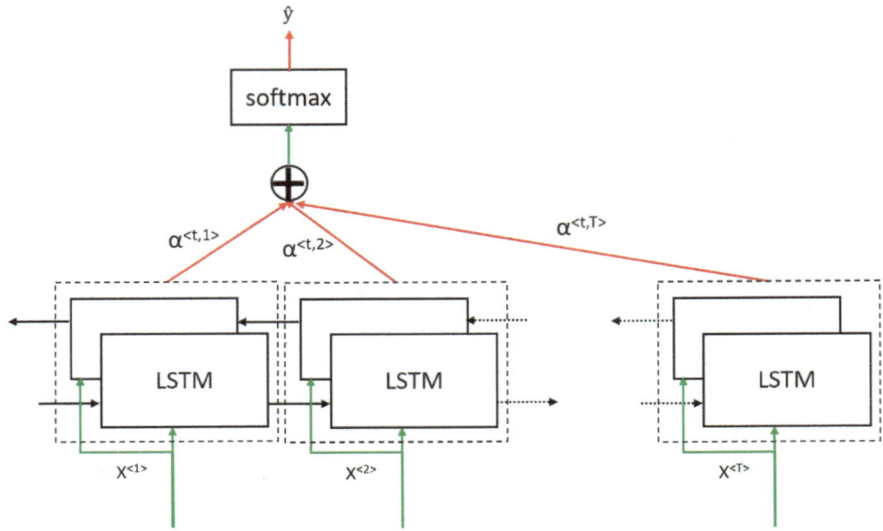

Fig. 3 Bi-LSTM with attention network

where g is a nonlinear function that outputs the probability of $y^{<t>}$ using a single directional RNN with state $s^{<t>}$. Context vector $c^{<t>}$ for each time-step will depend on features at different time-step $h^{<1>}, h^{<2>}, \ldots, h^{<T_x>}$ and attention parameter α that tells how much attention should pay on different features. Context vector $c^{<t>}$ is computed as weighted sum of features at different time-step $h^{<i>}$.

$$c^{<t>} = \sum_{t'=1}^{T_x} \alpha^{<t,t'>} h^{<t'>} \qquad (6)$$

Attention weight should satisfy this

$$\sum_{t'=1}^{T_x} \alpha^{<t,t'>} = 1 \qquad (7)$$

The weight $\alpha^{<t,t'>}$ of each annotation $h^{<t'>}$ is computed by

$$\alpha^{<t,t'>} = \frac{\exp(e^{<t,t'>})}{\sum_{t'=1}^{T_x} \exp(e^{<t,t'>})} \qquad (8)$$

where $e^{<t,t'>} = a(s^{<t-1>}, h^{<t'>})$.

A simple way is to utilize a small neural network, to parameterize the alignment model a. Instead of generating a common fixed-length vector for each step of output. At each step of output generation, decoder focuses on various sections of the sentence. Notably, the model grasps "what to attend" depending on the current input and previously generated outputs. Figure 3 shows a graphical illusion of attention-based network.

5.5 *Convolution Neutral Network (CNN)*

CNN has been used for the image classification task. More recently, with the development of word-embeddings, CNN produces state-of-the-art results in various NLP tasks also [14, 33]. Each sentence can be converted into a sentence matrix with the help of word-embeddings (word2vec, glove, etc.). If sentence length is s and word-embeddings dimension is d, then sentence matrix S dimension will become $s \times d$. Now sentence can be treated like an image a CNN can be easily applied over it. In CNN, a filter is convolved over the image an produce a feature map. If the image dimension is $n \times n$, filter size is $f \times f$ and feature map will be $(n - f + 1) \times (n - f + 1)$. Same in the case of text, convolution procedure comprises a filter $w \in \mathbb{R}^{hd}$, which is imposed on a window of h words to produce a new feature.

Following output sequence will be obtained by applying convolution operator over the sentence matrix S.

$$o_i = \mathbf{w} \cdot \mathbf{S}[i : i + h - 1]; \tag{9}$$

where $(\cdot)$ denotes dot-product and i has interval $[1, s - h + 1]$. After that feature map $c \in \mathbb{R}^{s-h+1}$ is introduced by adding a bias b and a nonlinear activation function f (e.g tanh).

$$c_i = f(o_i + b) \tag{10}$$

Finally, each feature map goes through a pooling function to generate a fixed-length vector. That vector can be used as an input in other neural network architecture. An illusion of sentence classification using CNN architecture is depicted in Fig. 4.

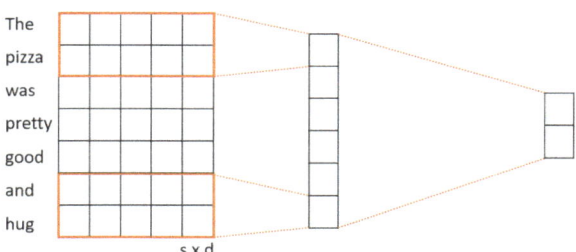

Fig. 4 CNN architecture for text

6 Deep Learning Architectures Used in ABSA

In this section summarizes the applications of deep learning approaches on different tasks of ABSA.

6.1 Sentiment Analysis

For sentence-level sentiment classification, [29] used basic CNN architecture in their sentiment model. The CNN model performs a convolution operation, using a filter, over the words of a sentence within a window size of h. This operation results in a feature map for the whole sentence. In next step, maximum value is extracted from the feature map using max-over-time pooling. The drawback of this approach is that it is aspect-agnostic. This approach works well only for sentences containing uni-sentiment.

In the approach by [4], authors also have done sentence-level sentiment analysis using CNN. In their approach, sentences were categorized into various groups depending on the total count of aspect-terms presented in the review sentence. Then, different CNN classifiers were trained separately on each sentence groups. Authors observed a significant increase in performance by separating sentences into different groups.

6.2 Aspect-Term Extraction

Chen et al. [4] used a neural network model consisting of bi-directional LSTM with CRF (Bi-LSTM-CRF) for extraction of aspect-terms presented in the reviews sentences.

Similar to the aforesaid architecture, to extract aspect-terms, a Bi-LSTM-CRF model was developed by Al-Smadi et al. [1]. This model used character-level features along with word-level features in the form of embeddings while predicting aspect-terms. Use of character-level embeddings benefited in analyzing of affixes without morphology analysis.

Giannakopoulos et al. [7] employed a Bi-LSTM architecture to extract features from the inputs. Randomly initialized character-embeddings and word-embeddings were passed as input to a Bi-LSTM layer. These character-embeddings were learned during the training process. Indirectly, a feature vector was created as a combination of pre-trained word-embeddings provided by fastText and character-based word-embeddings of each token of a sentence. This feature vector acts as input to main Bi-LSTM layer, which is utilized to extract features for the CRF layer. These features were generated by exploiting word morphology and sentence structure.

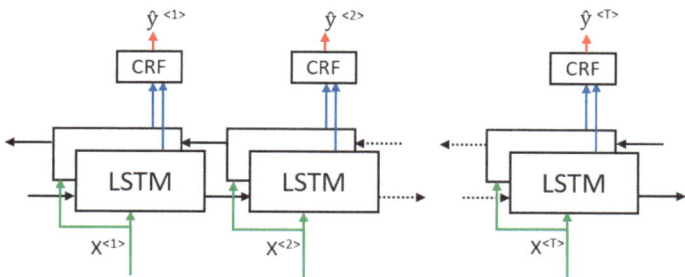

Fig. 5 Bi-LSTM-CRF network

Jebbara and Cimiano [13] used a hybrid neural architecture consisting of CNN and GRU for extraction of aspect-terms and opinion-terms. Deep CNN was able to capture local dependencies near a word of interest. Using the extracted features by CNN, GRU preserved the information over a long distance.

Wu et al. [31] have used a set of dependency relation-based rules for extraction of NP chunks. These NP chunks were treated as candidate aspect-terms and used to generate the noisy labeled dataset. Finally, a deep GRU network was trained to predict aspect-terms.

The basic Bi-LSTM-CRF model architecture is presented in Fig. 5.

6.3 Aspect-Category Extraction

In aspect model proposed by Wang and Liu [29], input was generated by averaging of word vectors of sentence and output was a probability distribution over aspects (where aspects were entity and attribute pair, E#A pair). They used a basic two-layer fully connected neural network for detection of aspects (aspect-categories).

Zhou et al. [34] also represented input sentence same as [29]. Then two separate neural networks were trained to learn shared features and aspect-specific feature. The first two-layer neural network was trained to classify the aspect-categories, and a hidden layer of this network was used as shared feature. The second neural network was trained for one specific category only (different neural networks were trained for each category). The hidden layer of this network was described as an aspect-specific feature. To form hybrid features, aspect-specific and shared features were concatenated. Finally, aspect-categories ware predicted using the hybrid features by a logistic regression classifier.

6.4 Aspect-Based Sentiment Detection

Wang and Liu [29] re-scaled word-embeddings in proportion to the aspect distribution in the review sentence. The idea behind this is to encode the relation between word and aspect associated with it. For this process, top n-aspects were filtered out based on the aspect distribution. Then filter the probabilistic mass with respect to each aspect. Aspect-specific probabilistic mass was propagated based on the parse tree. Finally, each word vector was re-scaled (weight distribution) and new word vectors were used in the CNN-based sentiment model.

Tang et al. [26, 27] claimed that standard LSTM is incapable to produce good results by incorporating only sentence representation. Along with sentence representation, if target information is used then the performance of sentiment classification will boost significantly. So they developed two models, target-dependent LSTM (TD-LSTM) and target-connection LSTM (TC-LSTM). In TD-LSTM, the selection of the relevant part of the context is based on the relatedness of it with the target word. They used two LSTMs, one for preceding context plus target word and another for target word plus following context. Last hidden vectors of both LSTMs after concatenation were passed to a softmax layer. Finally, target-specific sentiment polarity was inferred by the softmax layer. In this way, they include the context in both directions. An extension of the TD-LSTM was also introduced, named TC-LSTM. To expressly use the relationship between target word and context word, a target-connection component was used. Target vector was acquired by averaging the vectors of words it contains.

For first LSTM, they used word-embeddings of preceding context words concatenated with target vector, and for second LSTM, they used word-embeddings of following context words concatenated with target vector. This architecture works best for positive and negative examples but misclassifies examples belonging to the neutral category.

Neural network methods presented by [27] cannot efficiently identify which word in the sentence is more important. Existing work ignores or does not explicitly model the position information of the aspect-term in a sentence, which has been studied for improving performance in information retrieval (IR). Fortunately, attention mechanisms are an effective way to solve this problem. When an aspect-term occurs in a sentence, its neighboring words in the sentence should be given more attention than other words with long distance. Gu et al. [9] proposed a position-aware bidirectional attention network (PBAN) based on bidirectional Gated Recurrent Units (Bi-GRU). This model consists of three components: (1) Obtain position information of each word based on current aspect-term, then convert position information into position embedding. (2) The PBAN consists of two Bi-GRU for extracting aspect-level and sentence-level features respectively. (3) Use the bi-directional attention mechanism to model the mutual relation between aspect-terms and its corresponding sentence.

Wang et al. [30] explored the interrelation between an aspect and a sentence. To emphasize the crucial part of a sentence provided the aspect, aspect-to-sentence attention mechanism was used. There were two approaches by which this can be achieved.

One was the concatenation of aspect vector with sentence hidden representation. Another was to appending the aspect vector with the input vectors. So they proposed two models, attention-based LSTM (AT-LSTM) and attention-based LSTM with aspect-embedding (ATAE-LSTM). In AT-LSTM, aspect-embeddings were concatenated with the hidden representation of sentence to calculate the attention of different context words with respect to an aspect. On another hand, aspect-embedding was also appended into each input word vector in AT-LSTM to built ATAE-LSTM model. These models were designed to discover the aspect-based sentiment polarity. These models have the capabilities to deals with different parts of a sentence based on the current aspects.

Similar to aforementioned ATAE-LSTM model, Al-Smadi et al. [1] also developed an attention-based model named (AB-LSTM-PC) for aspect-based sentiment polarity classification. In this model, attention weights were calculated using word and aspect-embeddings. This model was trained on Arabic review dataset.

A sentence could contain multiple sentiment-target pairs. To isolate diverse opinion circumstances for different targets, He et al. [11] proposed an approach for improving the effectiveness of attention. To acquire the semantic significance of the opinion target, a better target representation method was proposed by authors. The computed attention weights rely entirely on the semantic associations between context words and the target representation. However, this may not be sufficient for differentiating opinions words for different targets. Apart from semantic information, syntactic information was also incorporated into the attention mechanism. Their syntax-based attention mechanism selectively focuses on a small subset of context words that are close to the target on the syntactic path which was obtained by applying a dependency parser on the review sentence.

Li et al. [15] model contains two key components, namely Truncated History-Attention (THA) and Selective Transformation Network (STN), for capturing aspect detection history and opinion summary, respectively. THA and STN were built on two LSTMs that generate the initial word representations for the primary ATE task and the auxiliary opinion detection task, respectively. THA was designed to integrate the information of aspect detection history into the current aspect feature to generate a new history-aware aspect representation. STN first calculates a new opinion representation conditioned on the current aspect candidate. Then, a bi-linear attention network was used to calculate the opinion summary as the weighted sum of the new opinion representations, according to their associations with the current aspect representation. Finally, the history-aware aspect representation and the opinion summary were concatenated as features for aspect prediction of the current time-step.

Inspired by the work of [24], Tang et al. [28] uses memory networks to capture the importance of context words for aspect-level sentiment classification. An external memory was created by stacking the context word vectors. A computational layer called hop took aspect vector as an input and focused on memory using attention mechanism. Each hop summed up the linearly transformed aspect vector and attention layer output, which was passed as an input to next hop. Last hop output was treated as sentence representation with respect to aspect and passed to softmax classifier to sentiment classification. Apart from performance enhancement, they observed

Table 2 Summary of state-of-the-art ABSA methods

Reference	Dataset	Domain	Task	Embeddings	Performance	Method
[1]	SemEval-16 Arabic	Hotel	ATE AC-PD	Word2vec + FastText	F1-score: 69.98 Accuracy: 82.70	Bi-LSTM-CRF LSTM + Attention
[4]	SemEval-16 Task-5	Restaurant	ATE	Word2vec	F1-score: 72.44	Bi-LSTM-CRF
[7]	SemEval-14 Task-4	Restaurant Laptop	ATE	FastText	F1-score: 84.12 F1-score: 77.96	2 layer Bi-LSTM-CRF 2 layer Bi-LSTM-CRF
[34]	SemEval-14 Task-4	Restaurant	ACD	–	F1-score: 90.10	2-layer NN + Logistic R
[26]	Dong et al. 2014 [5]	Twitter	AT-PD	SSWE	F1-score: 69.00 F1-score: 69.50	TD-LSTM TC-LSTM
[30]	SemEval-14 Task-4	Restaurant Laptop Restaurant	AT-PD AC-PD	Glove	Accuracy: 76.60 Accuracy: 77.20 Accuracy: 68.90 Accuracy: 68.70 Accuracy: 82.50 Accuracy: 83.10 Accuracy: 84.10	AE-LSTM ATAE-LSTM AE-LSTM ATAE-LSTM AE-LSTM AT-LSTM ATAE-LSTM
[9]	SemEval-14	Restaurant Laptop	AT-PD	Glove	Accuracy: 81.16 Accuracy: 74.12	Bi-GRU + Attention
[11]	SemEval-14 Task-4 SemEval-14 Task-4 SemEval-15 Task-12 SemEval-16 Task-5	Restaurant Laptop Restaurant Restaurant	AC-PD	Glove	Accuracy: 80.36 Accuracy: 71.94 Accuracy: 81.67 Accuracy: 84.64	LSTM + Attention
[29]	SemEval-15 Task-12 Laptop	Laptop Laptop	ACD AC-PD	Word2vec	F1-score: 51.30 Accuracy: 78.30	2-layer NN Parse Tree + CNN

(continued)

Table 2 (continued)

Reference	Dataset	Domain	Task	Embeddings	Performance	Method
[20]	SemEval-14 Task-4 + Dataset by Qiu et al. [21]	Restaurant Laptop	ATE	Word2vec + Amazon Reviews Embeddings	F1-score: 87.17 F1-score: 82.32	7-layer CNN + Linguistic Pattern
[28]	SemEval-14 Task-4	Restaurant Laptop	AT-PD	–	F1-score: 80.95 F1-score: 72.37	Memory Network + Attention
[3]	SemEval-14 Task-4 Dong et al. 2014	Restaurant Laptop Twitter	AT-PD	GloVe	F1-score: 70.80 F1-score: 71.35 F1-score: 67.30	Memory Network + Attention + GRU
[13]	SemEval-15 Task-12	Restaurant	ATE	Amazon Reviews Embeddings	F1-score: 65.90	CNN + GRU
[31]	SemEval-14 Task-4	Restaurant Laptop	ATE	Word2vec	F1-score: 76.15 F1-score: 60.75	Bi-GRU + POS

that deep memory network with 9 layers is 15 times faster than LSTM with CPU implementation.

Similarly, with described architecture Chen et al. [3] utilized multiple attention mechanism but differently, it uses Bi-LSTM to generate memory. The further relative position of words with respect to aspect target was used to generate location weighted memory. A recurrent network (GRU) was used to capture multiple attention on memory. Finally, softmax classifier was used to sentiment classification.

A summary of current state-of-the-art methods of ABSA is presented in Table 2.

7   Conclusion

In this chapter, we have presented some introduction about deep learning methodology used in Natural Language Processing especially for ABSA. This chapter discussed various tasks of ABSA and deals with the approaches used for these tasks. For sequence labeling task like aspect-term extraction, deep learning models consisting the Bi-LSTM-CRF are generally used. For the sentiment classification task, vanilla neural networks and CNN have shown state-of-the-art performance. For aspect sentiment detection task, exploiting the relationship between aspect and opinion, aspect and context words is beneficial. Researchers tried different approaches to extract this relationship and used in the deep neural networks. There is also a need to combine approaches to jointly perform two tasks, i.e., aspect detection and sentiment analysis. Overall, this chapter provides a starting point to dive into ABSA using deep learning approaches.

References

1. Al-Smadi, Mohammad, Bashar Talafha, Mahmoud Al-Ayyoub, and Yaser Jararweh. 2018. Using long short-term memory deep neural networks for aspect-based sentiment analysis of Arabic reviews. *International Journal of Machine Learning and Cybernetics*, pp. 1–13
2. Bahdanau, Dzmitry, Kyunghyun Cho, and Yoshua Bengio. 2014. Neural machine translation by jointly learning to align and translate. arXiv preprint arXiv:1409.0473
3. Chen, Peng, Zhongqian Sun, Lidong Bing, and Wei Yang. 2017. Recurrent attention network on memory for aspect sentiment analysis. In *Proceedings of the 2017 Conference on Empirical Methods in Natural Language Processing*, 452–461
4. Chen, Tao, Ruifeng Xu, Yulan He, Xuan Wang. 2017. Improving sentiment analysis via sentence type classification using BiLSTM-CRF and CNN. *Expert Systems with Applications* **72**, 221–230
5. Dong, Li, Furu Wei, Chuanqi Tan, Duyu Tang, Ming Zhou, and Ke Xu. 2014. Adaptive recursive neural network for target-dependent twitter sentiment classification. In *Proceedings of the 52nd Annual Meeting of the Association for Computational Linguistics (volume 2: Short papers)*, vol. 2, 49–54
6. Feng, Jinzhan, Shuqin Cai, and Xiaomeng Ma. 2018. Enhanced sentiment labeling and implicit aspect identification by integration of deep convolution neural network and sequential algorithm. *Cluster Computing*, 1–19

7. Giannakopoulos, Athanasios, Claudiu Musat, Andreea Hossmann, and Michael Baeriswyl. 2017. Unsupervised aspect term extraction with b-lstm & crf using automatically labelled datasets. arXiv preprint arXiv:1709.05094
8. Graves, Alex, Schmidhuber, Jürgen: Framewise phoneme classification with bidirectional LSTM and other neural network architectures. Neural Networks **18**(5–6), 602–610 (2005)
9. Gu, Shuqin, Lipeng Zhang, Yuexian Hou, and Yin Song. 2018. A position-aware bidirectional attention network for aspect-level sentiment analysis. In *Proceedings of the 27th International Conference on Computational Linguistics*, 774–784
10. Güngör, Onur, Güngör, Tunga, Üsküdarli, Suzan: The effect of morphology in named entity recognition with sequence tagging. Natural Language Engineering **25**(1), 147–169 (2019)
11. He, Ruidan, Wee Sun Lee, Hwee Tou Ng, and Daniel Dahlmeier. 2018. Effective attention modeling for aspect-level sentiment classification. In *Proceedings of the 27th International Conference on Computational Linguistics*, 1121–1131
12. Hochreiter, Sepp, Schmidhuber, Jürgen: Long short-term memory. Neural Computation **9**(8), 1735–1780 (1997)
13. Jebbara, Soufian, and Philipp Cimiano. 2016. Aspect-based relational sentiment analysis using a stacked neural network architecture. In *Proceedings of the Twenty-second European Conference on Artificial Intelligence*, 1123–1131. IOS Press
14. Kim, Yoon. 2014. Convolutional neural networks for sentence classification. arXiv preprint arXiv:1408.5882
15. Li, Xin, Lidong Bing, Piji Li, Wai Lam, and Zhimou Yang. 2018. Aspect term extraction with history attention and selective transformation. arXiv preprint arXiv:1805.00760
16. Liu, Bing: Sentiment analysis and opinion mining. Synthesis Lectures on Human Language Technologies **5**(1), 1–167 (2012)
17. Mikolov, Tomas, Kai Chen, Greg Corrado, and Jeffrey Dean. 2013. Efficient estimation of word representations in vector space. arXiv preprint arXiv:1301.3781
18. Mikolov, Tomas, Ilya Sutskever, Kai Chen, Gregory S. Corrado, and Jeffrey Dean. 2013. Distributed representations of words and phrases and their compositionality. *Advances in Neural Information Processing Systems*, 3111–3119
19. Pennington, Jeffrey, Richard Socher, and Christopher D. Manning. 2014. Glove: global vectors for word representation. In *Proceedings of the 2014 Conference on Empirical Methods in Natural Language Processing (EMNLP)*, 1532–1543. ACL
20. Poria, Soujanya, Cambria, Erik, Gelbukh, Alexander: Aspect extraction for opinion mining with a deep convolutional neural network. Knowledge-Based Systems **108**, 42–49 (2016)
21. Qiu, Guang, Liu, Bing, Jiajun, Bu, Chen, Chun: Opinion word expansion and target extraction through double propagation. Computational linguistics **37**(1), 9–27 (2011)
22. Ramshaw, Lance A., and Mitchell P. Marcus. 1999. Text chunking using transformation-based learning. In *Natural Language Processing Using Very Large Corpora*, 157–176. Berlin: Springer
23. Schuster, Mike, and Kuldip K. Paliwal. 1997. Bidirectional recurrent neural networks. *IEEE Transactions on Signal Processing*, 45(11), 2673–2681
24. Sukhbaatar, Sainbayar, Jason Weston, Rob Fergus, et al. 2015. End-to-end memory networks. In *Advances in Neural Information Processing Systems*, 2440–2448
25. Sutskever, Ilya, Oriol Vinyals, and Quoc V. Le. 2014. Sequence to sequence learning with neural networks. In *Advances in Neural Information Processing Systems*, 3104–3112
26. Tang, Duyu, Bing Qin, Xiaocheng Feng, and Ting Liu. 2015. Target-dependent sentiment classification with long short term memory. *CoRR*, abs/1512.01100
27. Tang, Duyu, Bing Qin, Xiaocheng Feng, and Ting Liu. Effective LSTMs for target-dependent sentiment classification. [9], 3298–3307
28. Tang, Duyu, Bing Qin, and Ting Liu. 2016. Aspect level sentiment classification with deep memory network. arXiv preprint arXiv:1605.08900
29. Wang, Bo, and Min Liu. 2015. Deep learning for aspect-based sentiment analysis. *Stanford University report*

30. Wang, Yequan, Minlie Huang, Li Zhao, et al. 2016. Attention-based LSTM for aspect-level sentiment classification. In *Proceedings of the 2016 Conference on Empirical Methods in Natural Language Processing*, 606–615
31. Wu, Chuhan, Fangzhao Wu, Sixing Wu, Zhigang Yuan, Yongfeng Huang. 2018. A hybrid unsupervised method for aspect term and opinion target extraction. *Knowledge-Based Systems* 148, 66–73
32. Xu, Hu, Bing Liu, Lei Shu, and Philip S. Yu. 2018. Double embeddings and cnn-based sequence labeling for aspect extraction. arXiv preprint arXiv:1805.04601
33. Zhang, Ye, and Byron Wallace. 2015. A sensitivity analysis of (and practitioners' guide to) convolutional neural networks for sentence classification. arXiv preprint arXiv:1510.03820
34. Zhou, Xinjie, Xiaojun Wan, and Jianguo Xiao. 2015. Representation learning for aspect category detection in online reviews. In *Twenty-Ninth AAAI Conference on Artificial Intelligence*, 417–424. AAAI Press

Transfer Learning for Detecting Hateful Sentiments in Code Switched Language

Kshitij Rajput, Raghav Kapoor, Puneet Mathur, Hitkul, Ponnurangam Kumaraguru and Rajiv Ratn Shah

Abstract With the phenomenal increase in the penetration of social media in linguistically diverse demographic regions, conversations have become more casual and multilingual. The rise of informal code-switched multilingual languages makes it tough for automated systems to monitor instances of hate speech, which are further intelligently disguised through the use of spelling variations, code-mixing, homophones, homonyms, and the absence of sophisticated grammar rules. Machine transliteration can be employed for converting the code-switched text into a singular script but poses the challenge of the semantical breakdown of the text. To overcome this drawback, this chapter investigates the application of transfer learning. The CNN-based neural models are trained on a large dataset of hateful tweets in a chosen primary language, followed by retraining on the small transliterated dataset in the same language. Since transfer learning can act as an effective strategy to reuse already learned features in learning a specialized task through cross-domain knowledge transfer, hate speech classification on a large English corpus can act as source tasks to help in obtaining pre-trained deep learning classifiers for the target task of classifying tweets translated in English from other code-switched languages. Effects

K. Rajput · R. Kapoor · P. Mathur · Hitkul · P. Kumaraguru · R. R. Shah (✉)
MIDAS Lab, IIIT Delhi, New Delhi, India
e-mail: rajivratn@iiitd.ac.in

K. Rajput
e-mail: rajput.kshitij97@gmail.com

R. Kapoor
e-mail: raghavk.co@nsit.net.in

P. Mathur
e-mail: pmathur3k6@gmail.com

Hitkul
e-mail: hitkuljangid@gmail.com

P. Kumaraguru
e-mail: pk@iiitd.ac.in

K. Rajput · R. Kapoor
NSIT, New Delhi, India

© Springer Nature Singapore Pte Ltd. 2020
B. Agarwal et al. (eds.), *Deep Learning-Based Approaches for Sentiment Analysis*, Algorithms for Intelligent Systems,
https://doi.org/10.1007/978-981-15-1216-2_7

159

of the different types of popular word embeddings and multiple supervised inputs such as the LIWC, the presence of profanities, and sentiment are carefully studied to derive the most representative combination of input settings that can help achieve state-of-the-art hate speech detection from code-switched multilingual short texts on Twitter.

Keywords Hate speech · Code-switching · Transfer learning · Multilingual · Social media · Offensive text classification

1 Introduction

The spark of natural language processing commonly known as NLP began around 60–70 years before, wherein the researchers worked upon large amounts of language data focusing on the different aspects of the natural or spoken language, analyzing their semantics, features, and other prominent attributes and determining how the machines interact with these human languages. Also, the growth of Internet in the last 25–30 years has led to collection of huge amount of data from online sources and has given rise to a completely new domain of research called data mining. Various tasks have been taken up by the researchers in the past few years in these domains like sentiment analysis, POS tagging, machine translation. However, with the advent of Internet and increasing popularity of social media among the masses, one of the most challenging problems in the field of natural language processing that has cropped up in the past few years is that of hate speech detection from social media content. We discuss the hate speech problem in the following subsection.

1.1 Hate Speech Problem

As the various social media platforms like Facebook and Twitter became famous among the users, they started expressing their feelings and views more freely in real time. This led to the problem of hate-inducing and abusive posts on the Internet. People would exert their anger on a government policy or somebody's views by writing a hateful or a abusive post against that person. Though the right to show dissent on any view is part of democracy, but venting out the anger in the form of a hateful or an abusive post online is not the correct way of expressing dissent. Moreover, as a recent trend, we see that certain people use social media as a platform to instill hate in the minds of the users against a particular religion or gender. Also, at the time of elections, it is to be seen that many new Facebook pages and Twitter handles crop up whose main aim is to write abusive and hate-inducing posts against the opponents which are mostly fake.

The flooding of social media by hateful posts is not good if we want the social media to be clean and fit to use for children and women. Hence, the task of hate speech detection becomes a very important task in the domain of natural language processing (NLP). In the next subsection we discuss the code switched and code mixed languages and why it is significant to consider code mixed languages when discussing the problem of hate speech detection on social media.

1.2   Code Switched and Code Mixed Languages

Though some work has been done in the field of hate speech detection in English and other languages like Hindi and Bengali separately, very little work has been done in the code switched and the code mixed version of the two languages. The alternation of languages across sentence boundaries is known as code-switching. For example, "Mausam pyaara hai. Let's g o t o play!" are two different sentences, one in Hinglish and the other in English. This translates to "The weather is great. Let's go to play!" The alternation of two or more languages within a sentence is known as code-mixing. Formally, code-switching is defined as juxtaposition within the same speech exchange of passages of speech belonging to two different grammatical systems or sub-systems, and code-mixing is defined as "The embedding of linguistic units such as phrases, words, and morphemes of one language into an utterance of another language." For example, "Aaj ek yaadgaar din hai because I finally stood first in class" meaning "Today is a memorable day because I finally stood first in class" is a text of Hindi-English code mixed language known as Hinglish.

According to the ICUBE 2018 report whose main task is to track digital adoption and usage trends in India, it was noted that the number of Internet users in India has registered an annual growth of 18% and is estimated at 566 million as of December 2018, a 40% overall Internet penetration among the population. Since most of the users in India know more than two languages, they generally use a code switched version of two languages in order to express their feelings on social media. This means that the major chunk of data that is produced on daily basis by the social media posts is of code switched languages. Hence, if we want to free the social media from hateful posts in order to make it a clean environment for everyone, we should consider hate speech detection in code switched languages. A code switched language that has gained popularity among the users on social media is Hinglish (code switched version of Hindi and English) used mainly in the northern part of India. Hence, in this chapter, we focus on the past researches of hate speech detection in the Hinglish language.

1.3 Challenges in Code Switched and Code Mixed Languages

The task of hate speech detection in a code mixed or a code switched language is not as straightforward as it is for a monolingual language. This is because of the following reasons: Firstly, code mixed language consists of no-fixed grammar and vocabulary. Hinglish, for example, derives a fraction of its semantics from Devnagari and another from the Roman script. Secondly, Hinglish speech and written text consist of a combination of words spoken in Hindi as well as English, but written in the Roman script. This makes the spellings variable and dependent on the author of the text. For instance, "haan" (Yes) can be written as "haa," "han," "haaan," or "hn," etc. Moreover, while writing on social media, we can have phonetic conversions of a single word like "yatra," "yatraa," "yaatraaa" [19]. Hence, because of these reasons, hate speech detection using code switched data provides a difficult challenge in terms of parsing and getting meaning out of the text. This has raised the need to use deep learning techniques to solve this complex problem. In the next subsection, we discuss deep learning techniques, which are an effective way to solve the intricate problem of hate speech detection in code switched languages.

1.4 Deep Learning

Deep learning (DL) is a sub-division of machine learning which mainly consists of techniques which are driven by the functioning and structure of the human brain called artificial neural networks (ANNs). Deep learning-based model architectures help us to learn complex and intricate encoded forms of the given information having multiple levels of abstraction, thus producing state-of-the-art prediction results.

Unlike traditional machine learning algorithms, deep learning techniques operate well even without domain expertise and hard core feature extraction [2]. The performance of deep learning models increases proportionately with the amount of training data available, and thus, advances in hardware and the exponential increase in the amount of training data made available are one of the main reasons of the widespread impact of deep learning. Deep learning-based techniques have produced very promising results in image recognition, speech recognition, image captioning, machine translation and video classification, among many other domains. In this work, we discuss some deep learning methods for hate speech detection.

1.5 Overview

Few researches have focused in the code mixed data in the past. Researches in the past have been done in the field of language identification, POS tagging, and named entity recognition (NER) of code-mixed data [4, 50, 51]. In the past decade, researchers have proposed deep neural network-based state-of-the-art models for sentiment analysis [37] for English text. For the problem of sentiment analysis of Hinglish code-mixed data, sub-word level representations in LSTM have shown promising results [37] and also some ensemble methods have been proposed like combining the outputs of character-trigrams-based LSTM model and word n-gram-based multinomial naive Bayes model to predict the sentiment of Hi-En code-mixed texts [18].

In this chapter, we aim to focus on the limited work that has been done in the field of hate speech detection using the Hinglish language. We aim to cover the methodology of each of the researches in the domain of hate speech detection in Hinglish language which is the most used code switched language on the Internet. Each research is discussed in detail showing the datasets which were considered, the features extracted and the classification models used by each of the research. We first explain the work by [5], who used feature extraction followed by SVM and random forest classifier for hate speech detection task. As we explore further, we will see there were many techniques that produce better results than the supervised models like the use of a ternary trans-CNN model [28]. This work shows the benefits significant benefits of using transfer learning. Another model was developed by [19] using LSTM with transfer learning. This model established itself as state-of-the-art for HEOT dataset [5]. Finally, we discuss the MIMCT model proposed by [27]. This model as we will see is a multi-input double-channeled, CNN-LSTM classification model which has produced the most optimum results for HOT dataset [27].

In this chapter, we will first discuss the background and related work followed by the discussion on datasets used in Sect. 3. Section 4 describes about the various methodologies and deep leaning techniques that have been used for hate speech detection task. In Sect. 5, we discuss the results that were obtained by the techniques discussed in Sect. 4. This is followed by conclusion and a brief description about the future work in Sects. 6 and 7.

2 Background and Related Work

In this section, we discuss the various researches that have taken place in the domain of natural language processing (NLP) on code switched data. Here, we will study about the previous work focusing on language identification, POS tagging, named entity recognition, and sentiment analysis which is the parent problem of hate speech detection. The primary emphasis will be on the work that involves the use of Hindi-English code mixed data.

2.1   Language Identification

Language identification is one of the initial tasks of the natural language processing (NLP). The task was performed at first on code mixed Hinglish data by [9] through his research which discussed word level language detection through N-gram language profiling and pruning, dictionary-based detection and SVM-based word language detection. In SVM-based word language detection, the task of identifying language is seen as a classification problem. The SVM classifier considered the following features:

(i) **N-gram with weights**: N-gram was implemented using the bag of words principle.
(ii) **Dictionary-based features**: This is a binary feature for each language.
(iii) **MED-based weight**: This is minimum edit distance which is calculated if a
word is not found in any of the dictionaries.
(iv) **Word context information**: A window of length 7 to determine the contextual
information is used as the feature.

The SVM model which considered all the above features was reported to produce a precision score of 90.84% for the language identification task. Post-processing, the model was shown to produce an even higher precision score of 94.84%.

Mave et al. [29], through his research compared CRF model, bidirectional LSTM model and a word character LSTM model for the task of language identification. For the Hi-En code mixed data, the CRF model with context window size of 2 was shown to produce a F1 weighted score of 96.84%.

Singh et al. [49] produced a language identification model (LIDF) based on the hypothesis that word or character sequences of different languages encode different structures. Hence, the aim was to capture the structure of the character for all the languages which is done by training a model for a token length n, which learns to predict the character at nth position given all the $(n - 1)$ characters. The model consisted of a RNN layer each containing 128 LSTM cells with ReLU activation. The output of the second RNN layer at the last time step is connected to a fully connected (FC) layer with softmax activation. The architecture of the model is shown in Fig. 1.

2.2   POS Tagging

According to the definition by Wikipedia, part-of-speech(POS) tagging is described as, "In corpus linguistics, part-of-speech tagging (POS tagging or PoS tagging or POST), also called grammatical tagging or word-category disambiguation, is the process of marking up a word in a text (corpus) as corresponding to a particular part of speech, based on both its definition and its context, i.e., its relationship with adjacent and related words in a phrase, sentence, or paragraph." POS tagging is one of the fundamental preprocessing steps for NLP which has various use cases like generating parse trees and named entity recognition (NERS).

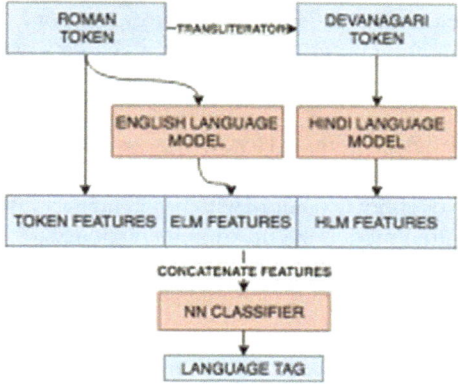

Fig. 1 Architecture of LIDF [49]

The POS tagging of Hi-En code mixed data was first provided by [51]. The corpus used for the task was created by extracting comments out of the Facebook posts of some famous personalities of India like Narendra Modi, Shahrukh Khan, Amitabh Bachchan, and BBC hindi news page. Every sentence was annotated using the following annotation scheme:

(i) Matrix
(ii) Word origin
(iii) Normalization/Transliteration
(iv) Part of speech.

The example of the annotation scheme is shown in Fig. 2. The system proposed by the authors tried to apply a pipelined approach in which the tasks of language

```
<s>
     <matrix name="Hindi">
     love_NOUN/E affection_NOUN/E lekar_VERB="ले कर" salose_NOUN=सालों
se_ADP=से sunday_NOUN/e ke_ADP=के din_NOUN=दिन chali_VERB=चली aarahi_VERB="आ
रही" divine_ADJ/e parampara_NOUN=परंपरा ko_ADP=को age_NOUN=आगे badhha_VERB=बढा
rahe_VERB=रहे ho_VERB=हो
     </matrix>
</s>
<s>
     <matrix name="Hindi">
     jindagi_NOUN=ज़िंदगी kaise_PRON=कैसी h_VERB=है paheli_NOUN=पहेली
haye_PRT=हाये
     </matrix>
     <matrix name="English">
     may_ADP his_PRON sol_NOUN=soul rest_VERB in_ADP peace_NOUN
     </matrix>
</s>
```

Fig. 2 Example annotation for POS tagging [51]

identification, normalization, and then POS tagging were done in a sequential order. The basic idea behind the POS tagging of code mixed data used was to divide the text into continuous maximum chunks of words whose language are the same. Then Hindi POS tagger was operated on the Hindi chunks, and the English POS tagger was operated on the English chunks. Hence, this task used language identification in order to separate the chunks of data on the basis of their language. CRF++-based POS tagger was used for Hindi text, and Twitter POS tagger was used for the English text. This being the initial model, the maximum chunk accuracy reported was 34% when the language identities and normalized forms were known. For the case where both of these factors were unknown, the maximum chunk accuracy reported was 25%.

Another approach for POS tagging was proposed by [13] which produced much better results than the previous approaches. The datasets used in the experiment were provided by the organizer of POS tagging tool contest at ICON-2016. Nine exhaustive set of features were used for the task of POS tagging. These features were Context word, Character n-gram, Word normalization, Prefix and suffix, Word class feature, Word position, Number of upper case characters, Word probability including Top@1-Probability and Top@2-Probability, Binary features which include *isSufficientLength, isAllCapital, isFirstCharacterUpper, isInitCap, isInitPunDigit, isDigit, isDigitAlpha,* and *isHashTag.*

These feature set were used to build a POS model. Conditional random field (CRF) is used was the underlying classifier. CRF ++3 , an implementation of CRF was used to perform the experiment. The paper reported a high precision score of 0.782 for the Hi-English code mixed data extracted from twitter. This was a major improvement from the previous approaches.

2.3 Named Entity Recognition

Named entity recognition (NER) is defined as, "The task of classifying named entities that are present in a text into pre-defined categories like individuals, companies, places, organization, cities, dates, product terminologies." Named entity recognition has several use cases in the field of natural language processing (NLP) like classifying contents from news providers, efficient search algorithms, and customer support. The few researches that have taken place for named entity recognition are described below.

The initial work on named entity recognition on code mixed Hi-En data was provided by [39]. The dataset was scraped using the twitter API, and the tag set was chosen which the Government of India standardized tag set was. In this tag set, named entity hierarchy was divided into three major classes, i.e., Entity Name, Time, and Numerical expressions. The Name hierarchy had eleven attributes, Numerical Expression had four attributes and Time had three attributes. The best results for this task were provided by the team Irshad-IIIT-Hyd as reported by the paper. This team used a simple feed forward neural network with activation function as rectifier,

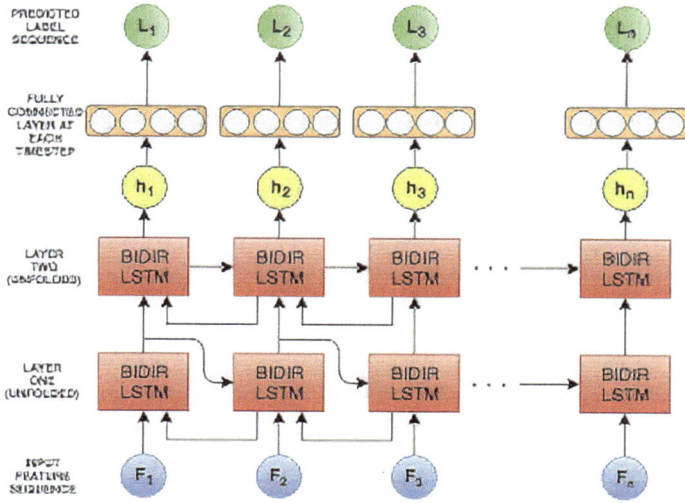

Fig. 3 LSTM NER architecture [49]

learning rate as 0.03, dropout as 0.5, and learning rule as *adagrad* with L2 regularization. English wiki corpus was used for developing word embeddings using Gensim Word2Vec. The team was able to secure a precision accuracy of 80.92%.

Other works have been performed for named entity recognition such as the [49], which considers that the named entities can be identified using the features extracted from the words surrounding it. The following features were used for the task: (i) Token-based features, (ii) Prefixes and suffixes, (iii) Character-based features, (iv) Language-based features, (v) Syntactic features, (vi) Tweet capitalization features.

A LSTM-based model comprising of two bidirectional RNN layers using LSTM cells and ReLU activation was used for the task of using the above features for named entity recognition. The architecture of the LSTM model is shown in Fig. 3. Also, a standard CRF model proposed by [21] added with a L1 and L2 regularization to prevent overfitting was also tested for the task. The CRF model was shown to have the highest precision accuracy of 84.95%, a major improvement from the previous work.

2.4 Sentiment Analysis

Sentiment analysis is the process of computationally identifying and categorizing opinions expressed in a piece of text, especially in order to determine whether the writer's attitude toward a particular topic or product is positive, negative, or neutral. Sentiment analysis (SA) on code-mixed data from social media has many use cases in opinion mining ranging from customer satisfaction to analyzing social media campaign in multilingual societies. The task of sentiment analysis in the code mixed data of Hindi-English was restrained due to the lack of a suitable annotated dataset.

The task of sentiment analysis on code mixed Hi-En (Hinglish) social media content was first performed by [37] who used sub-word level representations in LSTM architecture to perform the task. The dataset was created by extracting the comments from the Facebook pages of two of the most famous personalities of India, Salman Khan an Indian actor and Narendra Modi, the Prime Minister of India at the time. Then manual annotation of the dataset was performed. The comments were annotated by two annotators in a three-level polarity scale—positive, negative, or neutral. The size of the dataset was 3879 sentences which consisted of 15% negative, 50% neutral, and 35% positive comments owing to the nature of conversations in the selected pages. This research used intermediate sub-word feature representations learned by the filters during convolution operation. The relevant sub-word representation was circulated with LSTM using which the final sentiment of the sentence was calculated. The architecture used and the proposed methodology for sentiment analysis [37] is shown in Figs. 4 and 5.

The sub-word level representation using LSTM performed better than the system proposed by the [52] on this dataset with a accuracy of 69.7% and a F1 score of 0.658. The system also outperformed the [34] which used SVM unigram and unigram + bigram features and also the lexical lookup method proposed by [48]. The system performed better than the previous systems not only in the Hi-En code mixed dataset prepared but also on SemEval' 13 Twitter Sentiment analysis dataset with a accuracy of 60.57% and F1 score of 0.537. Hence, the sub-word level LSTM is performed significantly better than the baselines.

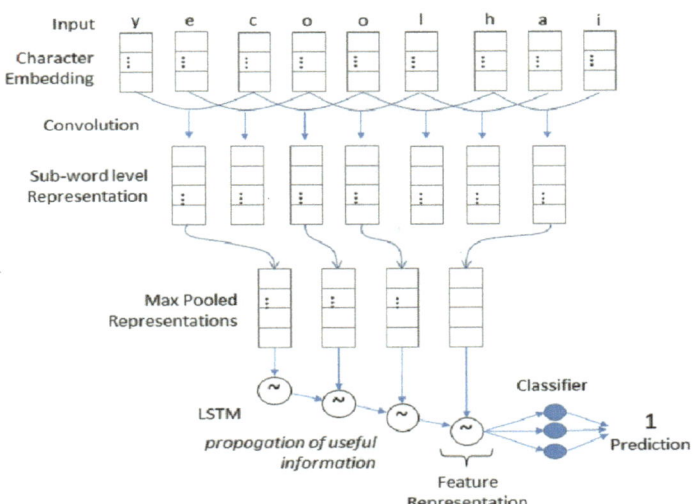

Fig. 4 Proposed methodology for sentiment analysis by [37]

Fig. 5 Model architecture used for sentiment analysis by [37]

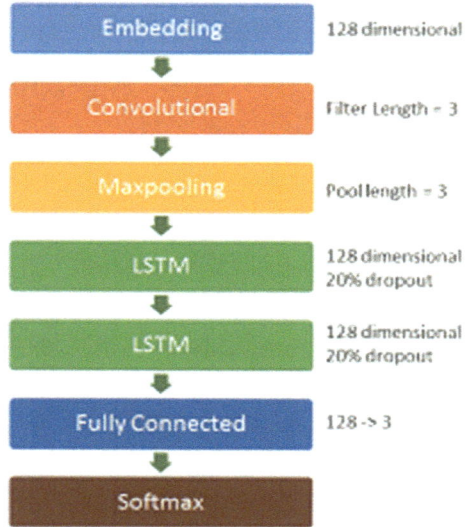

Jhanwar et al. [18] proposed an ensemble method for sentiment analysis on Hindi-English code switched dataset. The dataset used in this research was same as the dataset proposed by [37].

The ensemble model combined the outputs of character-trigrams-based LSTM model and word N-gram-based MNB model to predict the sentiment of Hi-En code-mixed texts. While the LSTM model encoded deep sequential patterns in the text, MNB captured low-level word combinations of keywords to compensate for the grammatical inconsistencies. This ensemble model produced a accuracy of 70.8% and a F1-score of 0.661 which was better than that of [37]. Figure 6 shows the architecture of the ensemble classifier used by [18].

Other methods have been proposed by [12] which performs the task of sentiment analysis using CNN which basically consists of a sentence representation matrix, convolution layer, pooling layer, and fully connected layer. The system was able to provide a high precision score 70.25% which outperformed the baseline models which used cosine similarities and SVM for the task of sentiment analysis. Text-based models such as [8, 16, 27, 28, 31, 40–43] also pave the way for sentence level classification over social media.

3 Dataset and Evaluation

In this section, we discuss the the various code mixed Hinglish datasets that have been used for the the task of hate speech detection in the various past researches. We focus our attention on the Hindi-English code switched datasets released in the public domain.

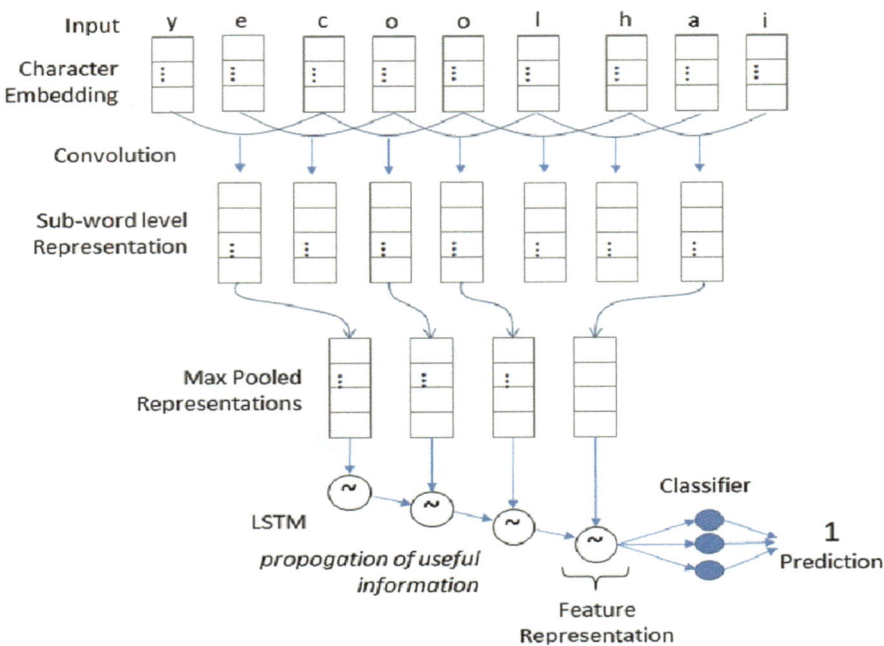

Fig. 6 Ensemble classifier architecture for sentiment analysis [18]

3.1 HOT Dataset

HOT dataset is a manually annotated dataset that was used by [27]. This dataset was created by scraping tweets from Twitter during the interval of 4 months between November 2017 to February 2018. Geo-location restrictions were imposed on the tweets so that the tweets hailing only from the Indian subcontinent were made part of the corpus. Since the tweets from famous personalities receive the maximum flak from the users, the tweets and their responses were crawled from the twitter handles of sports personals, political figures, movie stars, and news channels. From the initial corpus of 25,667 tweets, some of tweets containing only URLs, only images and videos, having less than three words, non-English, and non-Hinglish scripts and duplicates were removed.

The annotation was done in three categories: hate-inducing, abusive, and non-offensive, by three annotators having sufficient knowledge in the domain of NLP. The tweets were annotated as hate inducing if and only if they satisfied one or more of the following conditions: (i) tweet consisted of a sexist or racial barb to malign a minority, (ii) undignified stereotyping, or (iii) tweet consist of a hateful hashtag such as #HinduSc*m. The label which was in majority among the three annotators was chosen as the final label. In case of no majority was reached, the final annotation was decided by the help of a NLP expert. Finally, there were only 386 tweets that needed

Fig. 7 Examples of tweets in HOT dataset [27]

Tweet	Label
(i) Tum ussey pyar kyun nahin karti? (ii) Why don't you love him? (iii) you him love why no	Non-offensive
(i) Ch*d! Yeh sab ch*tiye hain! :/ (ii) F**k! They all are c*nts! :/ (iii) F**k they all c*nts are	Abusive
(i) M*d*rch*d Mus*lm**n sE nafrat (ii) m*therf*ck*r m*sl*m hate (iii) m*therf*ck*r m*s*l*m**n hate	Hate-inducing

expert annotation. The average value of Cohen's kappa inter-annotator agreement was $\kappa = 0.83$. Out of the total 3189 tweets, 2551 tweets were used for training purpose and 638 tweets were used for the testing purpose. An example of some of the tweets from the HOT dataset is shown in Fig. 7.

3.2   Bohra et al. [5] dataset

Another Hindi-English code mixed dataset was created by [5] in 2018 for hate speech classification as well as language identification. The dataset was created by scraping tweets from twitter, taking into account certain hashtags and keywords from politics, public protests, riots, etc., which have a good probability for the presence of hate speech. After all, the noisy tweets were manually removed, a dataset of 4575 code mixed tweets was created. The annotation was carried out in two schemes which are as follows:

(i) Language at Word Level: For each word in the tweet, a tag was associated with it. Tags are of three types namely "eng," "hin," and "other." These tags were assigned with the help of bilingual speakers.

(ii) Hate Speech or Normal Speech: Each tweet is annotated as hate speech or a normal tweet. An example of the annotation scheme is shown in Fig. 8.

Annotation of the dataset to detect the presence of hate speech was carried out by two human annotators having linguistic background and proficiency in both Hindi and English. Inter-annotator agreement is calculated by using Cohen's kappa coefficient. The kappa score was reported as 0.982 indicating the quality of annotation. Annotation instance of the dataset is shown in Fig. 8.

3.3   HEOT Dataset

HEOT dataset was also created by [28] for hate speech detection. HEOT dataset was created using the Twitter Streaming API by selecting tweets in Hindi-English code switched language which were mined using specific profane words. The dataset was

```
<tweet>
<id>954297321843433472<\id>
<word lang="eng">Congress</word>
<word lang="hin">ke</word>
<word lang="eng">agents</word>
<word lang="hin">ho</word>
<word lang="hin">ya</word>
<word lang="hin">maha</word>
<word lang="hin">murkh.</word>
<word lang="hin">rape</word>
<word lang="hin">rape</word>
<word lang="hin">hota</word>
<word lang="hin">hai</word>
<word lang="eng">use</word>
<word lang="hin">dalit</word>
<word lang="eng">or</word>
<word lang="hin">non</word>
<word lang="hin">dalit</word>
<word lang="eng">see</word>
<word lang="hin">Jo</word>
<word lang="hin">Kar</word>
<word lang="hin">mat</word>
<word lang="hin">dekho</word>
</tweet>
<class>
Hate Speech
</class>
```

Fig. 8 Annotation instance from Bohra et al. [5] dataset

Table 1 Examples of tweets in the dataset in HEOT dataset [19]

Category	Tweet	Translation
Benign	sache sapooto aap ka balidan hamesha yaad rahega	True sons, your sacrifice would be remembered
Hate-inducing	Bik gya Porkistan	Porkistan (Derogatory term for Pakistan) has been sold
Abusive	Kis m*darch*d ki he giri hui harkt	Which m*therf*cker has done this

compiled from November 2017 to December 2017 and was distributed to ten NLP researchers for annotation and verification. The dataset thus created consisted of 3679 tweets which consisted of 1414 non-offensive, 1942 abusive, and 323 hate-inducing tweets. The annotation scheme selected for this dataset was similar to that of HOT dataset, i.e., non-offensive, abusive, and hate-inducing. An example of some tweets from HEOT dataset is shown in Table 1.

3.4 Davidson Dataset

Davidson [10] provided one of the initial datasets for hate speech detection. Though this dataset is in English language and not code switched language, this dataset is used as the baseline dataset to check the accuracy of the classifying models. This dataset is used by both the primary researches in the field of hate speech detection in code switched language [19, 28] to validate the accuracy of their classifying models. The dataset was created by first collecting hate speech lexicons from Hatebase.org and then collecting the tweets that use those lexicons with the help of twitter API. After random sampling of about 25 K tweets, each tweet was annotated from three or more users. The inter-user agreement reached was reported to be 92%. The final dataset released consisted of 24,802 tweets each of which labeled as either hate-inducing, offensive, or neither.

4 Methodology

Over the period of time, different researchers have relied on various methods for the complex task of hate speech analysis. A few of the major techniques, some of which that have produced state-of-the-art results in this domain have been elaborately discussed.

4.1 SVM and Random Forest

One of the first works on hate speech recognition was done by "A Dataset of Hindi-English Code-Mixed Social Media Text for Hate Speech Detection" [5] in which the authors prepared an annotated dataset of 4575 tweets in code-mixed language pair of *Hinglish* and annotated them on two classes—normal speech and hate speech. This was one of the pioneer works in this domain.

This work was conducted in three major steps, i.e., preprocessing of the tweets, followed by feature extraction, and succeeded by using these feature vectors to train the supervised machine learning models (support vector machines and random forests).

4.1.1 Preprocessing of Tweets

The preprocessing of Hindi-English code-mixed tweets was done by removing the URLs present in the tweets and replacing by the word "URL." Similarly, all the user names were replaced by the term "USER." This was done since the authors believed that the user names and URL do not contribute to the sentiment of the tweet. Also, all the emoticons were replaced by the word "Emoticon." And finally, the punctuation was also removed and their count was stored separately since the punctuation were one of the features that were required later in the process of feature identification and extraction.

4.1.2 Feature Identification and Extraction

For the task of hate speech identification, the authors used five different features to train their supervised machine learning model. These features were as follows:

(i) **Character N-Grams (C)**: They are language independent features have therefore proven to be very proficient in text classification problems. It also has the capability to handle spelling errors [7, 15, 22]. In code-mixed languages like the Hindi and English pair, there is a lot of semantic variation and the words from English and Hindi vary remarkably. In such situations character, N-Grams help to capture the semantic meaning in the text. The authors have used this as a feature for hate speech classification task keeping the value of n from 1 to 3.

(ii) **Word N-Grams (W)**: The bag of words is a significant feature used in sentiment analysis task to capture the sentiment with the text [38]. This work uses the feature for hate speech detection as done by [53] where n is varied from 1 to 3 as a feature for training the supervised model.

(iii) **Punctuation (P)**: Punctuation is always used to depict the right sense of emotion and when punctuation is used repeatedly or multiple times in a single sentence, it might as well highlight the strong feelings of the person writing the tweet. For example, multiple question marks ("?") used together depict a feeling of annoyance and vexation. Before removing the punctuation marks from the tweets during preprocessing, the authors keep a count of every punctuation so as to capture the right sense of emotion depicted.

(iv) **Negation Words (N)**: The authors used a list of negation words from Christopher Potts sentiment tutorial. Also, the number of negation words was counted in each tweet by matching it with the list and this was used as a feature.

(v) **Lexicons (L)**: During the task, the authors have figured out 177 different Hindi and English words from the dataset which they have taken as a lexicon feature. Many previous researches have shown that people use a fixed set of words to express the feeling of animosity. And the use of lexicon as a feature for these sentiment analysis-based researches have been a common practice. Lexicon features when applied with other corpus-based features result in significant improve in the accuracy for classification tasks, especially if training and testing data are selected from the same domain. This was shown in a previous research by [32]. This is primarily because if the testing and training dataset have the same source, the set of words in those sentences or tweets will generally be of similar nature.

4.1.3 Training the Supervised Model

Subsequent to feature extraction, the paper addresses the use of supervised machine learning models for experimentation. In the work, two main approaches, i.e, support vector machines with radial basis kernel and random forest classifier have been used for the classification task.

The authors have used the above features one by one and later on used all the features at once to train the models. Since the size of feature vectors formed are very large, the chi-square feature selection algorithm was applied which reduces the size of our feature vector to 12,004. For training the system classifier, the authors have used scikit-learn [35] and a 10 fold classification was done for verification of results.

4.2 Ternary Trans-CNN Model

The problem of hate and abuse detection was also addressed by the paper "Detecting Offensive Tweets in Hindi-English Code-Switched Language" [28] where a CNN-based deep learning model was used by the authors for the classification task. The experiments were performed on HEOT dataset which contains the Hindi-English code switched language tweets classified on three categories, namely hate-inducing, abusive, and neutral. The ternary trans-CNN model is trained on a dataset provided by [10] and contains tweets during the US presidential elections in English language. The above experiments were performed with and without transfer learning to analyze the impact of transfer learning for the classification task.

4.2.1 Transfer Learning

Transfer learning [33] is a machine learning paradigm that refers to knowledge transfer from one domain of interest to another, with the aim to reuse already learned features in learning a specialized task. Transfer Learning is used in those situations where the distribution of input and the feature space are similar to each other. By this, we can obtain the maximum benefit from the transferred weights. Source task is defined as the task from which the system extracts the features and weights while the target task is one being benefited.

Transfer learning does not need lots of new data. A model that has been trained on a large corpus of annotated data will be able to handle a similar dataset with smaller data size as well. Also, using a pre-trained model for an identical task often accelerates the process of training the model on the new task and will also produce much better results.

Transfer learning is an increasingly being used to develop more sophisticated models such as in the domain of natural language processing. Hate speech detection is a great example of one such scenario where transfer learning can be befitting. This is due to the fact that there is semantic resemblance between English and the Hinglish languages. Transfer learning imparts an efficient performance with better accuracy.

Pan et al. [33] gave a mathematical definition of transfer learning. Let domain D consist of two components: a feature space X and a marginal probability distribution $P(X)$, where $X = x_1, x_2, \ldots, x_n \in X$, X is the space of all individual word vectors representing the input text, x_i is the ith vector corresponding to some tweet, and X is a particular learning sample. A task consists of two components: a label space Y and an objective predictive function $f()$, represented as $T = Y, f()$, which is not observed but can be learned from the training data, which consists of pairs (x_i, y_i), where $x_i \in X$ and $y_i \in Y$. In the experiments, Y is the set of all labels for a multi-class classification task, and y_i is one of three class labels.

4.2.2 Preprocessing of Dataset

Each of the tweets from the dataset was passed through the following steps to convert it into vectors. The intermediate steps involved the process were:

  (i)   Removal of user mentions, punctuations, URLs.
  (ii) Insertion of plain text in place of corresponding hashtags.
  (iii) The emoticons were replaced by their textual descriptions as described in the list by [1].
  (iv) Upper case characters were converted to lower case.
  (v)   Using gensim for removal of unnecessary stop words.
  (vi) Finally, the Hinglish words were converted to their English equivalents using a conversion dictionary.

4.2.3 Training Classification Model

The tweets were then converted into word vectors with the help of GloVe [36] pre-trained vector embeddings. The authors used the Twitter version of GloVe embeddings (2B tweets, 27B tokens, 1.2M vocab, uncased, 200d, 1.42 GB download). Finally, the word vector sequences were created which forms the input for the CNN model.

CNN models are first trained on an English dataset provided by [10] where the model learns low-level features of the English language. As we can see in Fig. 9, the last two dense layers are removed from the model and replaced with two new dense layers keeping the initial convolutional layers same as before and the model is retrained on HEOT dataset where it learns to extract high-level complex features due to syntactical variations of the Hinglish text.

This paper focuses on obtaining high accuracy in predicting the hate inducing tweets in Hinglish language using transfer learning with CNN model as shown in the architecture diagram in Fig. 10. The authors have chosen the word embedding dimension to be 200. The proposed CNN architecture consists of three *convolution1D*

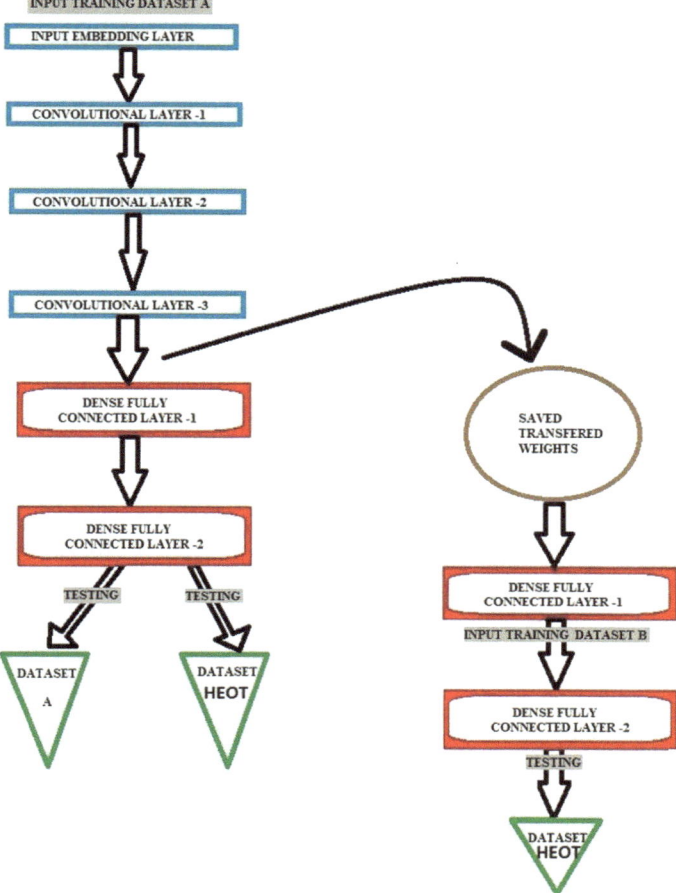

Fig. 9 Transfer learning technique used for ternary trans-CNN model [28]

layers with filter sizes being 15, 12, and 10, respectively, and kernel size is fixed to 3. Two dense layers have been added in the end of size 64 and 3 since the model aims to converge the vast set of features into 3 distinct classes. The activation function is selected as rectified linear unit (ReLU) [23] and Softmax, respectively, for the two dense layers. The authors have selected categorical cross-entropy as the loss function with Adam optimizer [20]. Using the grid search, batch size was examined from size 8 to 256 to procure the best results possible. Also, the number of epochs was chosen by experimenting from 10 to 50. Dropout layers were also added in the model after the dense layers to prevent overfitting and also to enhance the generalization of the systems. The model was initially trained on a data of size 11,509 and tested on data of size 3000 that were randomly split from the English dataset by [10]. To obtain the most favorable results, the batch size was set to 128 and epochs were set to 25

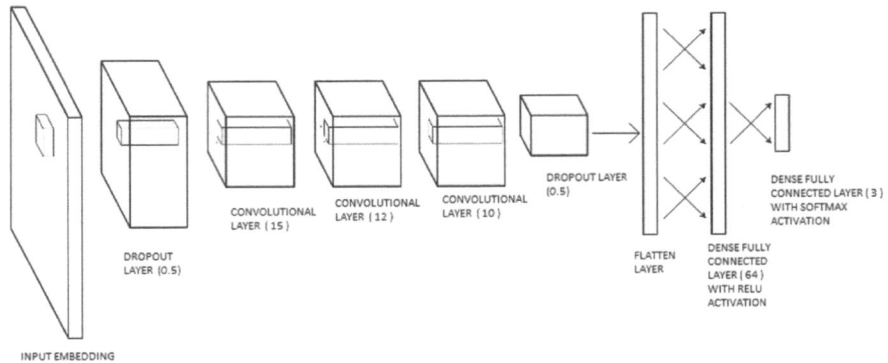

Fig. 10 Convolutional neural network (CNN) architecture used for ternary trans-CNN model [28]

with all layers as trainable. For applying transfer learning, the model is retrained keeping the complete model except the last two layers frozen. The two new dense layers were kept as trainable. The new model was trained on HEOT dataset which was split as 2679 tweets for training and the remaining 1000 for testing. The batch size was decreased to 64 with epochs reduced to 10 for minimum training loss and finer results.

4.3 LSTM-Based Model

The paper "Mind Your Language: Abuse and Offense Detection for code switched Languages" [19] uses an LSTM-based model with transfer learning for this task. This work deals with the problem of hate speech identification in a code-switched language pair of Hindi-English into, namely three categories—benign, abusive, and hate-inducing. This paper establishes itself in the field of hate speech detection as state-of-the-art work on HEOT dataset.

This involves the following steps: preprocessing, training of word embeddings, training of the classifier model, and then using that on HEOT dataset.

4.3.1 Preprocessing

In this work, two datasets were used. One of the dataset has been released by [10] which comprises of English language tweets and the other is HEOT dataset provided by [28].

The tweets obtained from datasets were channeled through a preprocessing pipeline with the final objective to transform them into feature vectors which involved the following steps:

Table 2 Examples of word pairs in Hinglish-English dictionary [19]

Hinglish	English	Hinglish	English
acha	good	gunda	thug
s**la	blo*dy	ra*di	h*oker

(i) Removal of punctuations, mentions, emoticons, URLs, stopwords, numbers, hashtags, etc.

(ii) This was followed by translation of each word to English using Hindi to English dictionary, i.e., Xlit-Crowd conversion dictionary.

(iii) Since Hinglish deals with a lot of spelling variations, the authors manually included some common variations of regularly used Hinglish words. The final dictionary comprised of 7200 word pairs.

(iv) Also, to deal with profane words, which are not present in Xlit-Crowd conversion dictionary, a profanity dictionary with 209 profane words was also created. An example is shown in Table 2.

4.3.2 Training Word Embeddings

After the tweets were preprocessed, different word embeddings were trained on the datasets. Firstly, the GloVe embeddings [36] were used. The version of word embedding used in this case was (2B tweets, 27B tokens, 1.2M vocab, uncased, 100d vectors, 1.42 GB download). Secondly, the Twitter word2vec embeddings [11], which is a 400 million tweets, word embedding model.

Both these were then trained on the datasets, namely HEOT and the one provided by [10]. These embeddings help to learn the distributed representations of tweets by creating word vectors. These word embeddings form the classifier model which then classifies the tweets in one of the three categories.

4.3.3 LSTM Model with Transfer Learning

After the creation of word embeddings, the final training and testing of the dataset was done on an LSTM-based ternary classification model which finally categorizes the tweets in one of the three categories—benign (non-offensive), abusive, and hate inducing. The model is depicted in Fig. 11.

The model takes the distributed tweets representation as the input which is the word embeddings. Next, a dropout layer is added to prevent overfitting of data. This has a value of 0.5. The model uses an LSTM layer and three dense layers. The LSTM layer has a dropout of 0.2, and the dense layer is of size 64, 32, 3, respectively. The authors have used categorical cross-entropy as the loss function in the last layer due to the existence of several classes. Also, to prevent overfitting, Adam optimizer along

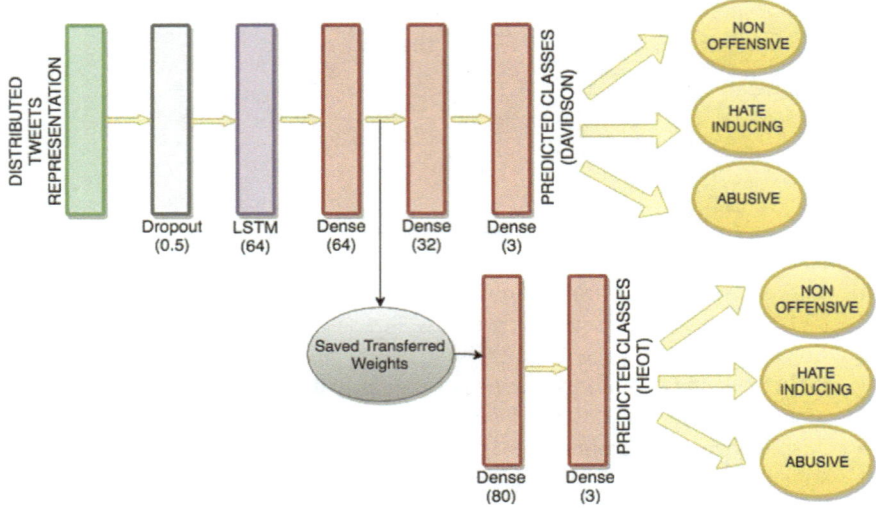

Fig. 11 LSTM-based model [19]

with L2 regularization has been used. To get the best set hyper-parameters and for the achievement of optimum results, extensive grid search was also done.

As we can note in Fig. 11, the LSTM model was trained on the dataset provided by [10]. After passing through the first dense layer, the trained model is saved and with the help of transfer learning, the model is retrained on HEOT dataset. The use of transfer learning helps to improve the score by making use of the saved weights from the model trained previously on Davidson tweets dataset. The retrained model is then passed through a series of dense layers and to finally distinguishes the tweet in the one of the three categories.

4.4 MIMCT Model

One of the recent works titled, "Did you offend me? Classification of Offensive Tweets in Hinglish Language" by [27] proposes a Hinglish offensive tweets (HOT) dataset which consists of tweets in Hindi-English code-switched language. The work also proposes an MIMCT model which inherits its benefits jointly from LSTM, CNN, and transfer learning. The main contributions of the work were to build an annotated Hinglish Offensive Tweet (HOT) dataset which is classified in three classes, namely abusive, offensive, and non-hate-inducing. Also, the paper demonstrates the usefulness of transfer learning for classifying offensive Hinglish tweets and finally building a novel model that outperforms the baseline models on HOT.

The methodology consists of the following subtasks: Preprocessing of the dataset followed by the training of the MIMCT model on HOT dataset.

4.4.1 Preprocessing the Dataset

As in the previous scenarios, the punctuation, URLs, user mentions (starting with "@"), stop words, and numbers were removed as they do not convey any sentiments. Hashtags and emoticons were converted by their textual counterparts, and finally, all the tweets were converted into lower case. The next task was the translation of each word in Hinglish into their corresponding English counterpart the Hinglish-English Xlit-Crowd Conversion dictionary. The syntax and grammatical notions of the converted tweets were not taken into consideration so that the resultant tweet can be converted to their corresponding word vectors. Once this was done, several word embedding models like FastText [6] embeddings, Twitter word2vec [11], and GloVe [36] were used for creating the word embeddings as the input to the model and to obtain the word sequence vector representations of the tweets. The train–test data split was 80:20.

4.4.2 Training MIMCT Model

Figure 12 depicts the architecture of the MIMCT model. It mainly consists of two components, namely inputs (primary and secondary) and a binary channel of neural networks consisting of CNN and LSTM layers.
Primary and Secondary Inputs

Fig. 12 MIMCT model [27]

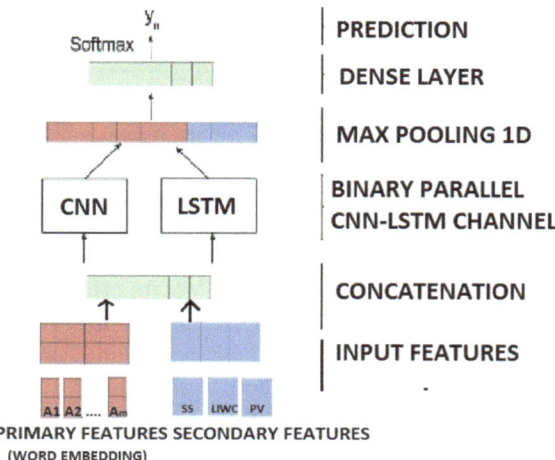

Word embeddings help the model to comprehend the low dimensional representations of tweets, and processing steps encode different aspects of the language. Word associations are stressed by the bag of word statistics. Word embeddings with their dimensions as $d_1, d_2, \ldots, d_m$ were independently inputted into the MIMCT model as primary inputs. The dimension of the embeddings was kept constant with a value of 200. The model comprised of multiple sentence matrices $A_1, A_2, \ldots, A_m$ where each $A_l \epsilon R s * dl$ having s as zero-padded sentence length and dl as dimensionality of the embedding. Next, feature vectors were obtained for every set of word embeddings which were known as primary inputs. Apart from the embedding inputs, auxiliary hierarchical contextual features were also added for the classification task. These features focus on the sentiments and abusive language which is not present in the regular dictionary. This was identified as one of the most serious problems in the regular methods that were being used to perform this task.

The secondary input consists of the following:

(i) **Sentiment Score**: With the help of Sentiwordnet [14], the authors evaluated the sentiment score and this was used a secondary input to the MIMCT model. This was a 1 dimensional vector represented by $+1$ for positive sentiment, 0 for neutral and -1 for a negative sentiment. This was done to emphasize on the polarity of tweets.

(ii) **LIWC Features**: LIWC or Linguistic Inquiry and Word Count describes the various language modalities depicting the linguistic statistical formation of text. The authors have not considered the numbers and punctuation in LIWC features as these were removed during the time of preprocessing stage.

(iii) **Profanity Vector**: A profanity list consisting of a number of swear words was created. This list consisted of the profane words in Hinglish with their English translation. An integer vector of dimension 210 was constructed for each tweet. This was done such that the presence of a particular profane word was demarcated by its corresponding profanity score while its absence was demarcated by a null value to stress on the presence of contextually subjective swear words.

In the MIMCT model, many combinations of trainable and static layers were experimented to derive the finest combination that gives the most favorable results. Fine tuning of the model was done, and tenfold cross-validation was performed on LSTM layer to identify the best set of hyper-parameters. Grid search was also applied to find the most optimum hyper-parameters.

For CNN, the authors selected the filter size to be 15, 12, 10, respectively, followed by a dropout layer of 0.2 to prevent overfitting. Two dense layers were subsequently added of size 64 and 3 with the activation function being *ReLU* and *softmax* respectively. For the LSTM side of the model, an LSTM layer of size 64 and dropout as 0.25 was chosen which was followed by two dense layers of sizes 64 and 3, and their activation functions being *relu* and *sigmoid*, respectively. The final layer of both CNN and LSTM side of the models was a compile layer with categorical cross-entropy as the loss function, Adam as the optimizer, learning rate kept at 0.001 and L2 regularization with strength of 1E-6. The model was trained using three different sets of embeddings, namely GloVe, Twitter Word2Vec, and Fasttext. The dimension for each of them was chosen as 200. After several experimentation, the epochs and batch size were chosen to be 20 and 64 to maintain consistency in performance without compromising on the accuracy of the results.

The word embedding representation generated was combined with the secondary features and fed to the MIMCT model as independent inputs to both LSTM and CNN parts of the model. These features after passing through both the channels are merged again and then sent to the Max-pooling 1D layer. The vector now obtained is reshaped and finally inputted to a softmax layer. MIMCT model was trained on the EOT dataset and then the model is again trained on the HOT dataset so as to benefit from the transfer of learned features in the last stage.

5 Results

As we have moved from the basic machine learning models to the deep learning models, we can easily observe how the results have drastically improved by the use of deep learning techniques. Hate speech detection for code switched languages is indeed a complex problem involving great degree of intricacies in terms of the varied semantics that the code mixed languages has to offer.

5.1 SVM and Random Forest Classifier

As we discuss the progress that different models bring in, we observe that the work by [5] carries out experiments by training the supervised machine learning model. Accuracy was considered as the metric, and tenfold cross-validation was done, and the results were as depicted in Tables 3 and 4.

First each of the features was considered individually, and then all the five features, namely Character N-Grams, Word N-Grams, punctuation, negation words, lexicons, were taken together. The highest recorded accuracy in the case of support vector

Table 3 Accuracy of each feature using support vector machines [5]

Features	Accuracy
Character N-Grams	71.6
Word N-Grams	70.1
Punctuation	63.6
Lexicons	64.2
Negations	63.6
All features	**71.7**

Bold illustrate significant change in results, emphasis on outperformance

Table 4 Accuracy of each feature using random forest classifier [5]

Features	Accuracy
Character N-Grams	66.6
Word N-Grams	**69.9**
Punctuations	63.2
Lexicon	63.8
Negations	63.6
All features	66.7

Bold illustrate significant change in results, emphasis on outperformance

machine was 71.7% when all the features were considered simultaneously. In case of random forest classifier, optimum results were seen in case Word N-Grams as the sole feature which gave an accuracy of 69.9% over two classes, i.e, hate speech and normal speech. In general, the support vector machine model performs better than the random forest classifier. Another remarkable observation was that the best feature for detection of hate speech was the Character N-Grams in case of SVM and Word N-Grams in case of random forest classifier.

5.2 Ternary Trans-CNN Model

With the onset of deep learning as a technique for detection of abusive and hate-inducing speech, the ternary trans-CNN model proposed by [28] displayed significant results. The results were compiled in terms of F1 score, precision, accuracy, recall by choosing micro-metrics as the class imbalance is not severe enough to strongly bias the outcomes. The CNN model was initially trained on the Davidson tweets dataset and the model's performance on it was taken as the baseline. The model was tested on HEOT dataset without and with transfer learning to compare the results by the two methods. There was a decline in the results on HEOT dataset when compared to the Davidson dataset Hinglish language tweets in HEOT dataset suffer from syntactic degradation after preprocessing and translation which leads to a loss in the contextual structuring of the tweets. But the transfer learning helps to improve the performance of the model. The results are provided in Table 5. The highest accuracy recorded was 83.9% on the three classes of HEOT dataset by using transfer learning. There was a massive rise in the accuracy of about ∼25%.

5.3 LSTM Model with Transfer Learning

Inspired by the success of ternary trans-CNN model, the paper by [19] uses an LSTM model with word embeddings trained on GloVe [36]. Transfer learning was

Table 5 Results for ternary trans-CNN task: non-offensive, abusive and hate-inducing tweet classification on datasets A, HEOT without transfer learning (w/o TFL), and HEOT with transfer learning (TFL) [28]

Dataset	A	HEOT (w/o TFL)	HEOT (TFL)
Accuracy (%)	75.40	58.70	83.90
Precision	0.672	0.556	0.802
Recall	0.644	0.473	0.698
F1 Score	0.643	0.427	0.714

Table 6 Comparison of accuracy scores on HEOT dataset [19]

Model	Accuracy
Davidson et al.	0.57
Our Model with embeddings trained on GloVe	**0.87**
Our Model with embeddings trained on Word2Vec	0.82
Our Model with pre-trained Word2Vec embeddings	0.59
Mathur et al.	0.83

Bold illustrate significant change in results, emphasis on outperformance

Table 7 Comparison of accuracy scores on davidson dataset [19]

Model	Accuracy
Davidson et al.	**0.90**
Our Model with embeddings trained on GloVe	0.89
Our Model with embeddings trained on Word2Vec	0.86
Mathur et al.	0.75

Bold illustrate significant change in results, emphasis on outperformance

also used to benefit from the transferred weights that were trained on the previous dataset. Tables 6 and 7 show the performance of LSTM model after getting trained on [10] with two types of embeddings in comparison to the models by [28] and [10] on the HEOT dataset. The results were averaged over three runs. The authors also compared the results on pre-trained word2vec embeddings. As depicted in the table, the custom trained GloVe embeddings performs better than all other word embeddings. The results were also compared for the Davidson model which produces great accuracy for English language tweets. The model by [10] gives an accuracy of 90% of Davidson English tweets dataset while giving an accuracy of 57%, which indicates that model finds it hard to deal with the intricacies of Hinglish semantics. The LSTM model with transfer learning on the hand gives a comparable accuracy of 89% on Davidson dataset and also performs well on the HEOT dataset [28] resulting in an accuracy of 87% which is the state-of-the-art results on HEOT dataset.

Table 8 Baseline results for non-offensive, abusive, hate-inducing tweet classification on HOT [27]

Feature	Char N-grams		BoWV		TF-IDF	
Classifier	SVM	RF	SVM	RF	SVM	RF
Precision	0.679	0.565	0.688	0.579	**0.721**	0.655
Recall	0.708	0.587	**0.731**	0.664	0.724	0.678
F1-Score	0.688	0.574	0.703	0.639	**0.723**	0.666

Bold illustrate significant change in results, emphasis on outperformance

5.4 MIMCT Model

One of the most recent works is by [27] using MIMCT model, which uses a combination of CNN and LSTM to extract the benefits of both and combine them into a single unit. The experiments are performed on HOT dataset. Firstly, the experiments were conducted on supervised machine learning models—SVM and random forests. Table 8 shows that SVM produces better results than random forest classifier on HOT dataset. Also, we can conclude that TF-IDF is the most effective feature for semantically representing Hindi-English code switched language text and performs better than other two features, i.e., Character N-grams and Bag of Words Vector (BoWV) on respective classifiers.

Next, the MIMCT model was tried for the HOT dataset with F1 score, recall, and precision as the metrics for evaluation. Table 9 shows results for the classification task. Macro-metrics are often preferred in evaluation because the class imbalance is not severe enough to skew the outcomes. Since the Hinglish tweets suffer from syntactic degradation after translation, there is a sharp decline in the model's performance without the use of transfer learning. However, the use of transfer learning enhances the results further strengthening the argument that there was a positive transfer of features from English to Hinglish tweet data. The several experiments suggest the fact that the performance of LSTM model is better than its CNN counterpart. Also, the Twitter word2vec outperforms its contemporary embeddings in many cases. Lastly, we observe the effect of multiple inputs in MIMCT model. The combination of Twitter word2vec and FastText shows superior performance of 0.861 as the precision score which is much better than other embedding combinations. Also, we observe that the inclusion of sentiment score barely affects the overall performance of the classification task. On the other hand, the features such as profanity vector and LIWC boost the metric values and the best results are obtained when all the features are used simultaneously in combination with FastText and Twitter word2vec embeddings. As of today, MIMCT model establishes itself as the state-of-the-art model in the domain of hate speech identification and classification. MIMCT model (Tw + Ft + SS + PV + LIWC) outperforms SVM supplemented with TF-IDF features and the Twitter–LSTM transfer learning model by 0.166 and 0.165 F1 points, respectively. The best result obtained HOT dataset is recall of 0.928 and F1 score as 0.895 when Tw + Ft + SS + PV + LIWC configuration is considered. With these results, this model establishes itself as the state-of-the-art model for detecting hate speech in code switched Hinglish language.

6 Conclusion

The problem of hate speech detection had started off as a child problem to sentiment analysis, but in today's world, it has become a standalone problem and is in itself, a completely new research domain. As the problem of hate speech in code switched languages keeps on rising with the increasing reach of the Internet, the need of the hour is to filter out all the illicit content on the social media so as to make it a cleaner environment for everyone. The task of handling code mixed data is a complex problem in itself involving great deal of intricacies in terms of varying semantics that code mixed language offers. Through this work, we tried to discuss all the researches that have taken place in this domain for the Hindi-English code switched language (the most used code switched language on the Internet).

We discussed various datasets that have been used for this task which was followed by the discussion on the methodologies used by the various researchers for hate speech detection.We also briefly discussed the various different advancements that can take place in this domain so as to increase the scope and precision of the hate speech problem. Hence through this work, we hope to encourage the readers to take up this task of hate speech detection and hence make social media a cleaner place to exhibit one's views. This chapter is inspired by the works of Zhang et al. [54], Chowdhury et al. [8], Mahata et al. [25], Shah et al. [44], Meghawat et al. [30], and Shah et al. [45].

7 Future Work

Though there have been few researches that have focused on the task of detecting hate speech on code switched languages, there still seems to be some scope for improvement in terms of scalability to other code switched languages and also better accuracy. Some of the enhancements that can be looked upon are as follows:

(i) Implementing feature selection methods to choose the most pronounced features among all the known features. This is similar to the research provided by [42].
(ii) Using GRU-based models for the task of hate speech detection.
(iii) Exploring a model based on stacked ensemble of shallow convolutional neural networks for Twitter data similar to the work provided by [24].
(iv) Extending the models to other code switched and code mixed languages.
(v) Tuning the neural network models using boosting methods such as gradient boosting [3].

Table 9 Results for non-offensive, abusive, hate-inducing tweet classification on EOT, HOT, and the HOT dataset with transfer learning (TFL) for GloVe, Twitter Word2vec, and FastText embeddings

Embedding	GloVe						Twitter Word2vec						FastText					
Model	CNN			LSTM			CNN			LSTM			CNN			LSTM		
Data	EOT	HOT	TFL	EOT	HOT	TFL	EOT	HOT	TFL	EOT	HOT	TFL	EOT	HOT	TFL	EOT	HOT	TFL
Precision	0.843	0.734	0.789	0.819	0.753	0.802	0.856	0.762	0.793	0.821	0.756	0.810	0.800	0.730	0.758	0.799	0.746	**0.823**
Recall	0.841	0.804	0.820	0.834	0.764	0.819	0.861	0.811	0.817	0.835	0.779	**0.846**	0.820	0.805	0.827	0.807	0.677	0.838
F1-score	0.841	0.755	0.801	0.816	0.752	0.813	0.857	0.799	0.815	0.835	0.765	**0.830**	0.811	0.772	0.793	0.800	0.755	0.823

Bold illustrate significant change in results, emphasis on outperformance

(vi) Extending the work to code mixed languages that contain more than two languages for multilingual societies.

(vii) Exploit multimodal information in hate speech detection.

(viii) Try several advanced neural network architectures as done by [26] using multi-headed attention models in the context of NLP.

(ix) Develop an offensive video segmentation system [47] so as to get rid off abusive and hate-inducing videos on the Internet.

(x) Look over aspects [17], tag relevance [44], and events [46] for this task as hate speech is closely related with sentiments, keywords , and some associated events.

(xi) Exploring relative positions of words for analyzing Hindi data.

References

1. Agarwal, Apoorv, Boyi Xie, Ilia Vovsha, Owen Rambow, and Rebecca Passonneau. 2011. Sentiment analysis of Twitter data. In *Proceedings of the Workshop on Language in Social Media (LSM 2011)*, 30–38

2. Ayyar, Meghna, Puneet Mathur, Rajiv Ratn Shah, and Shree G. Sharma. 2018. Harnessing AI for kidney Glomeruli classification. In *2018 IEEE International Symposium on Multimedia (ISM)*, 17–20. New York: IEEE

3. Badjatiya, Pinkesh, Shashank Gupta, Manish Gupta, and Vasudeva Varma. 2017. Deep learning for hate speech detection in tweets. In *Proceedings of the 26th International Conference on World Wide Web Companion*, 759–760. International World Wide Web Conferences Steering Committee

4. Bali, Kalika, Jatin Sharma, Monojit Choudhury, and Yogarshi Vyas. 2014. I am borrowing ya mixing? An analysis of English-Hindi code mixing in Facebook. In *Proceedings of the First Workshop on Computational Approaches to Code Switching*, 116–126

5. Bohra, Aditya, Deepanshu Vijay, Vinay Singh, Syed Sarfaraz Akhtar, and Manish Shrivastava. 2018. A dataset of Hindi-English code-mixed social media text for hate speech detection. In *Proceedings of the Second Workshop on Computational Modeling of Peoples Opinions, Personality, and Emotions in Social Media*, 36–41

6. Bojanowski, Piotr, Edouard Grave, Armand Joulin, and Tomas Mikolov. 2017. Enriching word vectors with subword information. *Transactions of the Association for Computational Linguistics* 5: 135–146

7. Cavnar, William B., John M. Trenkle, et al. 1994. N-gram-based text categorization. In *Proceedings of SDAIR-94, 3rd Annual Symposium on Document Analysis and Information Retrieval*, vol. 161175. Citeseer

8. Chowdhury, Arijit Ghosh, Ramit Sawhney, Puneet Mathur, Debanjan Mahata, and Rajiv Ratn Shah. 2019. Speak up, fight back! detection of social media disclosures of sexual harassment. In *Proceedings of the 2019 Conference of the North American Chapter of the Association for Computational Linguistics: Student Research Workshop*, 136–146

9. Das, Amitava, and Björn Gambäck. 2014. Identifying languages at the word level in code-mixed indian social media text. In *Proceedings of the 11th International Conference on Natural Language Processing*, 378–387

10. Davidson, Thomas, Dana Warmsley, Michael Macy, and Ingmar Weber. 2017. Automated hate speech detection and the problem of offensive language. In *Eleventh International AAAI Conference on Web and Social Media*

11. Godin, Fréderic, Baptist Vandersmissen, Wesley De Neve, and Rik Van de Walle. 2015. Multimedia Lab @ ACL W-NUT NER shared task: Named entity recognition for twitter microposts using distributed word representations. In *Proceedings of the Workshop on Noisy User-Generated Text*, 146–153
12. Gupta, Deepak, Ankit Lamba, Asif Ekbal, and Pushpak Bhattacharyya. 2016. Opinion mining in a code-mixed environment: A case study with government portals. In *Proceedings of the 13th International Conference on Natural Language Processing*, 249–258
13. Gupta, Deepak, Shubham Tripathi, Asif Ekbal, and Pushpak Bhattacharyya. 2017. SMPOST: Parts of speech tagger for code-mixed Indic social media text. arXiv preprint arXiv:1702.00167
14. Haccianella, S., A. Esuli, and F. Sebastiani. 2010. SentiWordNet 3.0: An enhanced lexical resource for sentiment analysis and opinion mining. In *Proceedings of the Seventh Conference on International Language Resources and Evaluation*
15. Huffman, Stephen. 1995. Acquaintance: Language-independent document categorization by n-grams. Technical report, Department of Defense Fort George G Meade MD
16. Jain, Roopal, Ramit Sawhney, and Puneet Mathur. 2018. Feature selection for cryotherapy and immunotherapy treatment methods based on gravitational search algorithm. In *2018 International Conference on Current Trends Towards Converging Technologies (ICCTCT)*, 1–7. New York: IEEE
17. Jangid, Hitkul, Shivangi Singhal, Rajiv Ratn Shah, and Roger Zimmermann. 2018. Aspect-based financial sentiment analysis using deep learning. In *Companion of the The Web Conference 2018 on The Web Conference 2018*, 1961–1966. International World Wide Web Conferences Steering Committee
18. Jhanwar, Madan Gopal, and Arpita Das. 2018. An ensemble model for sentiment analysis of Hindi-English code-mixed data. arXiv preprint arXiv:1806.04450
19. Kapoor, Raghav, Yaman Kumar, Kshitij Rajput, Rajiv Ratn Shah, Ponnurangam Kumaraguru, and Roger Zimmermann. 2018. Mind your language: Abuse and offense detection for code-switched languages. arXiv preprint arXiv:1809.08652
20. Kingma, Diederik P., and Jimmy Ba. 2014. Adam: A method for stochastic optimization. arXiv preprint arXiv:1412.6980
21. Lafferty, John, Andrew McCallum, and Fernando C.N. Pereira. 2001. Conditional random fields: Probabilistic models for segmenting and labeling sequence data
22. Lodhi, Huma, Craig Saunders, John Shawe-Taylor, Nello Cristianini, and Chris Watkins. 2002. Text classification using string kernels. *Journal of Machine Learning Research* 2: 419–444
23. Maas, Andrew L., Awni Y. Hannun, and Andrew Y. Ng. 2013. Rectifier nonlinearities improve neural network acoustic models. In *Proceedings of ICML*, vol. 30, 3
24. Mahata, Debanjan, Jasper Friedrichs, Rajiv Ratn Shah, et al. 2018. # phramacovigilance-exploring deep learning techniques for identifying mentions of medication intake from twitter. arXiv preprint arXiv:1805.06375
25. Mahata, Debanjan, Haimin Zhang, Karan Uppal, Yaman Kumar, Rajiv Shah, Simra Shahid, Laiba Mehnaz, and Sarthak Anand. 2019. MIDAS at SemEval-2019 task 6: Identifying offensive posts and targeted offense from twitter. In *Proceedings of the 13th International Workshop on Semantic Evaluation*, 683–690
26. Mathur, Puneet, Meghna Ayyar, Rajiv Ratn Shah, and Sg Sharma. 2019. Exploring classification of histological disease biomarkers from renal biopsy images. In *2019 IEEE Winter Conference on Applications of Computer Vision (WACV)*, 81–90. New York: IEEE
27. Mathur, Puneet, Ramit Sawhney, Meghna Ayyar, and Rajiv Shah. 2018. Did you offend me? classification of offensive tweets in Hinglish language. In *Proceedings of the 2nd Workshop on Abusive Language Online (ALW2)*, 138–148
28. Mathur, Puneet, Rajiv Shah, Ramit Sawhney, and Debanjan Mahata. 2018. Detecting offensive tweets in Hindi-English code-switched language. In *Proceedings of the Sixth International Workshop on Natural Language Processing for Social Media*, 18–26
29. Mave, Deepthi, Suraj Maharjan, and Thamar Solorio. 2018. Language identification and analysis of code-switched social media text. In *Proceedings of the Third Workshop on Computational Approaches to Linguistic Code-Switching*, 51–61

30. Meghawat, Mayank, Satyendra Yadav, Debanjan Mahata, Yifang Yin, Rajiv Ratn Shah, and Roger Zimmermann. 2018. A multimodal approach to predict social media popularity. In *2018 IEEE Conference on Multimedia Information Processing and Retrieval (MIPR)*, 190–195. New York: IEEE

31. Mishra, Rohan, Pradyumn Prakhar Sinha, Ramit Sawhney, Debanjan Mahata, Puneet Mathur, and Rajiv Ratn Shah. 2019. SNAP-BATNET: Cascading author profiling and social network graphs for suicide ideation detection on social media. In *Proceedings of the 2019 Conference of the North American Chapter of the Association for Computational Linguistics: Student Research Workshop*, 147–156

32. Mohammad, Saif. 2012. Portable features for classifying emotional text. In *Proceedings of the 2012 Conference of the North American Chapter of the Association for Computational Linguistics: Human Language Technologies*, 587–591. Association for Computational Linguistics

33. Pan, Sinno Jialin and Qiang Yang. 2009. A survey on transfer learning. *IEEE Transactions on Knowledge and Data Engineering* 22 (10): 1345–1359

34. Pang, Bo, Lillian Lee, et al. 2008. Opinion mining and sentiment analysis. *Foundations and Trends® in Information Retrieval*, 2(1–2): 1–135

35. Pedregosa, Fabian, Gaël Varoquaux, Alexandre Gramfort, Vincent Michel, Bertrand Thirion, Olivier Grisel, Mathieu Blondel, Peter Prettenhofer, Ron Weiss, Vincent Dubourg, et al. 2011. Scikit-learn: Machine learning in python. *Journal of Machine Learning Research*, 12: 2825–2830

36. Pennington, Jeffrey, Richard Socher, and Christopher Manning. 2014. Glove: Global vectors for word representation. In *Proceedings of the 2014 Conference on Empirical Methods in Natural Language Processing (EMNLP)*, 1532–1543

37. Prabhu, Ameya, Aditya Joshi, Manish Shrivastava, and Vasudeva Varma. 2016. Towards subword level compositions for sentiment analysis of Hindi-English code mixed text. arXiv preprint arXiv:1611.00472

38. Purver, Matthew, and Stuart Battersby. 2012. Experimenting with distant supervision for emotion classification. In *Proceedings of the 13th Conference of the European Chapter of the Association for Computational Linguistics*, 482–491. Association for Computational Linguistics

39. Rao, Pattabhi R.K., and Sobha Lalitha Devi. 2016. CMEE-IL: Code mix entity extraction in Indian languages from social media text@ fire 2016-an overview. In *FIRE (Working Notes)*, 289–295

40. Sawhney, Ramit, Prachi Manchanda, Puneet Mathur, Rajiv Shah, and Raj Singh. 2018. Exploring and learning suicidal ideation connotations on social media with deep learning. In *Proceedings of the 9th Workshop on Computational Approaches to Subjectivity, Sentiment and Social Media Analysis*, 167–175

41. Sawhney, Ramit, Prachi Manchanda, Raj Singh, and Swati Aggarwal. 2018. A computational approach to feature extraction for identification of suicidal ideation in tweets. In *Proceedings of ACL 2018, Student Research Workshop*, 91–98

42. Sawhney, Ramit, Puneet Mathur, and Ravi Shankar. 2018. A firefly algorithm based wrapper-penalty feature selection method for cancer diagnosis. In *International Conference on Computational Science and Its Applications*, 438–449. Berlin: Springer

43. Sawhney, Ramit, Ravi Shankar, and Roopal Jain. 2018. A comparative study of transfer functions in binary evolutionary algorithms for single objective optimization. In *International Symposium on Distributed Computing and Artificial Intelligence*, 27–35. Berlin: Springer

44. Shah, Rajiv Ratn. 2016. Multimodal analysis of user-generated content in support of social media applications. In *Proceedings of the 2016 ACM on International Conference on Multimedia Retrieval*, 423–426. New York: ACM

45. Shah, Rajiv Ratn, Debanjan Mahata, Vishal Choudhary, and Rajiv Bajpai. 2018. Multimodal semantics and affective computing from multimedia content. In *Intelligent Multidimensional Data and Image Processing*, 359–382. IGI Global

46. Shah, Rajiv Ratn, Anwar Dilawar Shaikh, Yi Yu, Wenjing Geng, Roger Zimmermann, and Gangshan Wu. 2015. Eventbuilder: Real-time multimedia event summarization by visualizing social media. In *Proceedings of the 23rd ACM International Conference on Multimedia*, 185–188. New York: ACM

47. Shah, Rajiv Ratn, Yi Yu, Anwar Dilawar Shaikh, Suhua Tang, and Roger Zimmermann. ATLAS: automatic temporal segmentation and annotation of lecture videos based on modelling transition time. In *Proceedings of the 22nd ACM International Conference on Multimedia*, 209–212. New York: ACM

48. Sharma, Shashank, P.Y.K.L. Srinivas, and Rakesh Chandra Balabantaray. 2015. Text normalization of code mix and sentiment analysis. In *2015 International Conference on Advances in Computing, Communications and Informatics (ICACCI)*, 1468–1473. New York: IEEE

49. Singh, Kushagra, Indira Sen, and Ponnurangam Kumaraguru. 2018. Language identification and named entity recognition in Hinglish code mixed tweets. In *Proceedings of ACL 2018, Student Research Workshop*, 52–58

50. Solorio, Thamar, Melissa Sherman, Yang Liu, Lisa M. Bedore, Elisabeth D. Peña, and Aquiles Iglesias. 2011. Analyzing language samples of Spanish–English bilingual children for the automated prediction of language dominance. *Natural Language Engineering*, 17(3): 367–395

51. Vyas, Yogarshi, Spandana Gella, Jatin Sharma, Kalika Bali, and Monojit Choudhury. 2014. POS tagging of English-Hindi code-mixed social media content. In *Proceedings of the 2014 Conference on Empirical Methods in Natural Language Processing (EMNLP)*, 974–979

52. Wang, Sida, and Christopher D. Manning. Baselines and bigrams: Simple, good sentiment and topic classification. In *Proceedings of the 50th Annual Meeting of the Association for Computational Linguistics: Short Papers, vol. 2*, 90–94. Association for Computational Linguistics

53. Warner, William, and Julia Hirschberg. 2012. Detecting hate speech on the world wide web. In *Proceedings of the Second Workshop on Language in Social Media*, 19–26. Association for Computational Linguistics

54. Zhang, Haimin, Debanjan Mahata, Simra Shahid, Laiba Mehnaz, Sarthak Anand, Yaman Singla, Rajiv Ratn Shah, and Karan Uppal. 2019. Identifying offensive posts and targeted offense from twitter. arXiv preprint arXiv:1904.09072

Multilingual Sentiment Analysis

Hitesh Nankani, Hritwik Dutta, Harsh Shrivastava, P. V. N. S. Rama Krishna, Debanjan Mahata and Rajiv Ratn Shah

Abstract Sentiment analysis has empowered researchers and analysts to extract opinions of people regarding various products, services, events and other entities. This has been made possible due to an astronomical rise in the amount of text data being made available on the Internet, not only in English but also in many regional languages around the world as well, along with the recent advancements in the field of machine learning and deep learning. It has been observed that deep learning models produce the state-of-the-art prediction results without the need for domain expertise or handcrafted feature engineering, unlike traditional machine learning-based algorithms. In this chapter, we wish to focus on sentiment analysis of various low resource languages having limited sentiment analysis resources such as annotated datasets,

H. Nankani · H. Dutta · H. Shrivastava · P. V. N. S. Rama Krishna · D. Mahata · R. R. Shah (✉)
MIDAS Lab, IIIT Delhi, New Delhi, India
e-mail: rajivratn@iiitd.ac.in

H. Nankani
e-mail: hiteshnankani97@gmail.com

H. Dutta
e-mail: duttahritwik@gmail.com

H. Shrivastava
e-mail: harsh.vardhan.shri@gmail.com

P. V. N. S. Rama Krishna
e-mail: ramakrishnapinnimty@gmail.com

D. Mahata
e-mail: dmahata@bloomberg.net

H. Nankani · H. Dutta
NSIT, New Delhi, India

H. Shrivastava
Rustamji Institute of Technology, Gwalior, India

P. V. N. S. Rama Krishna
IIIT, Bhubaneswar, India

D. Mahata
Bloomberg, New York, USA

© Springer Nature Singapore Pte Ltd. 2020
B. Agarwal et al. (eds.), *Deep Learning-Based Approaches for Sentiment Analysis*, Algorithms for Intelligent Systems, https://doi.org/10.1007/978-981-15-1216-2_8

193

word embeddings and sentiment lexicons, along with English. Techniques to refine word embeddings for sentiment analysis and improve word embedding coverage in low resource languages are also covered. Finally, we discuss the major challenges involved in multilingual sentiment analysis and explain novel deep learning-based solutions to overcome them.

Keywords Sentiment analysis · Multilingual · Machine learning · Low resource languages · Deep learning

1 Introduction

Sentiment analysis is the process of classifying and determining the overall opinion and sentiment of people toward goods, services, current affairs and other entities. How other people think about some goods and services often influence our own purchasing choices. Moreover, companies and businesses are always keen on knowing public opinion of their goods and services, and this is becoming increasingly important to them as this knowledge can help them greatly in improving the quality of their goods and services and ultimately make more profits. Therefore, it not only is applicable to businesses and corporations, but also finds uses in areas such as e-commerce, financial field, politics, current events, health and medicinal sciences and even history [1].

With the advent of Web 2.0, the amount of review data being made available online has been steadily on the rise. All of such data is made available in public discussion forums, micro-blogging Web sites, such as *Twitter* and *Reddit*, and newspapers and magazines among other sources. The sheer volume of such online content calls for the need for automated sentiment analysis. Earlier, much of this data used to be in English, but now the content in other languages such as Turkish, Dutch, Spanish, Russian, Arabic, and in languages originating from such as Hindi, Marathi, Bengali and Tamil has been increasing.[1] Thus, multilingual sentiment analysis is becoming a research area of increasing importance.

1.1 Low Resource Language

Informally, any language in which we do not have large, well-maintained and annotated public datasets for Natural Language Processing (NLP) is known as a low resource language. Challenges lie in making best use of the low resources available and to develop robust and reliable sentiment analysis tools. For low resource languages, we need to make changes in both our traditional machine learning (ML) techniques and more recent deep learning (DL) techniques for better extraction of

---

[1]https://en.wikipedia.org/wiki/Languages_used_on_the_Internet.

relevant knowledge from these scarce datasets. In this chapter, we will discuss sentiment analysis in both English and low resource languages, but focus primarily on low resource languages.

1.2   Challenges of Sentiment Analysis

Sentiment analysis surely is one of the most researched topics in NLP, but it does come with its fair share of challenges. Here, we describe some of the challenges in both English and other languages. According to a study conducted in the work [2], there are three types of sentiment containing texts which one would encounter—structured sentiment, which is found in formal sentiment reviews and is usually written by professionals; unstructured sentiment, which is written informally without any grammatical or spelling restrictions and is usually the most difficult to determine the polarity of; and semi-structured sentiment, which lies between structured and unstructured reviews and might often require domain knowledge. One of the most researched challenges is negation handling, in which the appearance of negative words can change the polarity of a sentence only partially or completely. The scope of negation is not fixed and varies with different linguistic attributes. In the work [3], the authors propose a negation handling in the sentence level by identifying the negation scopes through consideration of features such as conjunctions, punctuation marks and heuristics based on parts of speech (POS) of negation terms, alongside a few exceptions and a static window. Some of the other challenges of sentiment analysis include domain dependence [4], spam and fake detection [5, 6], bipolar words [7] and sarcasm detection [8, 9].

Now, we discuss some of the challenges in the multilingual scenario. Let us first begin with Hindi, which has established itself as one of the most spoken languages worldwide. According to a comprehensive study conducted on Hindi sentiment analysis as reported by work [10], the main difficulties include the free word orderings of the Hindi language, its morphological variations, spelling variations and perhaps most importantly the lack of resources. A free order language like Hindi is structured in which the subject and verb could appear without any ordering restrictions, which can significantly alter the text sentiment. Moreover, multiple Hindi expressions originate from the same root word, but differ in their gender, tense, sense and other factors. Lastly, the lack of resources such as annotated and labeled corpora, standard sentiment lexicons and word embeddings in the Hindi language poses major challenges for sentiment analysis of Hindi. Apart from the lack of word embeddings, lack of standard and proven evaluation metrics for evaluating the word embeddings generated also poses a major challenge to sentiment analysis. El-Masri et al. [11] conducted an extensive literature review on the successes and challenges of Arabic sentiment analysis. The difficulties of determining text polarity of Arabic include lack of corpora and datasets, named entity recognition, spelling errors and transliterations in blogging Web sites, lack of parsers, negation handling, sarcasm detection, lack of sentiment lexicons and the use of local Arabic dialects.

A common trend in multilingual communities today is code mixing, which is incorporating the native language along with a prominent language like English in written texts to express sentiment on any particular entity. Some of the popular examples of code mixing are German–English and Hindi–English, which is specially becoming very popular in India. Code-mixed text comes with its own set of challenges such as word and spelling variations, multiple number of possible transliterations of a given word when mapping from one language to another and the practice of not following the rules of formal grammar in either of the two languages involved.

1.3 Deep Learning

Deep learning (DL), a subfield of machine learning, is mainly comprised of techniques motivated by the functioning and structure of the human brain called artificial neural networks (ANNs). DL-based model architectures enable us to learn complex and intricate encoded forms of the given information having multiple abstraction levels, thus producing the state-of-the-art prediction results [12]. Unlike traditional machine learning algorithms, DL techniques operate well even without domain expertise and hard core feature extraction. The performance of DL models increases proportionately with the amount of training data available; thus, advances in hardware and the exponential increase in the amount of training data made available are the main reasons of the widespread impact of DL. DL-based techniques have produced very promising results in image recognition [13], speech recognition [14], image captioning [15], machine translation [16] and video classification [17], among many other domains. For a comprehensive description and application of DL, the reader is referred to the work [18]. In this work, we describe some DL model architectures for the sentiment analysis in English and more specifically for low resource languages.

We commence our discussion by introducing word embeddings, and the fact that they are connected related representations of words of the input corpora in a vector space. These word vectors grasp syntax-related and meaning concerning patterns of the words from the various contexts in which they occur, and serve as input features to many of the DL models, following which we discuss techniques on how to encode explicitly sentiment-related information in word embeddings. We also mention how to improve the coverage of word embeddings in low resource languages, which typically have lesser number of annotated datasets to train our DL models with. Moving on, we introduce convolutional neural networks (CNNs) and discuss character-level and word-level CNNs, among other variants employed in the literature. Then, we proceed to introduce recurrent neural networks (RNNs) and an improvement over RNNs, namely long short-term memory (LSTM) networks. Both of these have achieved tremendous success in the literature, owing to the fact that they learn complex and intricate sequential patterns of the input corpora and also model long-term dependencies in the input text. We explain character-level and subword-level LSTM models, along with other alternatives such as combination of LSTMs and CNNs, LSTMs at three different lexical granularity levels and gated recurrent units (GRUs). The tremendous success of LSTMs is also because of incorporating the attention mechanism, which is basically learning of a set of weights that enable the model to attend and focus on the important parts of the input. We mention application of the attention mechanism for the aspect-based sentiment analysis of English and Chinese reviews. Then, we move our discussion toward autoencoders and their usage in bilingually constrained settings. Since autoencoders cannot work effectively for morphologically rich languages like Arabic, we will also discuss AROMA model which handles the problems using various novel methods. Finally, we conclude by giving a brief description of Siamese neural networks and their application in performing polarity detection of text in which some Hindi transliterated expressions are mixed with predominantly English text.

The work is therefore formulated as mentioned. Section 2 reviews non-DL approaches for sentiment analysis of English and resource-scarce languages. Section 3 begins by introducing importance and role of word embeddings in sentiment analysis and also describes how to refine word embeddings for sentiment analysis in English followed by techniques for improving word embedding coverage in low resource languages. Section 4 describes the novel and state-of-the-art DL-related approaches for multilingual sentiment analysis. Section 5 presents the conclusion.

2   Literature Survey

Here, we present a brief literature survey in the field of sentiment analysis concerning both high resource languages and low resource languages. We start by explaining the lexicon-based approaches, where a pre-defined set of rules determined by experts in the field are used to evaluate the polarity in sentence. Then, we survey the popular ML-based techniques which have been found dominant usage in academia.

2.1   *High Resource Languages*

Let us first cover the high resource languages. Since these languages, by definition, have plenty of annotated datasets, we can use a variety of models on them and get competitive accuracy. The English language, for example, has a large number of annotated datasets for textual sentiment analysis. These datasets when combined provide us with millions of sentences to train our model. Also, there are datasets from various sources pertaining to various different topics. Thus, if there are some words which express different polarities in different contexts, the abundance of training data results in all such scenarios being taken care of. Let us now discuss the various techniques which have been employed for sentiment analysis of high resource languages.

2.2 Lexicon-Based Approaches

Lexicon-based approaches are the ones where the rules and information about words are hard-coded in the program. Lexical approaches vary in accordance with the context they were created in. Lexicon-based approaches can essentially be broken down into two sub-parts—first one is the feature extraction, and second one is the application of the suitable sentiment analysis method. Both of the steps will first be explained keeping languages in mind. There are four popular traditional feature extraction methods proposed by Bakliwal [19], namely:

1. **Term presence versus Frequency**: Term presence is used to denote whether the term is present or not, while term frequency takes into account the presence of a term along with its recurring frequency.
2. **Opinion Words and Phrases**: The opinion words and phrases which express sentiment are extracted. These words can include *love, bad, amazed, I'd suggest that*, etc.
3. **Part-of-Speech Tags**: Parts of speech can be extracted to see the role words play in these sentences.
4. **Negation**: Negation is applied to reverse sentiment in the sentence. For example, *I was not happy*.

2.2.1 Sentiment Analysis Methods

Before starting the discussion, it is essential that we discuss coverage. Since it is known that in lexicon-based approaches some sort of information is hard-coded in the program, it is possible that the sentence being evaluated cannot be evaluated by the method in consideration. One possible reason is that the words or symbols we seek are not present in the sentence. The sentences which can be judged will certainly be less than or equal to total sentences present, and the percentage of sentences which can be judged using that lexicon-based approach will denote its coverage. The following are sentiment analysis methods. Although the coverage varies from dataset to dataset, we will still try to give an idea of the coverage that one can expect when scraping tweets from Twitter as done by Gonçalves et al. [20]:

1. **Emoticons**—Emoticons are graphical depictions of various facial expressions that express human emotions and are constructed using punctuation marks and other characters. Analyzing emoticons is the simplest way to detect polarity. The method is fairly straightforward and gives high accuracy. The idea is to extract and classify the tweets on the basis of emoticons they have used even though the coverage for this method is very low (around 14% on the dataset taken into consideration).
2. **SentiStrength**—SentiStrength is a tool created after an extensive survey of on-line social networks (OSNs). It takes into consideration a lot of factors like *very* to enhance the sentiment while *somewhat* to reduce its degree. Various learn-

ing techniques were used while making SentiStrength dataset, and best results, empirically obtained training model, were produced.

3. **SentiWordNet**—This is a widely used tool in lexicon-based sentiment classification inspired from WordNet, an English lexical dictionary. Adjectives, nouns, verbs, etc, are grouped together to form synonym sets called synsets. These sets in synset have positive score, negative score and objective score. The scores with their individual value in the range[0,1] cumulatively add up to one.

4. **SenticNet**—SenticNet is a prevalent concept-based resource. There have also been localization toolkits built for SenticNet. Concept disambiguation algorithms which can discover context were used in building SenticNet. In simple terms, this tool used NLP to figure out the major topics being talked about, which can be people, actions, events or real-world objects. After identifying the important concepts, their individual polarities are calculated and added. For example in *Boring, it's Monday morning*, *boring* and *Monday morning* are extracted. Sentiment score with *boring* is −0.383 while with *Monday morning* is +0.228. Thus, we have a final score of −0.077 which happens to be the average of these values.

Some other sentiment analysis methods are LIWC, PANAS-t and Happiness Index. Linguistic Inquiry and Word Count (LIWC) is a text analysis tool that makes use of a dictionary containing words and their polarities to assess the emotional and lexical properties of a given piece of text. The PANAS-t is a psychometric scale which is an extension of the Positive and Negative Affect Schedule (PANAS) scale used in psychology that is based on a large vocabulary of words expressing eleven emotions, and it is used to observe mood swings of users on *Twitter*. Happiness Index is a sentiment scale that is based on the popular resource Affective Norms for English Words (ANEW), and it scores texts on a scale of 1–9 where higher the score, the higher amount of positive emotion in the text. Till now, we have discussed the types of features which can be extracted from a text document and have also discussed the sentiment analysis methods which are widely used (Fig. 1).

2.3 Traditional Machine Learning-Based Approaches

1. **Naive Bayes Classifier**: Naive Bayes is among the simplest models of machine learning which is used in binary classification problem. Here, in sentiment analysis if we consider only two classes, positive and negative, then we can apply Naive Bayes model as in [21]. Although it is often said Naive Bayes works well when all the input words are conditionally (conditioned on the classes) independent of each other, which is rarely the case in sentiment analysis, even then it shows considerable accuracy. The Naive Bayes classifier is expressed as

$$P(c|x) = \frac{P(x|c)P(c)}{P(x)} \tag{1}$$

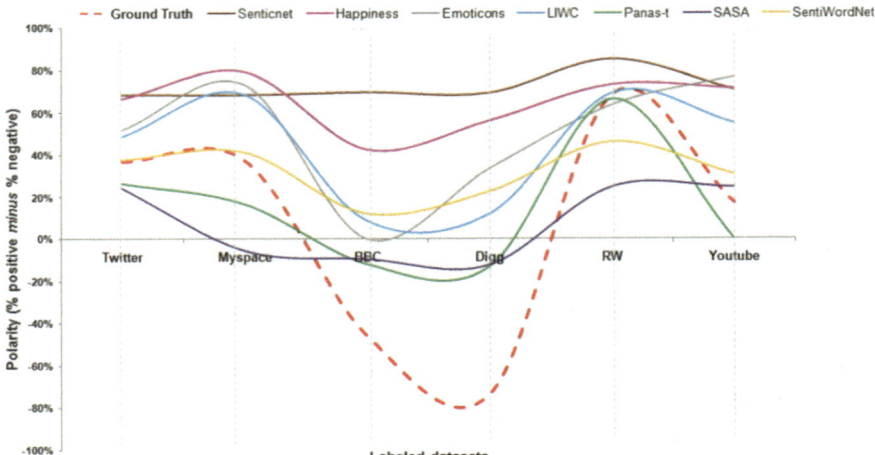

Fig. 1 Graph representing accuracy of various techniques over different datasets as given in work [20]

where $P(c|x)$ stands for posterior probability of class (target) given predictor attributes, $P(x|c)$ stands for the likelihood which is the probability of predictor given class, $P(c)$ stands for the prior probability of class and $P(x)$ stands for prior probability of predictor.

2. **Support Vector Machines**: Support vector machines (SVMs) [22] project the data points into a higher-dimensional space which makes it easier to find a hyper-dimensional plane that separates the two classes. They learn to classify the data points based on the training dataset provided. This technique generally shows the best result among all the classifiers when considered individually.

2.4 Low Resource Languages

2.4.1 Lexical Analysis

As discussed above, performing sentiment analysis using lexical approaches is a two fold process that involves feature extraction, followed by the application of an appropriate sentiment analysis technique. For low resource languages, even if feature extraction is not an issue, sentiment analysis methods can certainly become one. These sentiment analysis methods require the opinions of words to be taken care of and already having enough information about the word in every context to be able to predict the polarity of given sentence. This is a resource which generally lacks for low resource languages because an investment of spending Leviathan man-hours for creation of such dictionaries is not very profitable and also creating dictionaries for each language individually will certainly take a lot of time as pointed out by Altowayan et al. [23]. This is a dilemma researchers faced, of low resources available to analyze the text. One unanimously held approach is to translate the dataset that we have into high resource language (preferably English) and perform sentiment analysis on the new data received after translation. It is true that machine translation (MT) systems are not efficient enough to produce the exact translation, but it has been observed that for a lot of languages the translated sentence is reasonable enough to give significant accuracy increase in polarity detection of low resource languages. For such languages for which even a machine translation engine is not available, it is suggested to spend some time doing the manual work and make a MT system, or effective dictionaries to perform sentiment analysis. Among the two, the former is usually preferred.

2.4.2   Traditional Machine Learning Techniques

In this category as well, a widely used approach is to translate the present dataset to English and use pre-trained classifiers for sentiment analysis in English. Although SVMs have been observed to yield high accuracy, one is free to experiment with suitable ML algorithms for the dataset they are working with.

For now, we did state that it is generally more profitable to spend some man-hours on creating effective translation engines or resources similar to SentiWordNet for low resource languages. But still there are some languages like Arabic for which both of the options become a drudgery. Arabic is a morphologically complex language and has various nonstandard dialects. We will discuss the ways to tackle these challenges in great detail in coming sections.

3   Word Embeddings for Sentiment Analysis

We begin this section by briefly mentioning the importance of word embeddings in sentiment analysis and techniques on how to improve them specifically to perform sentiment analysis. We further discuss various methods to improve the coverage of word embeddings in low resource languages and some related work. The representation of words in a form which can be manipulated mathematically and also capture relevant domain information has always been a wide research field in NLP. Research efforts in this direction primarily focused on representing words as dense, high-dimensional representations capturing the syntax and semantic properties of contexts appearing in the corpora provided, called word embeddings. Latent semantic analysis (LSA) proposed by Landauer et al. [24] was one of the first word embedding models introduced and produced satisfactory results in word meaning similarity tasks. The more recent word embedding models include `Word2vec` [25, 26], `GloVe` [27] and `fastText` [28, 29]. `Word2vec` is a shallow network consisting of a couple of layers which either predicts the middle word(s) given a context

(CBOW) or predicts the context given the input word. `fastText` is an improvement over the `Word2vec` model which takes into account subword information and also handles the word similarities for words not present in the training data. The main idea behind `GloVe` is exploring the statistical properties of word occurrences in large text corpora.

3.1 Refining Word Embeddings for Sentiment Analysis

We now discuss techniques to refine and improvise word embeddings for sentiment analysis. Traditionally, word embeddings trained on labeled data in an unsupervised manner usually capture semantic and syntactic information very well, but they typically are unable to capture adequate sentiment-related information. This leads to semantically similar words with conflicting sentiments being mapped much nearer to one another, affecting sentiment classification performance. Yu et al. [30] provide a novel solution to this challenge. Initially, the top-k similar words to a given sentiment expressing word are determined using cosine similarity, following which they are ranked according to the decreasing order of sentiment similarity in accordance with a sentiment lexicon. The final stage is the improvement module which modifies the already obtained word embeddings such that they are mapped closer to both words having comparable meanings and sentiment information and further away from words expressing conflicting sentiments, while making sure that the modified embeddings are not mapped in very differing locations as compared to the initial embeddings. The refining process is based on the iterative reduction in the distances among the target words and their k closest neighboring words. The objective function $\Phi(V)$ for refining n vectors is defined as

$$\Phi(V) = \sum_{i=1}^{n} \sum_{j=1}^{k} w_{ij} dist(v_i, v_j) \tag{2}$$

where $V = \{v1, v2, \ldots, vn\}$ are the initially trained embeddings, v_i represents vector of the word under consideration, v_j is the vector of one of its closest words, $dist(v_i, v_j)$ is the distance between v_j and v_i and w_{ij} is the weighted reciprocated representation of the closest neighbor of the target word. Variables α and β regulate the degree of movement of the original vector toward its closest neighbors by accounting for them in the updated objective function as

$$argmin\Phi(V) = argmin \sum_{i=1}^{n} [\alpha dist(v_i^{t+1}, v_i^{t}) + \beta \sum_{j=1}^{k} w_{ij} dist(v_i^{t+1}, v_j^{t})] \tag{3}$$

where $dist(v_i^{t+1}, v_i^{t})$ denotes the separation among the original embedding and the improved embedding. For calculation of partial derivative of $\Phi(V)$, $dist(v_i, v_j)$ is measured by Euclidean distance as

$$dist(v_i, v_j) = \sum_{d=1}^{D} (v_i^d - v_j^d)^2 \tag{4}$$

where the word vectors are D-dimensional. The global optimum solution for $\Phi(V)$ can be found by an iterative update procedure which is found by setting

$$\frac{\delta}{\delta v_i^t} \Phi(V) = 0 \tag{5}$$

and is given as

$$v_i^{t+1} = \frac{\gamma v_i^t + \beta \sum_{j=1}^{k} w_{ij} v_j^t}{\gamma + \beta \sum_{j=1}^{k} w_{ij}} \tag{6}$$

The refining process is halted when all target words are refined.

Ye et al. [31] come up with a technique to encode sentiment information into word embeddings during training by combining a feedforward neural network model SentiNet alongside a CNN classifier and also leveraging a sentiment lexicon. The architecture used by the authors is shown in Fig. 2. The input to the embedding component is the collection of pre-trained word embeddings for words comprising the input document $[w_1, w_2, \ldots, w_N]$, and it outputs a real-valued matrix $W = [w_1, w_2, \ldots, w_N]^T$ consisting of embeddings to represent the input document. The CNN layer takes this matrix and determines the polarity of the document. The loss function of the CNN is given by

$$\mathcal{L}_{CNN} = CE(f_m^h(W), f_m^g(W)) \tag{7}$$

where CE is a scalar representing the categorical cross-entropy between the prediction and the gold standard distribution, and $f_m^h(W)$ and $f_m^g(W)$ are the prediction and gold standard distribution of the document. The CNN parameters are learned in accordance with $\mathcal{L}_{CNN}$. Sampled word vectors $[w_{s1}, w_{s2}, \ldots, w_{sN}]$ obtained by sampling M words are used by SentiNet to predict the sentiment distribution according to SentiWordNet. The loss function of SentiNet is

$$\mathcal{L}_{SentiNet} = \sum_{k=1}^{M} CE(f_w^h(w_{sk}), f_w^g(w_{sk})) \tag{8}$$

where $f_w^h(w_{sk})$ and $f_w^g(w_{sk})$ are prediction and gold standard sentiment prediction of word w_{sk}. Parameters of SentiNet are learned according to $\mathcal{L}_{SentiNet}$, and the embedding parameters are updated according to cumulative loss

$$\mathcal{L} = \mathcal{L}_{SentiNet} + \mathcal{L}_{CNN} \tag{9}$$

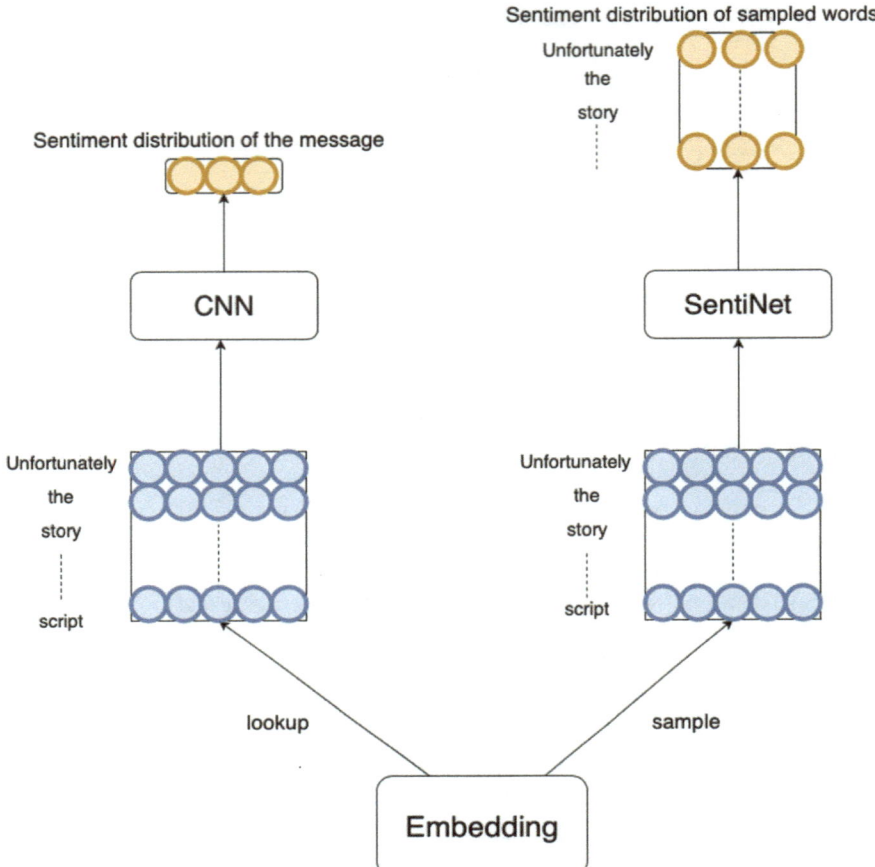

Fig. 2 Model architecture with SentiNet and non-static-based CNN as given in the work [31]

Hence, the word vectors are tuned with SentiNet model along with the CNN classifier, and the converging of the training of the former implies that sentiment information has successfully been encoded into the word vectors.

Cano et al. [32] conduct a comprehensive survey on the effect of factors like domain relevance, training data size and training method of word embeddings for the sentiment analysis of various datasets. Rezaeinia et al. [33] present a method for enhancing the text polarity prediction performance of already computed word embeddings by a combination of part-of-speech tagging methods, lexicon-based approaches and Word2vec/GloVe model.

3.2 Improving Word Embedding Coverage in Low Resource Languages

As mentioned before, sentiment analysis in low resource languages is challenging because of the limited availability of resources such as standard annotated datasets, sentiment lexicons and word embeddings. Therefore, increasing the coverage and enhancing the performance of word embeddings in these languages are the tasks of utmost importance. In this section, we discuss some novel approaches to improve word embedding coverage in low resource languages.

Yucesoy et al. [34] introduce an innovative weight selection method for co-occurrence counting. The motivation behind this line of thought is that sometimes two closely semantically related words might actually appear quite far away from each other in a corpus, and thus such pairs of words are interpreted as being semantically distant during co-occurrence counting. The problem is exacerbated in a low resource language scenario, owing to the lower frequencies of co-occurrences of such word pairs as compared to resource-rich languages. Therefore, the authors propose a weighting strategy on the basis of a polynomial fitting process to favor such word pairs, while other weights are left untouched. The original counting method proposed by Pennington et al. [27] was

$$M_{ab} = \sum_{n \in \kappa} f(x_{abn}) \tag{10}$$

where x_{ij} expresses the relative spacing among words a and b, $f(x) = 1/x$, M_{ab} is an element of the co-occurrence matrix representing the weighted co-occurrences of words a and b and κ is set of all the times words a and b occur together in the corpus. To this function, the authors fit a sixth-order polynomial function given by:

$$P(x) = p_6x^6 + p_5x^5 + p_4x^6 + p_3x^3 + p_2x^2 + p_1x + p_0 \tag{11}$$

The authors want to be able to cover different weighting strategies while keeping the value of the polynomial as close to $1/x$ for different values of x so they introduce two perturbing parameters α and β into the original polynomial and these parameters will not make any difference so long the distantly occurring word pairs are being accounted for. The resulting polynomial is given by

$$P(x) = (p_6 + \alpha)x^6 + p_5x^5 + p_4x^6 + (p_3 - \beta)x^3 + p_2x^2 + p_1x + p_0 \tag{12}$$

Owing to this weighting scheme, a considerable increase in performance is observed in small corpora, while it is almost unchanged in large corpora. The proposed framework enhances the performance of word analogy and similarity tasks in a low resource language like Turkish, along with a standard sentiment analysis task in the same language.

The concept of leveraging resource-rich languages to improve the word embedding coverage and text classification accuracy in resource-poor languages has gained significant attention from researchers. Akhtar et al. [35] employ bilingual word embeddings trained on a parallel English–Hindi corpora which are projected on a common vector space. If the vector representation of a Hindi word is unknown, it is translated into English and the embedding of that English word is used, thus leveraging contextual information in both sides and improving word embedding coverage in Hindi. They successfully apply this technique to perform aspect-based sentiment analysis (ABSA) in Hindi.

Barnes et al. [36] propose their model bilingual sentiment embedding (BLSE) which collectively expresses sentiment information in a resource-poor and resource-rich language. The proposed model architecture is depicted in Fig. 3. Their model leverages a relatively small bilingual lexicon, monolingual word embeddings in both languages and an annotated sentiment analysis corpus for the resource-rich language. For precomputed vector spaces S and R for the source and target languages and a bilingual lexicon of length n having word-to-word translation pairs $L = \{(s_1, t_1), (s_2, t_2), \ldots, (s_n, t_n)\}$ mapping from source to target, two linear projection matrices M and M' are used to map from the originally precomputed vector spaces to shared sentiment-aware bilingual spaces z and $\hat{z}$. Maintaining the quality of the projections is facilitated by minimization of the mean squared error

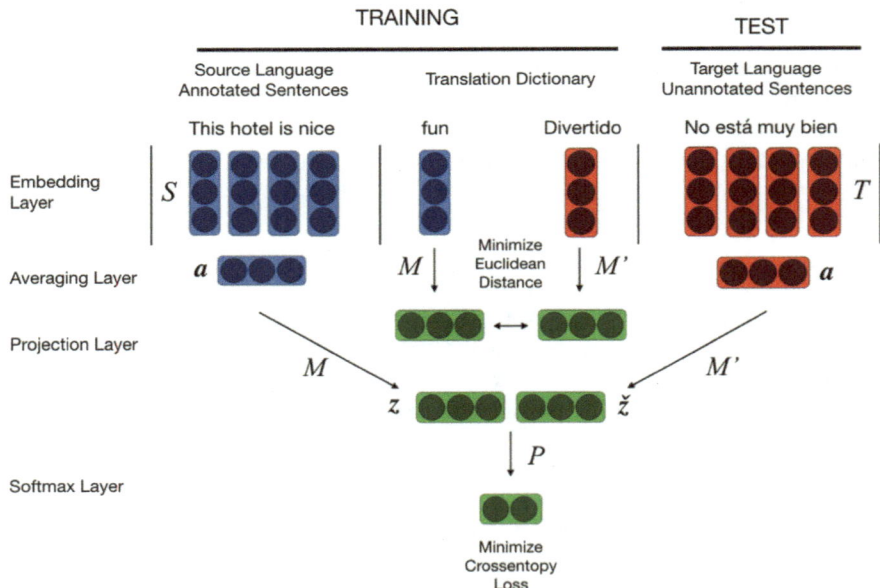

Fig. 3 Bilingual sentiment embedding (BLSE) model as given in work [36]

$$MSE = \frac{1}{n} \sum_{i=1}^{n} (z_i - \hat{z}_i)^2 \tag{13}$$

where $z_i = S_{s_i}.M$ is the embedding dot product for source word s_i and source projection matrix, and similarly for the target word t_i we have $\hat{z}_i = T_{t_i}.M'$. A source language corpus $C_{source} = \{(s_1, p_1), (s_2, p_2), \ldots, (s_n, p_n)\}$ where p_i is the polarity of each sentence s_i is required to train M' to capture sentiment. The average a_i of the word embeddings from S for sentence x_i is projected to the bilingual space $z_i = a_i.M$ following which prediction $\hat{y}_i = softmax(z_i.P)$ is obtained by passing z_i through softmax layer P. Cross-entropy error of the model is minimized to facilitate sentiment prediction as

$$H = -\sum_{i=1}^{n} \log \hat{y}_i - (1 - \hat{y}_i) \log(1 - \hat{y}_i) \tag{14}$$

To train the projection component and to capture sentiment, parameter matrices M, M' and P are optimized as

$$J = \sum_{(x,y) \in C_{source}} \sum_{(s,t) \in L} \alpha H(x, y) + (1 - \alpha).MSE(s, t) \tag{15}$$

The authors perform sentence-level crosslingual sentiment analysis on a combination of Spanish, Basque and Catalan with promising results and provide a detailed analysis which highlights the fact that their model produces word embeddings in the low language setting without any annotated corpora present while also reflecting sentiment information.

Ruder et al. [37] present a comprehensive survey of crosslingual word embedding models in the literature, comparing their data requirements and objective functions, while also elaborating on the methods of evaluating these models and further providing research directions and challenges. Akhtar [38] presented a detailed work for generating reliable word vectors in low resource languages and also released word similarity datasets for six Indian languages which have limited resources for sentiment analysis.

Duong et al. [39] propose a method for learning crosslingual word embeddings with the requirements being a bilingual dictionary and monolingual corpora in both languages, thus eliminating the need for bilingual corpora. Their model improves over the CBOW model by Mikolov et al. [25] and relies on using the context in one language to estimate the translation of the central term of the other language. Let $\bar{w}_i$ be the translation of a middle word w_i, h_i be a vector representing the context over a k size window, p be the number of negative points arbitrarily extracted from a noise distribution $P_n(w)$, D_e and D_f be the monolingual corpora in both languages, and V and U be the context embedding and word embedding matrices, respectively. Learning to predict a word and its translation from monolingual context ensures that both of them appear in very similar contexts, which gives rise to the following optimization function:

$$\mathcal{O} = \sum_{i \in D_e \cup D_f} (\alpha \log \sigma (u_{w_i}^T h_i) + (1 - \alpha) \log \sigma (u_{\bar{w}_i}^T h_i) + \sum_{j=1}^{p} E_{w_j \tilde{P}_n(w)} \log \sigma (-u_{w_j}^T h_i)) \tag{16}$$

where α determines the contribution of the two terms. The final step is the fusion of the two embedding spaces V and U to produce better crosslingual representations. To facilitate this, a regularization term is added in the above objective function to make the model capable of learning similar representations for every word in the combined dictionary $V_e \cup V_f$. This produces the updated optimization function

$$\mathcal{O}' = \mathcal{O} + \delta \sum_{w \in V_e \cup V_f} ||u_w - v_w||_2^2 \tag{17}$$

where δ denotes the extent to which the two spaces should be bound together. The proposed model produces competitive results on the monolingual word similarity and crosslingual document classification tasks and achieves excellent outcomes on a bilingual lexicon induction problem.

Jiang et al. [40] propose a framework to learn word vectors for resource-scarce languages by positive unlabeled (PU) learning. Often in such low resource languages with small training corpora, majority of the entries in the co-occurrence matrices are zeros because such combination of words either cannot appear together or do not happen to appear in the related corpus. The authors argue that such entries in the co-occurrence matrices also provide valuable insights and incorporate this information for learning vector representation of words. Their process involves construction of a co-occurrence matrix, followed by the administering of a PU learning algorithm to facilitate factorization of the constructed matrix to generate word and context vectors, and finally generating the vector for each word by combining the context and word vectors through a post-processing step.

4 Deep Learning Techniques for Multilingual Sentiment Analysis

This section describes the DL-based approaches for sentiment analysis, with more emphasis on the models and techniques for low resource languages. We begin by giving a brief description and basic model architectures of some of the popular neural network model architectures, namely convolutional neural networks (CNNs), recurrent neural networks (RNNs), autoencoders and Siamese neural networks. Following this, we describe some novel and state-of-the-art multilingual sentiment analysis approaches with these networks.

4.1 Convolutional Neural Networks

Convolutional neural networks [41] are a special class of neural network model architectures to facilitate processing data having predominantly grid-like arrangement. These networks employ a mathematical operation known as *convolution* in atleast one of their layers in place of matrix multiplication. The motivation behind using convolutional networks in place of traditional neural networks is reducing intra-modular network interactions, sharing of hyperparameters and learning of non-comparable depictions as mentioned in the work [42]. Unlike traditional neural networks in which every unit in one layer interacts with every other unit in the next layer, CNNs typically contain sporadic interactions. Convolving of a filter or a kernel initialized with random numbers across the entire input image makes this possible. Moreover, each kernel is replicated across the entire input and all the kernels share the same set of parameters and weights throughout which eliminates the need for learning a new set of hyperparameters everywhere.

We now describe some work for sentiment analysis in English language using CNNs. Santos et al. [43] propose a CNN architecture that effectively exploits character-level and word-level features from tweets to perform sentiment analysis, along with incorporating the contextual information of word embeddings pre-trained on a Wikipedia corpus by using `Word2vec`. They thus overcome the limitation of the limited amount of context that these short texts usually contain. They achieve the state-of-the-art accuracy of 85.7% for sentence-level polarity detection on the Stanford Sentiment Treebank (SSTb). Using a similar approach, Zhang et al. [44] propose a character-level CNN architecture without considering any word features at all for performing sentiment analysis on multiple domains. Their model accepts encoded characters as inputs, a collection of m characters is assigned for an input language, and each character is encoded by 1-of-m encoding technique. A pair of CNN models each nine layers deep with six convolutional layers and three fully connected layers was proposed for the task. The authors constructed several extensive sentiment analysis datasets and demonstrated the effectiveness of their approach on such datasets. They also suggested that their approach can work for other languages as well because characters are important language constructs along with having the advantage of being able to learn naturally unusual character arrangements such as misspellings and emoticons.

Kim [45] employs several CNN variants trained on top of already trained word vectors released by Mikolov et al. [25] with slight hyperparameter tuning for sentence-level sentiment analysis. Initially, only static vectors are employed to train the model but it was suggested that learning precise vectors by fine tuning brings added performance gains. The model architecture is shown in Fig. 4. Each sentence containing n words (after padding wherever necessary) is expressed by the concatenation of k-dimensional word vectors x with x_i corresponding to the word vector of the ith word in the sentence. A feature y_i is obtained by convolving a window containing h words with a filter w given by

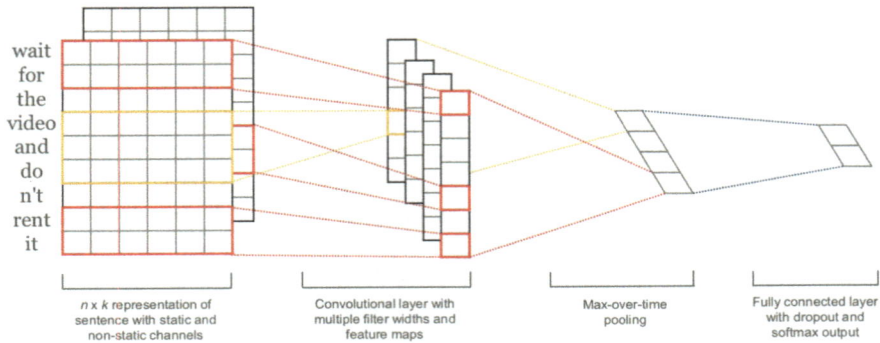

wait
for
the
video
and
do
n't
rent
it

$n \times k$ representation of
sentence with static and
non-static channels

Convolutional layer with
multiple filter widths and
feature maps

Max-over-time
pooling

Fully connected layer
with dropout and
softmax output

Fig. 4 Proposed architecture for an input sentence with two channels, as proposed in work [45]

$$y_i = f(w.x_{i:i+h-1} + b) \qquad (18)$$

where f is a nonlinear function like $\tan h(x)$ and b represents bias. A feature list f given by $f = [f_1, f_2, \ldots, f_{n-h+1}]$ is produced by convolving the filter w over the set of all possible word groupings of the input sentence. After this, the feature list f is subjected to max pooling and the highest value $\hat{f} = \max\{f\}$ is extracted, which is the most relevant feature for this feature list. The CNN model employs many kernels of differing coverage to obtain multiple features which are then sent to the fully connected softmax layer, the output of which is the probability distribution over the various polarities. Dropout is applied on the layer of the max pooled feature lists $z = [\hat{f}_1, \hat{f}_2, \ldots, \hat{f}_n]$ which is expressed by

$$o = w.(z \circ v) + b \qquad (19)$$

where o is the output unit, $\circ$ is the element-wise multiplication operation and v represents the vector of Bernoulli random variables with $P(r = 1) = p$. The proposed model outperforms the most popular techniques in four specific tasks.

Severyn et al. [46] introduce a CNN-related model architecture focusing on starting the training process by good initialization of the model parameters and weights to perform phrase-level and message-level sentiment analysis of tweets. The main steps they take to ensure initialization of their model in a beneficial way are as follows: (i) The authors trained Word2vec embeddings on 50M tweets collected by them by using skipgram model of window size 5 and minimum occurrences of word also as 5; (ii) to further incorporate sentiment information into the trained word embeddings, they use a distant supervision learning approach using CNNs as in the work [47]; (iii) the parameters obtained previously are used to train their network which is used to perform the sentiment analysis task. Wang et al. [48] propose a combined CNN- and RNN (LSTM and GRU)-based model architecture to perform sentiment analysis of short texts. Already trained Word2vec vectors are given to the model as input; following convolution and max pooling procedures, the feature maps are used as input to RNNs; finally, fully connected softmax layer produces the desired classification. For output y_i of the LSTM, the softmax output which is the probability distribution over all the labels is computed as

$$\hat{P}_i = \frac{exp(y_i)}{\sum_{j=1}^{X} exp(y_j)} \tag{20}$$

where X is the input sentence matrix. The training purpose is to optimize the cross-entropy loss that expresses the difference between the actual sentiment distribution $\hat{P}^t(X)$ and the output arrangement $\hat{P}(X)$ of the input sentences.

$$loss = -\sum_{s \in T} \sum_{i=1}^{V} \hat{P}_i^t(X) \log(\hat{P}_i(X)) \tag{21}$$

where training corpus T contains sentence s and V polarity labels, and $\hat{P}_i^t(X)$ is the one-hot encoded vector of V dimensionality in which element 1 corresponds to the actual polarity label of the sentence and element 0 otherwise. The CNN model extracts high-quality linguistic characteristics from the input corpus, whereas the RNN model learns long-term dependencies. This proposed architecture improves over the benchmark results on three benchmark sentiment analysis corpora.

We now describe some work for multilingual sentiment analysis using CNNs. Ruder et al. [49] come forward with a CNN-related model architecture to perform multilingual ABSA as a part of SemEval-2016 Task 5. Their model takes as input the sentence and the mentioned aspect, initially pads the input sequence to a length n and then extracts the word embeddings for every word in the processed input; for the English language, they use the pre-trained GloVe word vectors by Pennington et al. [27], while for the other languages randomly initialized word vectors are employed. To take into account aspect extraction, the authors define a probability distribution for aspect a and a sentence s as follows:

$$p(a|s) = \begin{cases} 1/n & \text{if s contains a and s contains n aspects} \\ 0 & \text{otherwise} \end{cases} \tag{22}$$

A threshold f is defined, and all aspects where $p(a|s) < f$ are removed. Then, an aspect vector is created by splitting the constituent tokens of the aspect and averaging the individual embeddings of the tokens in the vector space. The aspect vector thus formed is then linked with each word vector to form a sentence matrix. To this sentence matrix, convolution and max pooling operations are applied while a final softmax layer outputs the probability distributions over the output labels. The authors achieve competitive results across all domains and languages, and state that the performance in the low resource languages can be increased further if pre-trained embeddings on a monolingual corpora for that particular language are employed to train the model.

Singhal et al. [50] leverage vectors and sentiment labels of English words to perform multilingual sentiment analysis using variants of CNNs in Indian languages such as Hindi and Marathi, and seven other languages provided by Araújo et al. [51]. They initially perform word-to-word translation of review texts in other languages into English using Google Translate[2] and use the pre-trained Word2vec embeddings to initialize the embeddings for these translated words. Random vectors are allotted to those words which fail to get translated. This training data is augmented with English polar words along with their polarity labels before proceeding to perform sentiment analysis using static, non-static and multi-channel modes of CNN. All the CNN models consist of a convolutional layer with filters of sizes 3, 4 and 5, a feature map of size 50 for every filter and sigmoid activation function. The authors achieve excellent results in the low resource language scenario despite the paucity of annotated training corpora in these languages, and the mapping of unknown words to English word embeddings also contributes to the observed increase in the performance.

4.2 Recurrent Neural Networks

Recurrent neural networks (RNNs) are neural network architectures whose links among the various elements form a directed cycle and which are capable of processing data of a sequential nature such as text, audio and video. They contain internal memory states which enable them to not only process and consider the current input, but also remember relevant information about all the inputs they have encountered so far. However, RNNs fail to perform well specially when the distance between the relevant contextual information and the point where it is needed increases, as pointed out by Bengio et al. [52]. To overcome this issue, Hochreiter et al. [53] proposed LSTMs, which are RNN variants capable of learning long-term dependencies. For a detailed description of the fundamentals of RNNs and LSTMs, the interested reader is referred to the work [54].

The repeating modules in a traditional RNN have a rather simple structure, but this is not the case in LSTMs, as depicted in Fig. 5. AN LSTM has three gates and a memory cell state from which information can flow from one module to another with either some linear interactions or unchanged. The gates are means to remove or add information to the cell states, and they consist of a multiplication operation and a sigmoid layer. The three gates in a LSTM module are the input , forget and the output gates. The forget gate decides what portion of the previous data to discard, the input gate conjectures what new knowledge is to be updated in the current module, while the output gate reflects output of the present module.[3] Each component is calculated as follows:

$$M = \begin{bmatrix} h_{t-1} \\ x_t \end{bmatrix} \qquad (23)$$

---

[2]https://translate.google.com.
[3]http://colah.github.io/posts/2015-08-Understanding-LSTMs/.

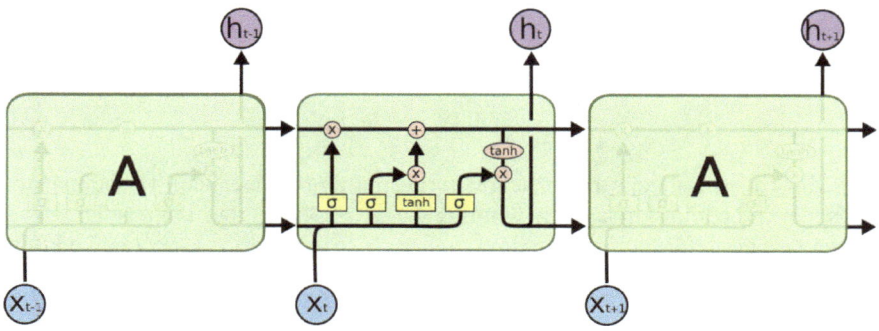

Fig. 5 Repeating module of an LSTM network which has four interacting units

$$f_t = \sigma(W_f . M + b_f) \tag{24}$$

$$i_t = \sigma(W_i . M + b_i) \tag{25}$$

$$o_t = \sigma(W_o . M + b_o) \tag{26}$$

$$C_t = f_t * C_{t-1} + i_t * \tanh(W_c.M + b_c) \tag{27}$$

$$h_t = o_t * \tanh(c_t) \tag{28}$$

where h_{t-1} is the output of the preceding repeating module, x_t represents present input, f_t is the output of the forget gate, i_t represents information to be updated at the current module, o_t is the output of the cell state, C_{t-1} represents previous cell state, C_t represents updated cell state, σ is the sigmoid function and h_t represents final output of the current module. For a description of some of the LSTM variants employed over the years, the interested reader is referred to [55–57].

We now describe some LSTM-based models employed for sentiment analysis in English. Wang et al. [58] propose an attention-based LSTM model for ABSA. They explore the correlation between the aspect and the polarity of a given review text and propose an attention technique that focuses on the various important portions of a line of text given the aspect. They represent various aspects of the input sentence by an aspect embedding v_{a_i} for aspect i. The proposed attention mechanism produces a vector α consisting of attention weights and a weighted representation r of the input sentence consisting of N words given an aspect, as follows:

$$\alpha = softmax(w^T M) \tag{29}$$

$$r = H\alpha^T \tag{30}$$

where H represents the matrix of outputs that the LSTM model generated and M stands for a projection parameter. The resulting feature depiction of the sentence given aspect h^* is

$$h^* = \tanh(W_p r + W_x h_N) \tag{31}$$

W_p and W_x are parameters that are learnt during training of the model. The authors use the aspect embeddings to assist in learning the attention weights, but a better way to leverage aspect information according to them is to append the aspect embeddings learned with the input word vectors, the model architecture of which is shown in Fig. 6.

The authors used the pre-trained `GloVe` vectors to train their model. They e-valuate their model on the dataset of SemEval 2014 Task 4 and achieve superior performance as compared to the baseline models. Chen et al. [59] introduce a divide and conquer method for sentiment analysis by first classifying different sentences on the basis of the quantity of different opinion targets or entities contained in them. They propose a bidirectional LSTM (BiLSTM) model stacked together with conditional random fields (CRFs) to classify sentences into nontarget, one-target and multi-target sentences following which they train one-dimensional CNNs to determine the polarity of each type of sentence. For training the BiLSTM-CRF model to learn features, they use the MPQA dataset. Apart from achieving the state-of-the-art results on sentiment analysis in various benchmark datasets in the literature, their approach also performs the best on SemEval16 Task 5 ABSA dataset subtask 1 and slot 2 on six languages.

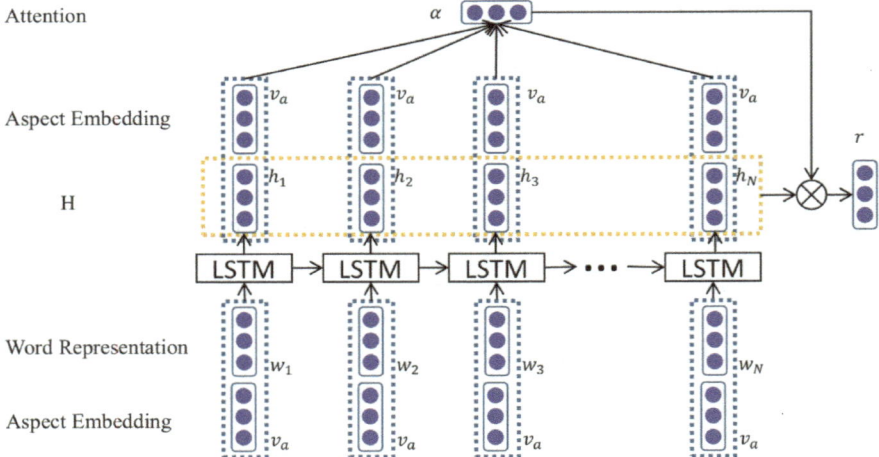

Fig. 6 Model architecture of attention-based LSTM with aspect embeddings as given in work [58]. The learned aspect embeddings have been taken as inputs to the model along with the word embeddings. $\{w_1, w_2, w_3, \ldots, w_N\}$ are the word embeddings for a sentence containing N words, $\{h_1, h_2, h_3, \ldots, h_N\}$ is the hidden vector, v_a is the aspect embedding, and α is the attention weight

We now describe some work done in multilingual sentiment analysis which uses LSTMs and their variants. Joshi et al. [60] introduce the concept of learning subword-level representations in LSTM (subword-LSTM) for performing sentiment analysis of Hindi–English code mixed text and also release a Hindi–English code mixed dataset for the same. The authors propose to learn subword-level representations of the input data which not only can generate meaningful linguistic features but also carry semantic information, thus overcoming the limitations of character-level and word-level representations. The ith element of the learned subword-level feature map is given by

$$f_i = g((Q[:, i : i + m - 1] * H) + b) \tag{32}$$

where Q is the matrix representing the input sentence s which is convolved with a filter H of length m, $Q[:, i : i + m - 1]$ represents matrix of ith to $(i + m - 1)^{\text{th}}$ character representation and g i s he ReLU activation function. The max pooling from p feature representations produces subword representations as

$$y_i = \max(f[p * (i : i + p - 1)]) \tag{33}$$

The learned representation f is provided to LSTM cell at time t, and the output representation is learned as

$$O_t = \sigma (X_{yt} + Yh(t - 1) + Z(C_t + b)) \tag{34}$$

$$h_t = O_t \tanh(C_t) \tag{35}$$

where X, Y and Z are the weight matrices and b reflects bias. The relevant information h_t is propagated to a fully connected module which then determines the polarity of the input sentence. The proposed approach attains higher accuracy than other classic methods on their dataset and also significantly outperforms the current tools for polarity detection of Hindi–English data. Jhanwar et al. [61] improve upon the above-proposed method for sentiment analysis of Hindi–English data which is subjected to code mixing by leveraging an ensemble of character trigram-associated LSTM model and word ngrams-based multinomial Naive Bayes (MNB). The LSTM model is able to successfully extract the rich and diversified sequential patterns from the input data, while the MNB model helps in determining the polarity of the sentences leveraging ngram-level features. The authors believe that the small size of such code-mixed datasets renders the deep learning models incapable of capturing all the intricacies from the textual information which is why they jointly employ the MNB model. The ngram features are passed on to the MNB classifier which computes the probability of a sentence belonging to either the *positive, negative* or *neutral* classes. Concurrently, each token of the input sentence is represented as a character trigram embedding matrix which is passed to the LSTM that outputs a feature map which is then fed to a fully connected layer and the polarity is assigned to the sentence accordingly. The final output is contributed to by the outputs of each of the individual modules.

Can et al. [62] leverage an RNN-based model trained using on English language datasets and use their model to perform sentiment analysis in other resource-scarce languages such as Turkish, Dutch, Spanish and Russian without the need for any additional resources in these languages. The model architecture consists of a pair of bidirectional gated recurrent unit (GRU) [55] layers with dropout. The authors use pre-trained GloVe embeddings to train their model, Google Translate API to translate the datasets in these languages to English, and use this model to evaluate the sentiments in other languages, outperforming all previous baselines. Alayba et al. [63] propose a mixed CNN-LSTM architecture to perform sentiment analysis in the Arabic language. Owing to the complex morphology of Arabic, the authors perform the sentiment analysis task in the character, character ngram and the word level to extract as many features as possible. The proposed framework is shown in Fig. 7.

The input data is encoded as a matrix where each sentence is encoded as a row of vectors and the vectors consist of either character-level, character ngram-level or word-level tokens. To this matrix, convolution operation is applied with filters of different lengths to generate suitable feature maps, following which the max pooling functionality grasps the features of utmost relevance and dropout is applied so that the model does not overfit . These feature maps are then used as inputs to the LSTM module, the outputs of which are sent to the fully connected layer, and then the polarity is determined using the sigmoid function.

Baly et al. [64] introduce a Twitter dataset which consists of tweets belonging to 12 nations of the Arab world, marked for dialect and sentiment. The dataset contains tweets from different Arabic dialects and is annotated for sentiment and dialect. They propose an LSTM model with universal- and dialect-specific embeddings to perform sentiment analysis on the dataset created. The universal embeddings are learned from all unlabeled tweets belonging to the dataset, while the dialect-specific embeddings were computed from the parts of the tweets that are applicable to the respective countries. Lemmatized and stemmed embeddings are also trained separately. Their

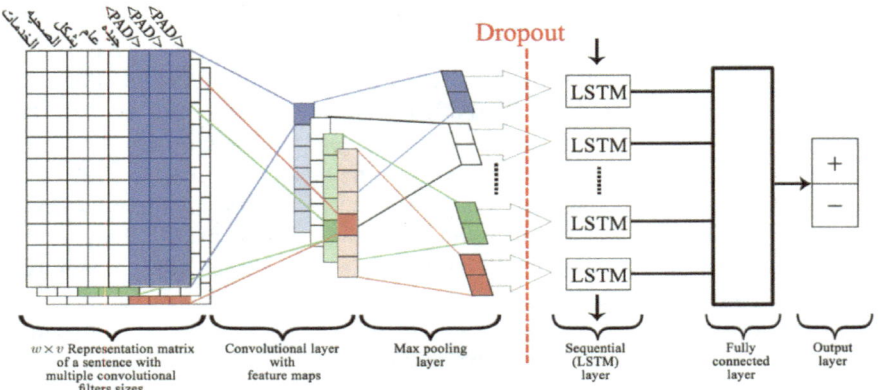

Fig. 7 Joint CNN-LSTM framework with an input sentence in Arabic, as given in work [63]

approach outperforms the other non-DL-based approaches dependent on manual feature engineering.

Peng et al. [65] perform aspect-based sentiment analysis in Chinese language at three different granularity levels with sentiment of the aspect target sequence (aspect term containing multiple words) level as the main focus. Earlier work on aspect-based sentiment analysis simply averaged word embeddings of aspect words, thus not only overlooking sequential information within the aspect target sequence but also perhaps misinterpreting the actual meaning of the aspect target sequence. The reason for considering the word-, character- and radical-level granularities is because linguistic features at the subword level in the Chinese language encode semantics and provide complementary context explanations, thus enhancing the meaning. In order to account for the above, the authors use RNN-based models to perform two subtasks—adaptive embedding learning and sequence learning of aspect target. Adaptive embedding learning is the learning and appending of rich intra-sentence contextual information to every unit of the aspect target sequence. This process leverages the attention mechanism to learn adaptive vectors, which are basically attention weights belonging to each aspect of the aspect target sequence with sequential sentence encoding outputs learned from an LSTM model. The adaptive vector V_{adapt} for aspect target unit u_i and its word vector v_i for a sentence containing n words is:

$$X = \begin{bmatrix} v_i \\ h_j \end{bmatrix} \tag{36}$$

$$V_{adapt} = \sum_{j=1}^{n} \alpha_j . X \tag{37}$$

where h_j is the jth output of LSTM sequential sentence encoding, and α_j is the weight for the jth memory in the sentence and reflects the semantic significance of the jth unit on u_i. α_j is computed from softmax as

$$\alpha_j = \frac{e^{g_j}}{\sum_{m=1}^{n} e^{g_m}} \tag{38}$$

where g_j is a score derived from an attention model given by:

$$g_j = \tanh(W_j . X + b) \tag{39}$$

Sequential learning of aspect target is the concatenation of adaptive vector of each unit in the aspect target sequence which is fed to the second LSTM module, and the hidden output of the last cell of the LSTM is the final representation of the aspect target sequence. The next stage of the sentiment analysis process is to merge the outputs from the three granularity levels to predict the final sentiment label of the input. The two fusion mechanisms proposed are early fusion, in which the joining occurs before the aspect target sequence learning and late fusion, where the fusion takes place at the classification step. The authors assess the performance of their framework on four Chinese datasets encompassing a variety of areas and conclude that their approach outperforms the best techniques for ABSA in the Chinese language. They also experimentally evaluate various aspect target sequence modeling techniques and improvements in performance brought about by fusing the granularities and provide a detailed analysis for the same. The model architecture with late fusion is given in Fig. 8.

Chung et al. [66] propose a character-level double embedding neural network model consisting of two independent CNN- and RNN-based model architectures to perform sentiment analysis on a Chinese news dataset. The CNN model extracts hierarchical information, while the RNN model extracts sequential information. The two models are merged in a concurrent manner to allow them to capture relevant information without interfering with each other instead of stacking them on top of one another. The framework is depicted in Fig. 9. The vocabulary consists of the k most frequently occurring characters during training, and each character is encoded by a vector of size n. The input sentences are subjected to character-level parsing. Two embedding matrices $E1$ and $E2$ of dimensions $l \times n$, which are initialized randomly and are updated during training, are computed for each document consisting of l characters. Matrix $E1$ is convolved over h words by filters of size $h \times n$ generating feature map m_i as

$$m_i = f(w.E_{i:i+h} + b) \tag{40}$$

where b represents bias factor and f is the ReLU function. Max pooling operation is applied which extracts the most relevant features and a feature map C is generated to which the one layer highway network performs the following operation to generate output C_h

$$C_h = t.f(W_H.C + b_H) + (1 - t).C \tag{41}$$

where t is the transform gate and $(1 - t)$ is the carry gate of the highway network. t is defined as

$$t = \sigma(W_T C + b_T) \tag{42}$$

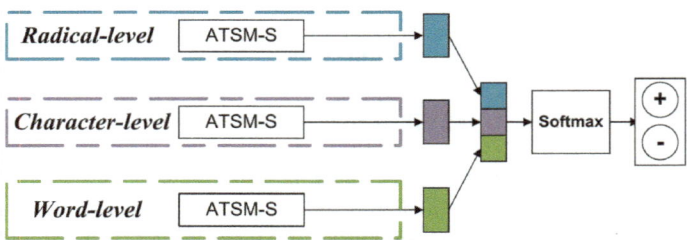

Fig. 8 Aspect target sequence model with single granularity (ATSM-S) incorporating late fusion of multi-granularity representations, as mentioned in work [65]

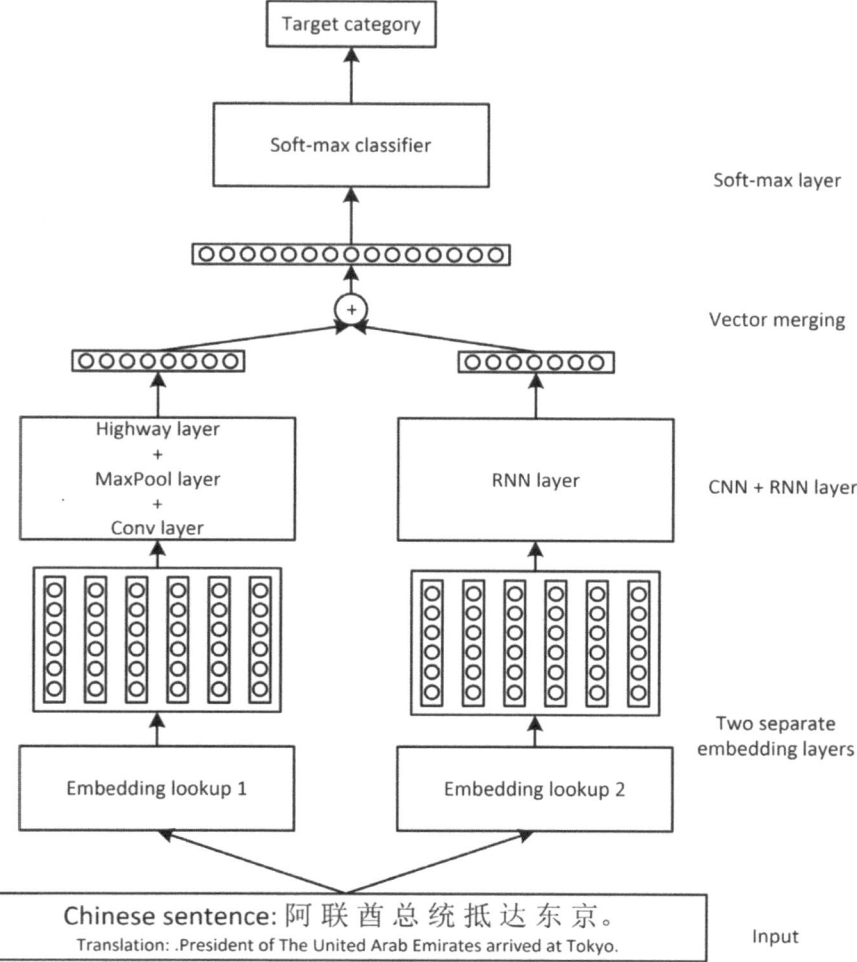

Fig. 9 Double character-level embedding CNN-LSTM model architecture, as in work [66]

The embedding matrix $E2$ is sent to the second LSTM module. The updated state hidden vector h_t and cell state vector c_t are computed as

$$f_t = \sigma(X_f.[e_t, h_{t-1}] + b_f) \tag{43}$$

$$i_t = \sigma(X_i.[e_t, h_{t-1}] + b_i) \tag{44}$$

$$g_t = \tanh(X_g.[e_t, h_{t-1}] + b_g) \tag{45}$$

$$c_t = f_t * c_{t-1} + i_t * g_t \tag{46}$$

$$o_t = \sigma(X_o.[e_t, h_{t-1}] + b_o) \tag{47}$$

$$h_t = o_t * \tanh(c_t) \tag{48}$$

where e_t represents embedding of the current character, h_{t-1} is the hidden vector of the preceding term and c_{t-1} is the preceding cell state embedding. Dropout operation is applied to both the neural network modules to prevent overfitting. Finally, outputs from both the modules are concatenated comprising the input of the softmax classifier which outputs a vector y of d dimensions containing real values between 0 and 1 computed as

$$P(y = k|x^s; W, b) = \frac{e^{W^k.x^s+b^k}}{\sum_{k \in K} e^{W^j.x^s+b^j}} \tag{49}$$

where class $k \in K$ has parameters W^k and b^k, and $x^s = [C_h, h_t]$ is input to the softmax layer. The authors evaluate the proposed model on the Fudan and Sougou news datasets, achieve better than all other methods and conclude that to increase performance, incorporating more pre-trained word embeddings and selecting convolution filters of appropriate sizes should be implemented.

4.3 Autoencoders

Autoencoders are feedforward, unsupervised neural network architectures introduced by Socher et al. [67]. They specialize in creating an output which is similar to the input. This is achieved in a recursive manner. The output created from the autoencoders is then decoded to create back the input. Mean squared error is taken from the difference between the input provided and the input generated. This means that if one is given the output, one can recreate the input, although the output and input need not be same. For a better representation, the reader is referred to Fig. 10. We have numbers 1–8, with each row representing one number. This is our input, and the representation immediately following it is our output. Basically, these autoencoders converted the integral numbers 1–8 into some other format. If you notice closely, then the output is actually a BCD representation x and that no significant data loss was experienced. The matrix size was reduced immensely (more than 50%). This can be achieved when autoencoders are used effectively. First, we will explain the autoencoder model, and then we will proceed to discuss distinguished work on multilingual sentiment analysis which uses autoencoders.

In recursive autoencoders (RAEs), the embedding forms a treelike structure, which is parsed in a bottom-up manner. The lowest embeddings stand for embeddings representing single words ($c1, c2$), and their parent embeddings are formed by combining the two child embeddings $y1$ given by

$$y_1 = f(W^{(1)}[c_1; c_2] + b^{(1)}) \tag{50}$$

Fig. 10 Sample autoencoder
input and output (*Source*
blog.wtf.sg)

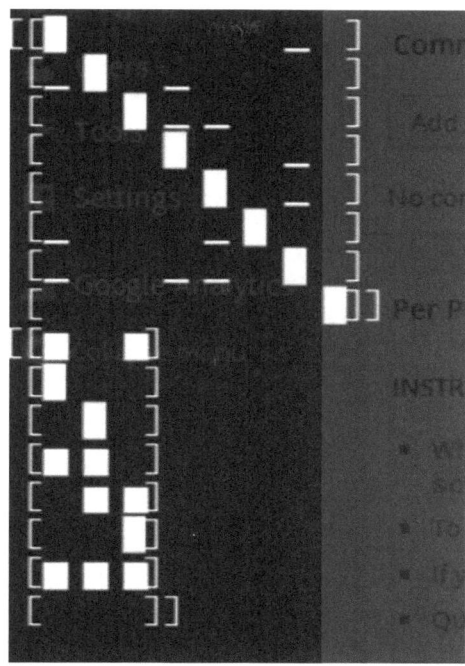

This is the equation for creating a parent embedding, where W is the weight matrix and b is the bias term. It is worth noting that the dimension of $y1$ is same as that of $c1$ and $c2$. As already stated before, autoencoders essentially recreate input from output. We try to recreate child embeddings (c_1, c_2) from parent embedding y_1 and back-propagate the squared error loss between the resultant child embeddings (n_1, n_2) and original child embeddings (c_1, c_2)

$$[n_1 : n_2] = f^{(2)}(W^{(2)}p + b^{(p)}) \tag{51}$$

$$E_{rec}(c_1; c_2) = \frac{1}{2}||[c_1; c_2] - [n_1; n_2]||^2 \tag{52}$$

The aim is to minimize the error at each embedding, which is given by:

$$Loss = argmin_{y \in A(x)} \sum_{s \in Y} E_{rec}([c_1; c_2]_s) \tag{53}$$

This was a basic explanation of RAEs. Let us move on to understanding how RAEs can be used in a multilingual setting, especially when two languages are taken into consideration.

4.4 Bilingual Constrained Recursive Autoencoders

Initially, recursive autoencoders were built by Zhang et al. [68] to perform superior machine translation, but they also found their application in bilingual sentiment analysis. The key reason is their ability to represent phrases instead of just words (Fig. 11).

The basic idea behind this is that bilingual constrained recursive autoencoders (BRAEs) employ training of two RAEs jointly, one for the source language and other for the target language. The BRAE model is depicted in Fig. 12. Let us assume the languages we are working with to be Japanese and English. First, RAE is trained on Japanese; that is, it can create back the input Japanese phrases from the output embeddings without any major loss on representation. We now use this as gold standard embeddings for the English phrase and use it to learn English phrase embeddings. A similar thing can be done by first creating embeddings for English phrase and then using it as gold standard embeddings for creating Japanese phrase embeddings. The main ideology represented here is that a phrase and its translation should have similar semantics.

We discussed the method of creating RAE above, but we trained only one RAE. In BRAEs, we are concerned with two different languages, which will have different phrases for the same meaning. Hence, we will train two different RAEs. One RAE

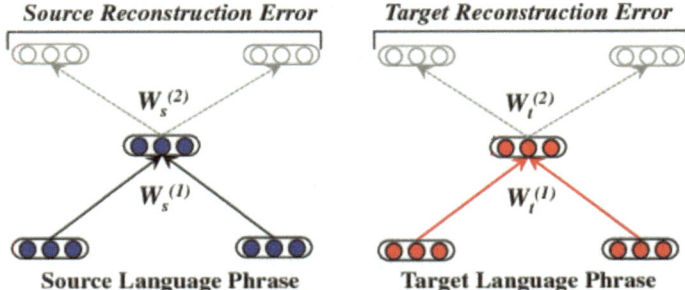

Fig. 11 Functioning as two individual RAE models

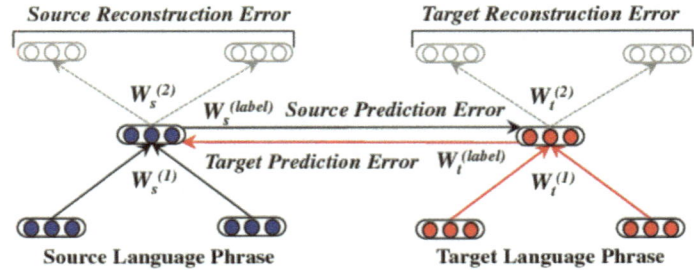

Fig. 12 BRAE model as given in work [68]

is used with source language, while the second RAE is used with target language. This ends our RAE training phase. We then proceed to our cross-training phase in which we first use target-side embeddings to improve our source-side phrase representations, followed by using the source-side embeddings to improve target-side phrase representations. This is done until we reach local minima in the error function. The two different error functions which are computed are

1. **Reconstruction Error**: It is the error which denotes the efficacy of the output embeddings to remake the input embeddings.
2. **Semantic Error**: This represents the semantic distance between the two embeddings formed for phrases in different languages but having the same meaning. The semantic error between source embeddings s and target embeddings t can be represented as

$$E_{sem}(s, t; \theta) = E_{sem}(s|t, \theta) + E_{sem}(t|s, \theta) \qquad (54)$$

where $\theta = (W, b)$. After training BRAE model which connects similar meaning phrases of two languages, we will discuss in depth an approach to use BRAE for bilingual sentiment analysis. This method given by Jain et al. [69] takes Hindi as the resource-scarce language and English as the resource-rich language.

We disconnect the two RAEs from each other since their semantic similarity has already been established and proceed toward adding sentiment information in these embeddings. This part is straightforward. We take monolingual corpora in both languages (corpora for the resource-scarce language would be fine as well) and make a sigmoid classifier in the last layer representing the classification we are to make. The output embeddings from the BRAE network will serve as input embeddings for this network. The rest of the process which follows is similar to a feedforward neural network. The model will train itself based on the text inputs and desired outputs, but with one major difference. One might ask if we use our output embeddings from BRAE and train them for sentiment analysis task, it is possible that these embeddings might change because of backpropagation, thus rendering our previous efforts futile. To tackle this very issue, the authors introduced a penalty on embeddings for movement in semantic vector space. This penalty was the squared error loss between new location and previous location of embeddings whenever they were forced to change. Thus, the final error is expressed as

$$E_{rec}^{*}([c_1; c_2]; \theta) = E_{rec}([c_1; c_2]; \theta) + \frac{\lambda p}{2} ||p - p^*||^2 \qquad (55)$$

This has two advantages: One is that sentiment information is added, and second is that semantic information is also not lost. Also, it is worth noticing that if two phrases are in same vector space semantically, it makes sense for them to be in similar vector space in sentiment model.

Thus, to summarize, we have produced phrase embeddings using RAEs which can recreate input embeddings without much loss of data and storing all their essential signatures. After that, we extracted phrase embeddings for a phrase and its translation in sync using BRAE. We stopped training after local minima were reached in total error and disconnect the two RAEs. These extracted phrase embeddings were used as inputs for a feedforward neural network which trains itself to classify sentences based on sentiment. To ensure that the semantic information in embeddings is not lost, we add a square error loss between the new position of embeddings and original position of embeddings.

After following through the above processes, we have achieved embeddings which captured both sentiment and semantic information of phrases. Also since phrases that are semantically similar, i.e., have similar meanings, it will essentially share same sentiment. Hence, we have developed embeddings which represent the words from resource-scarce language in an expressive way using corpora of resource-rich language.

4.5 AROMA

We have described how RAEs can be used to create embeddings which include both semantic and sentiment information of the phrase, and also discussed BRAEs. But there are some languages in which RAE is not able to reconstruct the input phrases from output phrases. This happens in languages which have rich and complex morphology. Essentially, the way the words are formed and their relationship with other words of the same language can be challenging for computers to imitate. One such language is Arabic. To tackle these challenges, Sallab et al. [70] proposed the AROMA model. So, let us first look at the challenges we faced when using RAE over Arabic, followed with their techniques to tackle them (Fig. 13).

Fig. 13 RAE over Arabic as mentioned in [70]

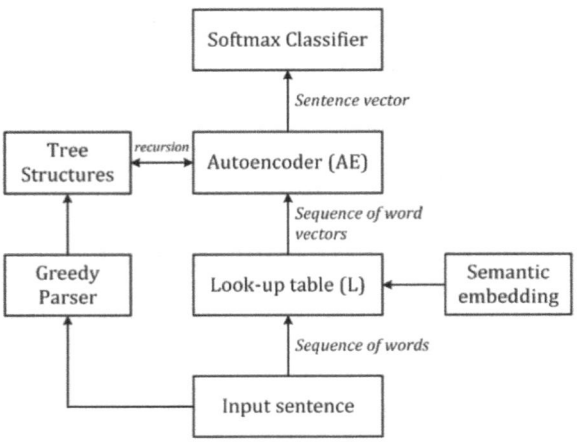

The basic RAE model in consideration suffers in various scenarios when used with morphologically complex languages. As described above, it is not able to imitate the input phrases from output embeddings. This leads to lexical ambiguity, because of which the model's generalization ability is circumscribed . Also, the baseline RAE does not take into consideration the sentiment aspects. Hence, it becomes unreasonable to expect decent accuracy in sentence-level sentiment analysis using these embeddings. Finally, the recursion tree which was formed when deriving phrase embeddings $y1$ from children embeddings ($c1$, $c2$) was formed by greedy method. The way a sentence was expressed by this greedy method may not be the accurate representation of the sentence in consideration. How AROMA model tackles this is rather interesting. The AROMA model is shown in Fig. 14. We will look up at all the changes introduced by the authors and the issues they address in detail.

Tokenization: Tokenization of words, as the authors noted, helped them overcome several Arabic-related challenges. RAEs fail in modeling Arabic, the main reason being that the morphology of the language is fairly complex, which can be further broken down into lexical sparsity and lexical ambiguity:

1. **Lexical Sparsity**: In Arabic, we can form complex vocabulary using a small set of roots via morphology. We are permitted to concatenate morphemes using derivational and inflectional morphology. Thus when the model is trained on the phrase *he wins* which differs from *they win* with just one morpheme inserted in between, the model will not be able to understand the relationship between the above two words. Also, machine translation-based techniques will not be effective here as it has been noticed that the vocabulary of Arabic, as observed in very large parallel corpora as given in the work [71], is two times to that of

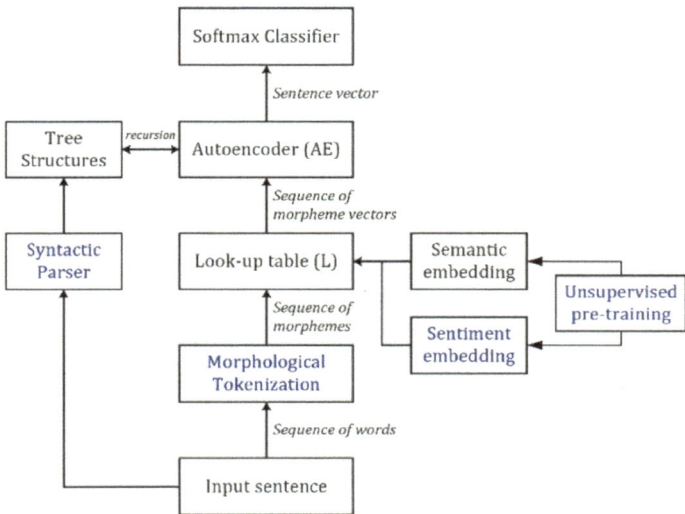

Fig. 14 AROMA model as given in [70]

English. Moreover, it is known that some Arabic words are complex enough to be translated into lengthy English phrases. One such example is that there is an Arabic word which translates to the English phrase *and they will correspond with her.*

2. **Lexical Ambiguity**: Some Arabic words pose an additional problem in sentiment analysis, as they can express different meanings and sentiments. Lexical ambiguity is said to be present in a word when it expresses more than two meanings or sentiments. For example, *sweet* and *suicide* in Arabic look very similar, and it is just a few morphemes which make all the difference. This further creates problems for RAEs in trying to create meaningful output embeddings.

To tackle these challenges, the authors performed morphological tokenization and provided this data to their final model. The tool MADAMIRA, a morphological analyzer and disambiguator provided by Pasha et al. [72], was used to perform tokenization. Words were split into morphemes using the Penn Arabic Tree Bank (ATB) scheme. Both base words and clitics were used to split the texts provided. This method of reducing words to their basic morphemes provided a more robust dataset for the model to be trained on. Evidently, this is because now the model will be trained using morphemes, which have basic meanings, rather than complex words that were built after combining these morphemes.

Syntactic Parsing to improve Composition: One of the problems mentioned above was that the tree which RAE generates is based on a greedy algorithm. This algorithm aims at reducing the reconstruction error of the RAE, but in turn compromises on understanding the syntax of the language and its grammatical structure. Hence, it was found out that for Arabic, the generated trees were not producing optimal embeddings which incorporated both semantic and sentiment information. Alternative to this greedy approach, the authors used the Stanford lexicalized phrase structure parser given by Green and Manning in the work [73]. This automatically generates parse trees over which the autoencoder can train itself recursively. So, the advantage of using the parse trees generated by this method is twofold. The first one is that this method is consistent with our choice of tokenization explained above. Secondly, the sentence representation by autoencoders improves since grammatical rules are taken into consideration.

There is one problem though, the trees generated by Stanford parser are not necessarily binary. Hence, it is not possible to train a recursive model using RAEs due to inconsistencies in input and output dimensions. For this, an ingenious step is used. The grammar of the parse trees was converted to the Chomsky Normal Form (CNF) in accordance with the work [74]. Left factoring was chosen over right factoring to account for the direction the readers follow to combine words while reading Arabic text. Since only unary and binary production rules are contained in CNF, we collapse the unary production. The thus obtained binary parse tree is used to train recursive model. Significant accuracy in sentiment analysis tasks was observed using this method.

4.6   Siamese Neural Networks

Siamese neural networks are a special type of neural network models comprised of two or more equivalent subnetworks which are joined at their outputs. This essentially means that both the subnetworks have the same model architecture and the same weights, and their hyperparameters are mutually shared. This sharing of parameters results in the training of fewer parameters, leading to the fact that lesser data is required and the chances of the model to overfit are reduced considerably.

Siamese neural networks are being widely used in tasks where one needs to find out the similarity between two comparable types of inputs. One of the earlier applications in the literature used Siamese networks for a signature verification task as mentioned in the work [75]. They feed in the pair of signatures to be compared to the model, which in turn elicits features from the input and forms two feature vectors. Then, the cosine similarity of these two feature vectors is calculated to establish the extent of relatedness between the inputs.

Perhaps, one of the most important capabilities of Siamese networks is their applicability in one-shot learning, which aims to learn attributes about object categories, from one or only very limited training instances. Koch [76] uses them for identifying images in one attempt, typically when the size of the input datasets is very small. Other application areas of Siamese networks include Pedestrian tracking for video surveillance [77], matching resumes to jobs [78] and facial recognition systems [79].

We now proceed to describe an application of such a model architecture to determine the polarity of text which incorporates elements of Hindi in an otherwise majorly English text by Choudhary et al. [80]. The authors propose a Siamese network consisting of a pair of bidirectional LSTM networks with shared parameters and a contrastive learning-based energy function at the top to improve sentiment analysis and emoji prediction accuracy. This property of having shared parameters ensures that sentences of both Hindi and English map into the same embedding space. The model consists of a pair of BiLSTM networks at the front of which there is a dense feedforward neural network model. The proposed arrangement is shown in Fig. 15. To the output of the LSTM model, ReLU function is applied, and in turn, the feedforward layer converts the output of the ReLU layer to a vector s given as

$$s = \max\{0, X[fw, bw] + b\} \tag{56}$$

where fw and bw are, respectively, the forward and rearward LSTM representation of the sentence, b represents bias and X is the learned matrix consisting of the hyperparameters. The loss function proposed given embeddings a_i and a_j of tweets and a polarity label $y \in \{-1, 1\}$

$$loss(a_i, a_j) = \begin{cases} 1 - \cos(a_i, a_j) & if\, y = 1; \\ \max(0, \cos(a_i, a_j) - m & if\, y = -1; \end{cases} \tag{57}$$

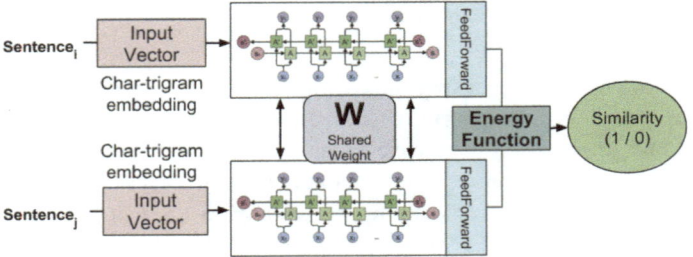

Fig. 15 Model architecture of SACMT, as mentioned in work [80]

where m is the degree by which unlike pairs should be moved away. The training purpose involves optimizing this function in such a way that sentences with label 1 are mapped closer to each other in the sentiment space while those with label -1 are mapped further away from each other. Formally, the training objective is to minimize

$$L(\lambda) = \sum_{(a_i,a_j) \in C \cup C'} loss(a_i, a_j) \qquad (58)$$

In code mixed text, some words which might seem misspelt might actually contain sentiment information because of the informal nature of writing such texts. For example, a term like *veerrryyy good* conveys a higher degree of positive sentiment than simply *very good*. Therefore, they used character-level trigram-based word embeddings of each Hindi–English code mixed tweet along with an English sentence as inputs to the Siamese networks to take these observations into account, along with the labels of the tweets. The model weights and the shared parameters are updated through backpropagation. To take into consideration all variations of transliterated Hindi words into English, they define a clustering-based similarity metric as follows:

$$f(v1, v2) = \begin{cases} \cos(vec(v1), vec(v2)) & \text{if } v1, v2 \text{ have same consonants} \\ 0 & \text{otherwise} \end{cases} \qquad (59)$$

where $v1$ and $v2$ are the variations of the transliterated versions of the word, $vec(v)$ returns the vector representation of v and $f(v1, v2)$ reflects the similarity between $v1$ and $v2$. All versions of the transliterated words occur in similar contexts, and they also have common consonants. Both these characteristics are captured by the above similarity metric. The authors state that their technique outperforms the best techniques in the literature for code mixed text by 7.6% in accuracy and 10.1% in F-score.

5   Discussion

In this section, we discuss some of the pros and cons of the architectures talked about in the chapter, along with instances when they can be applied suitably for the sentiment analysis task.

The model for refining word embeddings mentioned in work [30] can be applied to pre-trained word vectors trained using any model without the need for any labeled corpora. This technique improves the sentiment analysis accuracy on the Stanford Sentiment Treebank (SSTb) dataset, although this method remains to be tested for languages other than English. However, this model makes use of a sentiment lexicon of the language in consideration, and such resources might not be available for low resource languages. The method proposed in work [31] to encode sentiment information in word embeddings performs very well on the SSTb dataset and three other benchmark datasets for sentence-level polarity detection. But just like the previous technique, this method also leverages a sentiment lexicon and has only been tested in the English language. Therefore, the above two techniques can be employed when dealing with English data.

Now let us talk about the techniques discussed for improving the quality of word representations in low resource languages. For taking into account words with similar polarities that appear far from each other in the training corpora, the authors suggest an appropriate weighting scheme in work [34]. This method does not improve the performance in word analogy tasks for languages with abundant resources such as English and Italian. But a considerable increase in performance is observed on word similarity and analogy tasks in a low resource language like Turkish, it is important to mention that the model was trained on the entire Turkish Wikipedia corpus. This model also slightly increased the sentiment analysis accuracy for a Turkish dataset. So, this model seems to be applicable for a low resource language on instances with varying amounts of training data and can perhaps be experimented upon in other low resource language scenarios. The architecture described in work [36] collectively expresses sentiments of a source and target language in a common vector space. It surpasses the benchmarks on sentence-level polarity classification in a combination of three relatively resource-poor languages and is viable to be applied in other scenarios. The only feature of the model which might reduce its applicability when dealing with resource-scarce languages is that it makes use of a bilingual sentiment lexicon, which might not be always available. The framework described in work [40] improves the performance of the word embeddings as reported by the fact that the model outperforms the baselines for most corpora, in the word similarity and analogy tasks for low resource languages such as Czech, Danish and Dutch. Moreover, this model does not have stringent data requirements like other models in the literature as seen by its effectiveness for low resource languages. The fact that this model takes into account the zero entries in the co-occurrence matrices of the corpus in consideration, which is an issue addressed by rather few researchers in the literature, further enhances the prospects of achieving better accuracies for the task in hand.

We now proceed to discuss the main points about CNN-based model architectures talked about previously. People started using CNNs over traditional neural networks for tasks involving images and text. The motivation behind this transition is the reduced number of intra-modular network interactions, sharing of hyperparameters and learning of complex and non-comparable depictions in CNNs as mentioned in the work [42]. Another added advantage of using CNNs is that there is no need to learn a new set of hyperparameters everywhere in the network as all the kernels share the same set of parameters and weights. Let us discuss some notable works done on sentiment analysis in English language using CNNs. The work presented in [43] uses a CNN that extracts character-level and word-level features from tweets and uses it to perform sentiment analysis, along with incorporating the contextual information of word embeddings pre-trained on a Wikipedia corpus by using Word2vec. One advantage of this method is it overcomes the problem of limited information about context that the short texts usually contain. Another similar study [44] uses a character-level CNN which takes encoded characters as input with the exception of not using any word-level features as suggested in earlier method. The authors suggested that their approach can work for other languages as well with the advantage of being able to learn naturally unusual character arrangements such as misspellings and emoticons.

The work proposed in [46] introduces a CNN architecture with a primary focus on good initialization strategies for the weights and parameters of the network. Wang et al. [48] propose a hybrid network which is a combination of CNN- and RNN (LSTM and GRU)-based models to perform sentiment analysis on short texts. The job of the CNN model is to extract high-quality linguistic characteristics from the input corpus, whereas the job of the RNN model is to learn the long-term dependencies. This architecture improves over the benchmark results on three benchmark sentiment analysis corpora. Let us now move forward to using CNNs for multilingual sentiment analysis using CNNs. The authors in [49] introduce a CNN-related model architecture to perform multilingual ABSA. Their model takes as input the sentence and the mentioned aspect, initially pads the input sequence to a length n and then extracts the word embeddings for every word in the processed input; for the English language, they use the pre-trained GloVe word vectors as specified in [27], while for the other languages randomly initialized word vectors are employed. To take into account aspect extraction, the authors define a probability distribution. The work in [50] uses the vectors and sentiment labels of English words to perform multilingual sentiment analysis using variants of CNNs in Indian languages such as Hindi and Marathi, and seven other languages as given in [51]. They perform word-to-word translation of texts in other languages into English using Google Translate. However, one flaw with this method is that the model assigns random vectors for words that could not to get translated. The authors, however, achieve excellent results in the low resource language scenario despite the scarcity of annotated training corpora in these languages, and the mapping of unknown words to English word embeddings also contributes to the observed increase in the performance.

Let us now proceed to discuss the pros and cons LSTM-based model architectures talked about earlier. The fact that LSTMs are designed for handling data of sequential nature and can handle long-term dependencies in the input data makes them suitable for use in sentiment analysis. The attention mechanism-based LSTM model for ABSA given in work [58] takes into account various aspects of the input data by constructing aspect embeddings from the sentences and coupling them with the already trained word embeddings from the input corpus. This architecture is capable of discriminating different polarities of sentiment with different aspects. Moreover, it works particularly well even when the sentence structure is long and complex. However, the model cannot handle multiple aspects simultaneously. The BiLSTM model with CRF model architecture for performing ABSA from work [59] approaches the problem elegantly by first classifying sentences according to the number of opinion targets they contain. Each of these categories of sentences is then fed to a 1-D CNN network for polarity detection. Their model surpasses the state of the art on several benchmark datasets and also performs well for the Spanish, French, Russian, Dutch and Turkish languages. Moreover, extensive experiments were conducted to establish the significant improvements brought by this approach which provides further insights. However, it has low performance on the sentences containing a large number of pronouns such as "it" and "you." The subword-LSTM architecture proposed in work [60] focuses on the subword-level representations of the input data, which captures greater linguistic and sentiment information than word-level features. The fact that the model determines the polarity of code mixed data paves a whole new path for sentiment analysis research of code mixed text, which is becoming increasingly pervasive in social media content today. Despite the misspellings and the SMS language being used in the sentences, it is able to capture the word sections conveying the polarity of the sentiment. The LSTM model architecture proposed in [65] for performing ABSA in the Chinese language takes into account various granularity levels of the Chinese language. The final result of the polarity classification is dominated by the aspect target sequence. This is very advantageous for multi-grained representation nature texts as the ones which are found in the Chinese language. Furthermore, this model merges the outputs from the radical-level, character-level and word-level granularities through the use of appropriate fusion mechanisms. It outperformed the state of the art on various Chinese language datasets covering a wide range of domains. Therefore, this model can definitely be considered as one of the first choices for any sentiment analysis task in the Chinese language.

A fascinating research problem involving Hindi–English code mixed text is the detection of offensive content and hate speech. With the increasing use of code mixed text in multilingual societies today and also the increasing amount of such undesirable text content, it becomes essential to formulate deep learning models to filter out such content. Mathur et al. [81] propose the Multi-Input Multi-Channel Transfer Learning (MIMCT)-based model which leverages pre-trained word vectors alongside extracted features to train a multi-channel CNN-LSTM model architecture to classify offensive tweet using transfer learning. The work [82] introduces a Hindi–English Offensive Tweet (HEOT) dataset containing texts of three categories—non-offensive, hate speech and abusive speech. The authors classify tweets into one of the above three categories by training a CNN model architecture.

Moving on, we mention about another problem of social importance which has been addressed to great effect with DL and NLP techniques. Identification and classification of suicidal ideations in social media texts are the novel application areas in the literature. The increasing suicidal tendencies among youth today along with lack of consulting and awareness are one of the key contributing reasons which led to increased efforts by researchers to develop models to tackle this problem. The work [83] proposes a supervised learning approach for extraction of relevant features to train both linear and ensemble classifiers in order to identify such content in social media texts. Furthermore, Sawhney et al. [84] investigate the effectiveness of deep learning-based model architectures for the identification of suicidal content in social media in their work and demonstrate the capabilities of the C-LSTM model architecture for the same. For a considerable amount of time, work in this area was limited to making use of only textual features. However, Mishra et al. [85] make use of non-textual features such as information from social media networks in the form of condensed social media vectors and analyzing author profiles from historical data, and propose a deep learning framework NAP-BATNET for detection of suicidal content in social media texts.

Let us conclude this section by briefly summarizing autoencoders. The positive part about autoencoders is that they can reduce the size of the dataset considerably, the reasons of which were already explained before. They can help you convert your current dataset into a dataset which is of smaller size with minimal loss of information. They are ideal for removal of noise from data, data compression and dimensionality reduction for the purpose of visualization. BRAE has an advantage that it even gives you phrase embeddings instead of just word embeddings. Moreover, the phrases of the two languages that the model is being trained on would be related semantically. Cons of BRAE for sentiment analysis are that the embeddings from BRAE exploit the semantics of the corpus, not the sentiment information. To use them for a sentiment analysis task, we can further train these embeddings for sentiment. While training this, the semantic meaning of the embedding might get lost. To counter this, we add an additional penalty on embedding for movement from its semantic vector space. Another backdrop of BRAE is that it exploits a greedy algorithm internally to make phrases and the focus is primarily oriented toward minimizing the reconstruction error. When it comes to morphologically rich languages like Arabic and Sanskrit, this greedy approach of making phrases can lead to loss of vital information. As semantics of the languages get more challenging, the pipeline mentioned in the AROMA model is helpful. It has been designed keeping the morphologically rich languages in mind. The parse tree for autoencoder is provided externally using treebank, which ensures that the crucial information regarding the language is not lost. This feature is positive as well as negative as this enforces the demand of a treebank in the target language, which may not really exist.

6 Conclusion

In this chapter, we described a few novel and state-of-the-art deep learning-based approaches for performing sentiment analysis in English and more specifically in low resource languages. Apart from a discussion of the various challenges encountered while performing sentiment analysis in both English and low resource languages, we also described methods to improve word embedding coverage in low resource languages as word embeddings are an effective way to encode the features of the input corpora and often are the input to the various DL model architectures. Thus, we hope that after reading this chapter, the readers will be made aware of the more recent DL-based approaches in the field of multilingual sentiment analysis and will be able to make an informed choice when considering which model to use for their task or perhaps to use a desirable combination of the models discussed.

References

1. Pang, B., and L. Lee. 2008. *Opinion Mining and Sentiment Analysis*. Hanover, MA: Now.
2. Hussein, D.M.E.-D.M. 2016. A survey on sentiment analysis challenges. *Journal of King Saud University-Engineering Sciences*. https://doi.org/10.1016/j.jksues.2016.04.002.
3. Farooq, U., H. Mansoor, A. Nongaillard, Y. Ouzrout, and M.A. Qadir. 2016. Negation handling in sentiment analysis at sentence level. *JCP* 12: 470–478.
4. Xiang, B., and L. Zhou 2014. Improving twitter sentiment analysis with topic-based mixture modeling and semi-supervised training. In *ACL*.
5. Ott, M., Y. Choi, C. Cardie, and J. T. Hancock. 2011. Finding deceptive opinion spam by any stretch of the imagination. In *ACL*.
6. Li, F., M. Huang, Y. Yang, and X. Zhu. 2011. Learning to identify review spam. In *IJCAI*.
7. Flekova, L., D. Preotiuc-Pietro, and E. Ruppert. 2015. Analysing domain suitability of a sentiment lexicon by identifying distributionally bipolar words. In *WASSA@EMNLP*.
8. Felbo, B., A. Mislove, A. Søgaard, I. Rahwan, and S. Lehmann. 2017. Using millions of emoji occurrences to learn any-domain representations for detecting sentiment, emotion and sarcasm. In *EMNLP*.
9. Maynard, D., and M.A. Greenwood. 2014. *Who cares about sarcastic tweets?*. LREC: Investigating the impact of sarcasm on sentiment analysis.
10. Arora, P. 2013. *Sentiment Analysis For Hindi Language (MS thesis, International Institute of Information Technology Hyderabad, 2013)*. Hyderabad: International Institute of Information Technology Hyderabad.
11. El-Masri, M., N. Altrabsheh, and H. Mansour. 2017. Successes and challenges of Arabic sentiment analysis research: a literature review. *Social Network Analysis and Mining* 7: 1–22.
12. LeCun, Y., Y. Bengio, and G.E. Hinton. 2015. Deep learning. *Nature* 521: 436–444.
13. Krizhevsky, A., I. Sutskever, and G.E. Hinton. 2012. ImageNet classification with deep convolutional neural networks. *Communication of the ACM* 60: 84–90.
14. Graves, A., A. Mohamed, and G.E. Hinton. 2013. Speech recognition with deep recurrent neural networks. In *2013 IEEE International Conference on Acoustics, Speech and Signal Processing*, 6645–6649.
15. Xu, K., J. Ba, R. Kiros, K. Cho, A.C. Courville, R.R. Salakhutdinov, R.S. Zemel, and Y. Bengio. 2015. Show, attend and tell: Neural image caption generation with visual attention. In *ICML*.
16. Bahdanau, D., K. Cho, and Y. Bengio. 2015. Neural machine translation by jointly learning to align and translate. CoRR, abs/1409.0473.

17. Karpathy, A., G. Toderici, S. Shetty, T. Leung, R. Sukthankar, and L. Fei-Fei. 2014. Large-scale video classification with convolutional neural networks. In *2014 IEEE Conference on Computer Vision and Pattern Recognition*, 1725–1732.
18. Deng, L., and D. Yu. 2014. *Deep learning: Methods and applications*. Hanover, MA: Now.
19. Bakliwal, A. 2013. *Fine-Grained Opinion Mining from Different Genre of Social Media Content (MS thesis, International Institute of Information Technology Hyderabad, 2013)*. Hyderabad: International Institute of Information Technology Hyderabad.
20. Gonçalves, P., M. Araújo, F. Benevenuto, and M. Cha. 2013. Comparing and combining sentiment analysis methods. In *COSN*.
21. Rish, I. 2001. An empirical study of the naive Bayes classifier.
22. Joachims, T. 1998. Text categorization with support vector machines: Learning with many relevant features. In *ECML*.
23. Altowayan, A.A., and L. Tao. 2016. Word embeddings for Arabic sentiment analysis. In *2016 IEEE International Conference on Big Data (Big Data)*, 3820–3825.
24. Landauer, T.K., P.W. Foltz, and D. Laham. 1998. Introduction to latent semantic analysis. *Discourse Processes* 25: 259–284.
25. Mikolov, T., K. Chen, G.S. Corrado, and J. Dean. 2013. Efficient estimation of word representations in vector space. CoRR, abs/1301.3781.
26. Mikolov, T., I. Sutskever, K. Chen, G.S. Corrado, and J. Dean. 2013. Distributed representations of words and phrases and their compositionality. In *NIPS*.
27. Pennington, J., R. Socher, and C.D. Manning. 2014. Glove: Global vectors for word representation. In *EMNLP*.
28. Joulin, A., E. Grave, P. Bojanowski, and T. Mikolov. 2017. Bag of tricks for efficient text classification. In *EACL*.
29. Bojanowski, P., E. Grave, A. Joulin, and T. Mikolov. 2017. Enriching Word Vectors with Subword Information. *Transactions of the Association for Computational Linguistics* 5: 135–146.
30. Yu, L., J. Wang, K.R. Lai, and X. Zhang. 2017. Refining word embeddings for sentiment analysis. In *EMNLP*.
31. Ye, Z., F. Li, and T. Baldwin. 2018. Encoding sentiment information into word vectors for sentiment analysis. In *COLING*.
32. Çano, E., and M. Morisio. 2019. Word embeddings for sentiment analysis: A comprehensive empirical survey. CoRR, abs/1902.00753.
33. Rezaeinia, S.M., A. Ghodsi, and R. Rahmani. 2017. Improving the accuracy of pre-trained word embeddings for sentiment analysis. CoRR, abs/1711.08609.
34. Yücesoy, V., and A. Koç. 2019. Co-occurrence weight selection in generation of word embeddings for low resource languages. *TALLIP*.
35. Akhtar, M.S., P. Sawant, S. Sen, A. Ekbal, and P. Bhattacharyya. 2018. Improving word embedding coverage in less-resourced languages through multi-linguality and cross-linguality: A case study with aspect-based sentiment analysis. *ACM Transactions on Asian & Low-Resource Language Information Processing* 18: 15:1–15:22.
36. Barnes, J., R. Klinger, and S.S. Walde. 2018. Bilingual sentiment embeddings: Joint projection of sentiment across languages. In *ACL*.
37. Ruder, S., I. Vuli'c, and A. Sogaard. 2017. A survey of cross-lingual word embedding models.
38. Akhtar, S.S., M. Shrivastava, A. Gupta, and A. Vajpayee. 2018. *Robust Representation Learning for Low Resource Languages*.
39. Duong, L., H. Kanayama, T. Ma, S. Bird, and T. Cohn. 2016. Learning crosslingual word embeddings without bilingual corpora. In *EMNLP*.
40. Jiang, C., H. Yu, C. Hsieh, and K. Chang. 2018. Learning word embeddings for low-resource languages by PU learning. In *NAACL-HLT*.
41. LeCun, Y., L. Bottou, and P. Haffner. 2001. Gradient-based learning applied to document recognition.
42. Goodfellow, I., Y. Bengio, and A. Courville. 2017. *Deep Learning*. Cambridge, MA: The MIT Press.

43. Santos, C.N., and M.A. Gatti. 2014. Deep convolutional neural networks for sentiment analysis of short texts. In *COLING*.
44. Zhang, X., J.J. Zhao, and Y. LeCun. 2015. Character-level convolutional networks for text classification. In *NIPS*.
45. Kim, Y. 2014. Convolutional neural networks for sentence classification. In *EMNLP*.
46. Severyn, A., and A. Moschitti. 2015. Twitter sentiment analysis with deep convolutional neural networks. In *SIGIR*.
47. Sahni, T., C. Chandak, N.R. Chedeti, and M. Singh. 2017. Efficient Twitter sentiment classification using subjective distant supervision. In *2017 9th International Conference on Communication Systems and Networks (COMSNETS)*, 548–553.
48. Wang, X., W. Jiang, and Z. Luo. 2016. Combination of convolutional and recurrent neural network for sentiment analysis of short texts. In *COLING*.
49. Ruder, S., P. Ghaffari, and J.G. Breslin. 2016. INSIGHT-1 at SemEval-2016 task 5: Deep learning for multilingual aspect-based sentiment analysis. In *SemEval@NAACL-HLT*.
50. Singhal, P., and P. Bhattacharyya. 2016. Borrow a little from your rich cousin: Using embeddings and polarities of english words for multilingual sentiment classification. In *COLING*.
51. Araújo, M., J.C. Reis, A.M. Pereira, and F. Benevenuto. 2016. An evaluation of machine translation for multilingual sentence-level sentiment analysis. In *SAC*.
52. Bengio, Y., P.Y. Simard, and P. Frasconi. 1994. Learning long-term dependencies with gradient descent is difficult. *IEEE Transactions on Neural Networks* 5 (2): 157–66.
53. Hochreiter, S., and J. Schmidhuber. 1997. Long short-term memory. *Neural Computation* 9: 1735–1780.
54. Sherstinsky, A. 2018. Fundamentals of recurrent neural network (RNN) and long short-term memory (LSTM) network. CoRR, abs/1808.03314.
55. Cho, K., B.V. Merrienboer, Ç. Gülçehre, D. Bahdanau, F. Bougares, H. Schwenk, and Y. Bengio. 2014. Learning phrase representations using RNN encoder-decoder for statistical machine translation. In *EMNLP*.
56. Yao, K. 2015. Depth-gated recurrent neural networks.
57. Greff, K., R.K. Srivastava, J. Koutník, B.R. Steunebrink, and J. Schmidhuber. 2017. LSTM: a search space Odyssey. *IEEE Transactions on Neural Networks and Learning Systems* 28: 2222–2232.
58. Wang, Y., M. Huang, X. Zhu, and L. Zhao. 2016. Attention-based LSTM for aspect-level sentiment classification. In *EMNLP*.
59. Chen, T., R. Xu, Y. He, and X. Wang. 2017. Improving sentiment analysis via sentence type classification using BiLSTM-CRF and CNN. *Expert Systems with Applications* 72: 221–230.
60. Joshi, A., A. Prabhu, M. Shrivastava, and V. Varma. 2016. Towards sub-word level compositions for sentiment analysis of Hindi-English code mixed text. In *COLING*.
61. Jhanwar, M.G., and A. Das. 2018. An ensemble model for sentiment analysis of Hindi-English code-mixed data. CoRR, abs/1806.04450.
62. Can, E.F., A. Ezen-Can, and F. Can. 2018. Multilingual sentiment analysis: An RNN-based framework for limited data. CoRR, abs/1806.04511.
63. Alayba, A.M., V. Palade, M. England, and R. Iqbal. 2018. A combined CNN and LSTM model for arabic sentiment analysis. In *CD-MAKE*.
64. Baly, R., G.E. Khoury, R. Moukalled, R. Aoun, H.M. Hajj, K.B. Shaban, and W. El-Hajj. 2017. Comparative evaluation of sentiment analysis methods across Arabic dialects. In *ACLING*.
65. Peng, H., Y. Ma, Y. Li, and E. Cambria. 2018. Learning multi-grained aspect target sequence for Chinese sentiment analysis. *Knowledge-Based Systems* 148: 167–176.
66. Chung, T., B. Xu, Y. Liu, C. Ouyang, S. Li, and L. Luo. 2019. Empirical study on character level neural network classifier for Chinese text. *Engineering Applications of Artificial Intelligence* 80: 1–7. https://doi.org/10.1016/j.engappai.2019.01.009.
67. Socher, R., A. Perelygin, J. Wu, J. Chuang, C.D. Manning, A.Y. Ng, and C. Potts. 2013. Recursive deep models for semantic compositionality over a sentiment treebank. In *EMNLP*.
68. Zhang, J., S. Liu, M. Li, M. Zhou, and C. Zong. 2014. Bilingually-constrained phrase embeddings for machine translation. In *Proceedings of the 52nd Annual Meeting of the Association for Computational Linguistics* (Volume 1: Long Papers). https://doi.org/10.3115/v1/p14-1011

69. Jain, S., and S. Batra. 2015. Cross lingual sentiment analysis using modified BRAE. In *Proceedings of the 2015 Conference on Empirical Methods in Natural Language Processing*. https://doi.org/10.18653/v1/d15-1016

70. Al-Sallab, A., R. Baly, H. Hajj, K.B. Shaban, W. El-Hajj, and G. Badaro. 2017. Aroma. *ACM Transactions on Asian and Low-Resource Language Information Processing* 16 (4): 1–20. https://doi.org/10.1145/3086575.

71. Alotaiby, F.A., I.A. Alkharashi, and S.G. Foda. (2014). Processing large Arabic text corpora: Preliminary analysis and results. In *Language Resources and Evaluation Conference*.

72. Pasha, A., M. Al-Badrashiny, M. Diab, A.E. Kholy, R. Eskander, N. Habash, M. Pooleery, O. Rambow, R.M. Roth. 2014. MADAMIRA: A fast, comprehensive tool for morphological analysis and disambiguation of Arabic. In *Language Resources and Evaluation Conference*.

73. Green, S., and C.D. Manning. 2010. Better Arabic parsing: Baselines, evaluations, and analysis. In *COLING*.

74. Chomsky, N. 1959. On certain formal properties of grammars. *Information and Control* 2: 137–167.

75. Bromley, J., I. Guyon, Y. LeCun, E. Säckinger, and R. Shah. 1993. Signature verification using a Siamese time delay neural network. *IJPRAI* 7: 669–688.

76. Koch, G.R. 2015. Siamese neural networks for one-shot image recognition.

77. Leal-Taixé, L., C. Canton-Ferrer, and K. Schindler. 2016. Learning by tracking: Siamese C-NN for robust target association. In *2016 IEEE Conference on Computer Vision and Pattern Recognition Workshops (CVPRW)*, 418–425.

78. Maheshwary, S., and H. Misra. 2018. Matching resumes to jobs via deep siamese network. In *WWW*.

79. Schroff, F., D. Kalenichenko, and J. Philbin. 2015. FaceNet: A unified embedding for face recognition and clustering. In *2015 IEEE Conference on Computer Vision and Pattern Recognition (CVPR)*, 815–823.

80. Choudhary, N., R. Singh, I. Bindlish, and M. Shrivastava. 2018. Sentiment analysis of code-mixed languages leveraging resource rich languages. CoRR, abs/1804.00806.

81. Mathur, P., R. Sawhney, M. Ayyar, and R. Shah. 2018, October. Did you offend me? classification of offensive tweets in hinglish language. In *Proceedings of the 2nd Workshop on Abusive Language Online (ALW2)*, 138–148.

82. Mathur, P., R. Shah, R. Sawhney, and D. Mahata. 2018, July. Detecting offensive tweets in Hindi-English code-switched language. In *Proceedings of the Sixth International Workshop on Natural Language Processing for Social Media*, 18–26.

83. Sawhney, R., P. Manchanda, R. Singh, and S. Aggarwal. 2018, July. A computational approach to feature extraction for identification of suicidal ideation in tweets. In *Proceedings of ACL 2018, Student Research Workshop*, 91–98.

84. Sawhney, R., P. Manchanda, P. Mathur, R. Shah, and R. Singh. 2018, October. Exploring and learning suicidal ideation connotations on social media with deep learning. In *Proceedings of the 9th Workshop on Computational Approaches to Subjectivity, Sentiment and Social Media Analysis*, 167–175.

85. Mishra, R., P. Sinha, R. Sawhney, D. Mahata, P. Mathur, and R. Shah. 2019, June. SNAP-BATNET: Cascading author profiling and social network graphs for suicide ideation detection on social media. In *Proceedings of the 2019 Conference of the North American Chapter of the Association for Computational Linguistics: Student Research Workshop*.

Sarcasm Detection Using Deep Learning-Based Techniques

Niladri Chatterjee, Tanya Aggarwal and Rishabh Maheshwari

Abstract Sarcasm is a figure of speech in which the speaker says something that is outwardly unpleasant with an intention of insulting or deriding the hearer and/or a third person. Designing a model for successfully detecting sarcasm has been one of the most challenging task in the field of natural language processing (NLP) because sarcasm detection is heavily dependent on the context of the utterance/statement and sometimes, even human beings are not able to detect the underlying sarcasm in the utterance. In this chapter, we design features for detecting sarcasm using pragmatic features that take into account the context of the utterance. The approach is based on a linguistic model that describes how humans distinguish between different types of untruths. We then train various machine-learning-based classifiers and compare their accuracies.

Keywords Sarcasm detection · Natural language processing · Classifiers

1 Introduction

Sentiment analysis forms a crucial part for various natural language processing tasks such as movie reviews, product recommendations. Sentiment analysis is very intricately intertwined with detection of sarcasm. While doing sentiment analysis, it is of immense importance that the model is able to detect sarcastic sentences as they carry a sentiment which is opposite to the surface sentiment. For illustration, there are some sentences like "I love solving math problems all day", which might be sarcastic for one person while non-sarcastic for other. Thus, sarcasm detection is greatly influenced by the context in which an utterance is made and the difficulty

N. Chatterjee · T. Aggarwal · R. Maheshwari (✉)
Indian Institute of Technology, New Delhi, India
e-mail: Rishabh.Maheshwari.mt614@maths.iitd.ac.in

N. Chatterjee
e-mail: Niladri@maths.iitd.ac.in

T. Aggarwal
e-mail: Tanya.mt614@maths.iitd.ac.in

© Springer Nature Singapore Pte Ltd. 2020
B. Agarwal et al. (eds.), *Deep Learning-Based Approaches
for Sentiment Analysis*, Algorithms for Intelligent Systems,
https://doi.org/10.1007/978-981-15-1216-2_9

associated in capturing the context of the utterance is something that adds to the challenge associated with sarcasm detection.

In terms of definition, sarcasm is defined as a form of verbal irony that is intended to express contempt or ridicule, i.e. Sarcasm has an implied negative (generally) sentiment but may not have a negative surface sentiment. For example, the sentence "I love being ignored" has a positive surface sentiment but has a negative implied sentiment (thereby creating an incongruity) and hence is sarcastic. There are three important parts to the definition of sarcasm:

1. Sarcasm is a form of irony
2. It is mostly intended by the speaker and hence is not just an interpretation of the listener
3. It is used to express contempt or ridicule.

Any model that is designed to solve the problem of sentiment analysis must have some mechanism to differentiate between the sarcastic and non-sarcastic sentences, because if present and not identified correctly, sarcasm can completely change the underlying meaning of the sentence and the way in which it is comprehended. Sarcasm detection is greatly influenced by the context in which an utterance is made and the difficulty associated in capturing the context of the utterance is something that adds to the challenge associated with sarcasm detection. For illustration, there are some sentences like "I love cooking", there are different scenarios possible (assuming the conversation is taking place between two people):

1. If the listener knows the actual liking of the speaker towards cooking:

   a. If the speaker actually likes cooking, it will be perceived as non-sarcastic by the listener
   b. If the speaker does not actually like cooking, it will be sarcastic for the listener

2. If the listener has no knowledge of the liking of the speaker towards cooking, he/she might perceive as either sarcastic or non-sarcastic.

Thus, a same sentence can be perceived as sarcastic by one person and non-sarcastic by some other person, thereby making the context of the utterance extremely crucial to detect sarcasm.

Camp [1] classifies sarcasm into four categories:

1. **Propositional Sarcasm**: The sarcastic sentences that fall in this category would appear as ordinary propositions on the surface but they have a negative implied sentiment associated with them. For example, if you do not like a plan made by your friends and you say *"This plan sounds fantastic"*. Again, it must be noted that if we just look at this sentence, we would perceive that the sentence has a positive sentiment associated with it, we need to know the context and the manner in which the person saying this sentence says it to know that it is actually a sarcastic remark.
2. **Embedded Sarcasm**: In this types of sarcastic sentences, there is an incongruity in the sentence, that is, there would be positive phrases (or words) that are immediately followed by phrases (or words) that carry a negative sentiment and vice

versa. This type of sarcasm can generally be identified by checking if there is any incongruity present in the sentence or not. For example, "*I love being ignored*". Here, the word "love" carries a positive sentiment and it is immediately followed by the phrase "being ignored" which has a negative sentiment associated with it and hence, an incongruity is generated.

3. **Like-prefixed Sarcasm**: As the name of this category of sarcasms suggest, these are preceded by the "*Like*", which provides an implied denial of the argument being made. For example, "*Like you care*" is a common sarcastic retort.

4. **Illocutionary Sarcasm**: The sarcastic sentences that fall in this category would appear as non-sarcastic if we look only for the textual clues. Their sarcastic nature is attributed to some non-textual clues, like the body language, tone, gestures, etc., of the speaker that indicate an attitude opposite to that of a sincere utterance. For example, rolling one's eyes while saying "*Yeah right!*" The "*rolling of eyes*" is a gesture that indicates that the speaker does not literally means the statement he/she is saying and is being sarcastic.

The importance of detecting sarcasm correctly can be illustrated through the following examples:

1. Twitter is one of the places where the use of sarcasm is quite prevalent. When we try to do any sentiment analysis task on twitter data, our first task should be to segregate sarcastic and non-sarcastic tweets and then detect the sentiment. Some of the applications require very accurate sentiment analysis, predicting stock market behaviour using twitter sentiment analysis, being one of them. An inaccurate prediction by the model can lead to huge losses.

2. When we are dealing with product reviews on Amazon to find a rating of the product, many times the consumer writes sarcastic remarks. For example, a product on Flipkart has the following review "*Supercool..i just sold my 2nd kidney to buy this after i bought iphone6 s..now, i m in ventilation. Feels satisfied having this*". On the surface this review seems to be a positive one because of the presence of the words like "supercool", "satisfied", etc., but a human being on reading the entire sentence clearly gets to know that it is a sarcastic remark and hence should be considered as a negative review instead of positive one. An algorithm for review mining should be able to do the same kind of interpretation which is possible only if it is correctly able to identify sarcastic remarks.

3. Similar to the previous example, suppose that we are trying to summarize multiple reviews about a hotel or a movie. We must be able to identify the sarcastic ones as they can change the entire sentiment of the summary we generate. Consider the following two reviews about a same hotel:

   a. "*Very friendly service, continental breakfast was excellent (JUICEBOXES!) and the room was great. Very clean and the haunted sink and screaming toilet gave the bathroom personality!*"

   b. "*We spent 2 nights but received no hskg service. Carpet was filthy. Drapes were torn and dirty. Faucet was broken. Plumbing was noisy, especially at night when we were trying to sleep*".

Now, when a summarizer would try to give a combined review (summary) of the two reviews, it needs to correctly interpret the sarcasm in the first review to give an overall negative sentiment (in this case) to the summary, otherwise, it might take the first review as positive one and second as negative and get confused.

There are several other real-life natural language processing applications that require sarcasm to be detected correctly. We design some features of our model based on violations of Grice's Maxims which will be discussed in detail in the sections to follow. In our work, along with the lexical-, pragmatic- and polarity-based features, we devise four new features based on violations of Grice's Maxim of quality. Grice suggested that any conversation is based on shared principle of cooperation which describes how effective communication can be achieved in any conversation. He fleshed out the principle in a series of maxims. There are eight violations to the maxim of quality: Lies, White Lies, Hyperbole, Meiosis, Sarcasm, Euphemism, Metaphor and Paradox.

Based on these violations, we constructed four new features to detect sarcastic sentences as follows:

1. Overtness—How overt or obviously untrue a sentence is
2. Acceptability—Social acceptability of the sentence with help of number of unacceptable words
3. Exaggeration—Exaggeration in the sentence by evaluating intensity of words
4. Comparison—Similarity between the compared objects (if any) in the sentence using Wu–Palmer similarity [2] on Word-net.

The former two try to capture the semantic sense of a sentence while the latter two capture the implicit incongruity which is between the surface sentiment and the implied sentiment as in the example mentioned above. Mathematical formulas have been used to compute the above features from given text. Thus, the above-mentioned features have continuous values in their ranges. This allows us to use them in a machine-learning-based framework that we developed.

We train different machine-learning classifiers (random forest classifier, gradient boosted trees and SVM) as they are better than rules-based classifications. We worked on the Twitter dataset. For all semantic scoring purposes (positive intensity and negative intensity), we use Valence Aware Dictionary and sEntiment Reasoner (VADER) lexicon. The results show the effectiveness of ML algorithms in differentiating sarcastic and non-sarcastic statements.

2 Related Work

Sarcasm is one of the interesting subjects of language and has proven same for natural language processing as well as a perpetual challenge. There have been several approaches to this problem, from primitive rule-based, which needs close study of pattern of sarcastic sentences and its components to modern deep learning techniques, which need close study to create features to get a good detection model.

Tsur et al. [3] used semi-supervised learning for sarcasm detection. They used syntactic as well as pattern-based features. They gave labels to a sentence from 1 to 5, 5 being clear presence of sarcasm and 1 being absence.

They defined content words (CW) and high-frequency words (HFW). Words below some threshold were called content words, while words above some different threshold were called high-frequency words. Proper nouns were also considered high-frequency words. The patterns they used were which looked at a fixed-sized window of words in a sentence and then look at content words and high-frequency words, ordered sequence of high-frequency words and slots for content words.

For e.g. "I love waking up early in the morning" if considered a window of 2 CW or HFW, then the patterns will be as follows: "I CW CW up", "up CW in CW", "in the CW" and all other of such types.

Therefore, each sentence can have more than one pattern. Each observation (seed) was converted to feature vector, with each pattern having an entry in the feature vector. And this entry would be between 0 and 1 according to match (exact, sparse or incomplete) found in the sentence with the corresponding pattern. Punctuation-based features were also present in this vector, to represent length and frequency of different punctuation marks in the sentence.

For labelling test set, they used a k-nearest neighbour like algorithm. Euclidean distance to k matching vectors to a test observation t was calculated, where matching vectors are those observations which share at least one pattern feature with t. Then, the label is weighted average of the labels of the k vectors, weights being the frequency of label same as that of the vector.

Although the method has interestingly incorporated the use of patterns for sarcasm detection, it does not use sentiment analysis, which intuitively plays a larger part in sarcasm detection.

Riloff et al. [4] proposed a very interesting method to use sentiment analysis and syntactic analysis. They defined sarcastic sentence to have a positive sentiment followed by negative situation, which is intuitively true (although they do not consider the negative sentiment with positive situation). Then, they developed a bootstrap algorithm to learn negative phrases.

For e.g. if "I love exams" is sarcastic, then "love" is a positive verb phrase, and "exams" will then become negative phrase (trigram).

First, they decided a "seed" word or initial positive sentiment verb phrase, then in each sarcastic sentence that contains this word, they looked at immediate following n-grams (unigrams, bigrams or trigrams), because due to brevity of sarcastic tweets (which comprised their dataset), they assumed simple sentence structure, and considered these as negative situation phrases candidates. They further pruned these by parts-of-speech (POS) tagging the phrases, and considered only those which matched their manually developed structure for the n-grams. Further, negative phrase candidates were only added to the final list, if the conditional probability of the sentence being sarcastic, given the negative situation phrase comes after a positive sentiment verb phrase, is more than a threshold.

For e.g. If seed word is "love" and next we come across "We eagerly wait for exams", then as "exams" is a negative phrase, "eagerly wait" becomes positive verb phrase.

This generated a list of negative phrase candidates. Then, learning was done in the reverse direction, to learn positive verb phrases or positive predicate phrases using analogous conditional probability and corresponding threshold.

Therefore, if a sentence will contain one of the positive sentiment phrase and negative situation phrase, it will be predicted as sarcasm. This method looks not only at syntactic, but also sentimental behaviour of sarcasm, and defines a way to find negative situation phrases. But due to this, it depends a lot upon the versatility of training set in lexical sense also, rather than just syntactical.

Joshi et al. [5] use the same algorithm to look for implicit incongruity in a sentence, where implicit refers to the fact that negative situation is implied and not so apparent.

Knowledge of phrases implying negative sentiment was needed.

For e.g. "He loves this pant so much that he rarely wears it", here "rarely wears it" is a phrase with negative implicit implication.

Instead of using a rule-based approach, they created feature vector for each sentence (tweets) and fed them to machine-learning model. The presence of implicit incongruity was then used as a feature for learning model.

Presence of explicit incongruity was also used as a feature, where the incongruity can easily be detected by sentiment analysis.

For e.g. "I love bitter food", "bitter" is not known to be preferred taste.

Lexical features like bigram, unigram were used to contain properties of semantics.

As tweets were involved, pragmatic features like capitalized letters, punctuation marks, emojis' frequency and type were also used as they signify sentiment of the user too.

They specified in the paper that this method was based on world knowledge and may overlook individual-specific sarcasm.

For e.g. "I love solving maths problems" may not be sarcastic for some individuals.

Zhang et al. [6] suggested that neural networks would be better at performing, because of automatic feature induction, rather than manual feature feeding. That is, using embedding to represent words in a sentence and developing feature vector using simple feature templates, like, representing word as concatenation of word embeddings of a word before it, the word itself and a word after it, can then be used by the neural model to automatically gain contextual as well as other sentimental insights.

For modelling, they used gated recurrent neural network, which does not forget the context as well as not carry all of the historical data. For e.g. long-short-term-memory (LSTM) is a GRNN.

They developed feature of current tweet and also historical 80 tweets, using word embedding and feature templates (as mentioned before) and fed it to an LSTM.

Poria et al. [2017] use convolutional neural network to extract the features from a sarcastic sentence rather than handmade features because they thought convolution network will capture context better and may also learn the hierarchical structure if any would be present.

Their model can then be divided into 4 parts,

1. **Word embedding model**: They used word2vec embedding to represent a word and therefore a sentence by concatenating n words present in a sentence, n being the length of longest sentence. This vector is then fed to a convolution network till a fully connected layer (explained further after description of models) for feature extraction.
2. **Sentiment feature extraction**: They used a pre-trained CNN model for feature extraction of 100 dimensional vectors. The training was done on a benchmark dataset for sentiments, classifying sentences into positive, negative or neutral sentiments in the final layer.
3. **Emotion feature extraction**: This too was done using a CNN model trained on dataset to classify emotion of a sentence into six categories, namely anger, disgust, surprise, sadness, joy and fear. The feature vector obtained was 150 dimensional from fully connected layer of the model.
4. **Personality Feature extraction**: CNN models are used to extract features or traits for each personality, which are openness, conscientiousness, extraversion, agreeableness and neuroticism. So, for each personality, there is a CNN model, with each giving a feature vector or trait vector of 150 dimensions, which then are concatenated to form a 750 dimensional personality feature vector.

All these feature vectors are then concatenated, word embedding model's feature vector till the fully connected layer, and the features extracted from the other three models, and then fed into a CNN with softmax output layer or SVM classifier for sarcastic/non-sarcastic classification.

Hazarika et al. [7] also used Stylometric features which contain the information about author's writing style based on gender, age, diction, syntactic influence, etc., along with the features mentioned above.

3 Grice's Maxims

In simple words, Grice Maxims are a set of properties which when present in any kind of conversation can make it more meaningful and logical. Whenever we engage in any kind of vocal conversation, the things that we speak are often progressive remarks of related things. We usually do not make comments that are disconnected from what the conversation was about. To put it differently, one can say that any kind of *effective conversation* between people is a result of cooperative efforts of each participant who also is well aware of the purpose of the conversation. This can be labelled as the "*Cooperative Principle*". This was introduced by Paul Grice [8] in his pragmatic theory as:

'Make your contribution such as is required, at the stage at which it occurs, by the accepted purpose or direction of the talk exchange in which you are engaged'

Grice further suggested that this principal can be divided into four Maxims, which are popularly known as *Grice's Maxims* and these are:

1. Maxim of quantity
2. Maxim of quality
3. Maxim of relation
4. Maxim of manner

Maxim of Quantity: This maxim specifies the amount of information that a participant involved in the conversation should convey. So, according to this maxim:

1. Make your contribution as informative as is required (for the current purposes of the exchange).
2. Do not make your contribution more informative than is required.

Maxim of Quality: This maxim states how one can make a good quality contribution to the conversation. According to this maxim

1. Be truthful
2. Do not say what you believe is false
3. Do not say what you lack adequate evidence for.

Maxim of Relation: This maxim states that the contribution that one makes must *be relevant* to the topic of discussion going on

Maxim of Manner: The above three maxims focus only on the contribution made by a participant in terms of content. However, in any vocal conversation between people, it becomes of immense importance that how an utterance is made. This is precisely what this maxim lists, a set of points that dictate how we should make our contribution. We should

1. Be clear
2. Avoid difficult expressions
3. Avoid ambiguity
4. Be brief
5. Be orderly.

There can be different types of violations of the Grice's Maxims. For the purpose of this chapter, we would focus in particular on the violations of maxim of quality. Grice suggested that different violations of these maxims give rise to various figures of speech in discourse. Nair [7] proposes a model that describes how conversationalists across cultures differentiate systematically between different types of violations of the maxim of quality. There are eight different types of violations of the maxim of quality:

1. *Lie*: A lie is an outright untruth which is made in self interest. They tend to be harmful, malicious and to betray a person. For example, when you have stolen your friend's favourite watch and upon being asked you say that you have not seen it or do not know where it is.

2. **White Lie**: These are the lies that are spoken out of kindness and with the intention of not hurting anyone's feelings. They are selfless and harmless. For example, when you do not like a particular dish made by your mother but you do not want to hurt her feelings and say that the dish is tasty.
3. **Paradox**: A paradox is a statement that appears to be self-contradictory but has a hidden meaning attached to it. For example, Truth is honey, which is bitter.
4. **Metaphor**: A metaphor is an implicit comparison between two things that are not related to each other. To put it in other words, a metaphor draws a resemblance between two contradictory objects based on some shared characteristics. For example, Her voice is music to his ears. This statement is a metaphor because her voice is not literally music but it makes him feel happy and hence is compared to music.
5. **Meiosis**: Meiosis is an understatement. Thus, this figure of speech implies that something is less significant or small than it actually is. For example, "Don't worry, I'm fine. It's only a scratch" when you are actually experiencing pain from the injury.
6. **Hyperbole**: Hyperbole is an overstatement. It is basically an exaggerated or extravagant statement which is not meant to be taken literally. For example, "My grandmother is as old as the hills".
7. **Sarcasm**: In sarcasm, the underlying meaning is completely opposite to the surface meaning. For example, "I love being ignored".
8. **Euphemism**: Euphemism is a figure of speech, in which the speaker uses polite expressions instead of words or phrases that might otherwise be considered harsh or unpleasant. For example, saying "passed away" instead of "died".

Since all the above figures of speech involve some kind of misinterpretation of the reality (truth), they are violations to the maxim of quality. Having discussed the different types of violations to the maxim of quality, we need to find some way so that we can differentiate between them.

The above different types of violations can be distinguished from one another based on four features of **overtness**, **comparison**, **exaggeration** and **acceptability**. These features can be defined as:

1. **Overtness**: It is a measure of how obvious the untruth is, that is, how promptly can we identify the semantic or pragmatic violations of the literal truth
2. **Comparison**: As the name suggests, this features measures if there are two objects being compared in the statement being made. In case of violations of the maxim of quality, mostly we will find instances where two very different lexical items (words or phrases) are compared with each other (This in turn leads to the violation)
3. **Exaggeration**: This feature measures if the sentence that we are making, contains words that represent the entire situation or topic of conversation as better or worse than it actually is, hence leading to a violation of the literal truth.
4. **Acceptability**: It is a measure of how socially acceptable a statement is irrespective of the fact that whether it has been recognized by the participants in the conversation as a violation of the maxim of quality or not.

To get a better understanding of the above features, it is important to have an idea about how the hearer would know that a violation of the maxim of quality by the speaker. In other words, what triggers the hearer's beliefs that some kind of untruth has been uttered by the speaker. Again consider the sentence "I love cooking", a hearer can interpret this sentence as truth or untruth depending on the knowledge that he has about the speaker. Nair [7] explains that the mutual knowledge shared between the speaker and the hearer can be of two types:

1. **Pragma-linguistic knowledge**: This is related to the content of the sentence spoken by the speaker. It includes the entailments, presuppositions, etc., of the utterance along with the linguistic and lexical rules and the pragmatic norms.
2. **Encyclopedic knowledge**: This is related to the context of the utterance and background information of the utterance.

We term a shared knowledge between the speaker and hearer as 'mutual' if there is a match (assumed or created) between them. Since, any conversation have the possibility of becoming a part of the mutual knowledge, they have an equal probability of creating a mismatch between the knowledge of speaker and hearer, thereby creating a conflict and hence leading to an interpretation of the statement as untruth.

On the basis of the type mismatch with respect to the two types of mutual information mentioned above, one can think of two types of violations as explained below:

1. If there is a violation of the pragma-linguistic knowledge, then an obvious (overt) violation is made by the speaker. Also, it must be noted that when such kind of violation is made, the speaker has no intention of misleading the hearer as he believes that he has made an obvious violation which would straight away be identified by the hearer as an obvious untruth. Generally, the literary violations (hyperbole, metaphor, paradox, meiosis, sarcasm and euphemism) fall in this kind of violation.
2. In case of the violations of encyclopedic knowledge, it must be noted that the speaker has an intention of misleading the hearer. Such kind of violations is not obvious to identify (and hence are not overt). Lies and white lies fall in this category.

Now, coming to the feature of exaggeration, consider a set of sentences:

a. John is the worst cook in the world
b. John is a bad cook
c. John is not the best cook in the world.

The above examples, show a clear distinction in fact that how exaggerated is the fact that 'John is a bad cook'. Sentence a) is an overstatement (+ exaggeration) whereas sentence b) is an understatement (− exaggeration). Now again, if the hearer knows that the speaker has a limited experience and he makes a statement with phrases like "worst/best in the world", the hearer would immediately (thereby overt) know that this is an untruth.

The feature of comparison mainly just checks if there are two things that are being compared, irrespective of the fact that the two things being compared are actually similar, + comparison (metaphor) or not, − comparison (paradox).

The feature of acceptability, as the name suggests, checks if the statement would appear to be offensive to the hearer (− acceptability) or not (+ acceptability). That is, is the statement socially acceptable or not.

To exert the fact that these four features are enough to distinguish the eight kinds of violations from each other, we present a table with example sentences and corresponding symbols:

1. +: necessarily a criteria requirement of the violation
2. o: not necessarily a criteria requirement of the violation
3. −: necessarily not a criteria requirement of the violation.

Table 1 also helps to compare how similar and different the eight kinds of violations of the maxim of quality are to each other.

4   Challenges in Sarcasm Detection

The main reason why it is difficult to design a model for detecting sarcasm is that at times even human beings are not able to detect sarcastic sentences.

The main challenge faced while doing sarcasm detection is capturing the context of the utterance. As an example, let's say we are detecting sarcasm for twitter data (tweets), our model should be able to find the appropriate context in which the utterance i s made so that it can go about identify if it is sarcastic or not. Another thing that can be done is we can look at the kind of tweets that person has tweeted already, this would give us an idea about the manner in which the person writes tweets, that is whether he usually makes sarcastic remarks or not. If he/she usually makes sarcastic remarks then there are high chances that the tweet in consideration would also be sarcastic and vice versa. But again, our model should be able to identify the number of past tweets it will consider to get this detail. Also, some sarcastic sentences are sarcastic not because of their textual content but because of the manner in which they are spoken, which again is something which is difficult to deal with. In a nutshell, the challenges associated with sarcasm detection are:

1. Capturing the appropriate context of the utterance. Context in itself is a very vague term. There is no exact definition of what exactly is the context of the utterance. Some utterances would require the model to look at a very small context to say that it is sarcastic, whereas for others we might have to look at a very large context.
2. Capturing the tone and body language of the person who makes the statement, because at times the sarcastic nature of the sentence is attributed not to its content but to the manner in which it is spoken.

Table 1 Violations of Grice's Maxims

Violation	Example	Overtness	Exaggeration	Comparison	Acceptability
Lie	When you have stolen your friend's favourite watch and upon being asked you say you have not seen it or don't know where it is	–	o	o	–
White Lie	When you do not like a particular dish made by your mother but you do not want to hurt her feelings and say that the dish is tasty	–	o	o	+
Paradox	Truth is honey, which is bitter	+	o	–	O
Metaphor	Her voice is music to his ears	+	o	+	O
Meiosis	"Don't worry, I'm fine. It's only a scratch" when you are actually experiencing pain from the injury	+	–	o	O
Hyperbole	My grandmother is as old as the hills	+	+	o	O
Sarcasm	"I love being ignored"	+	o	o	–
Euphemism	"His father 'passed away'". instead of "His father is 'dead'"	+	o	o	+

5   Dataset Description

We used tweets to test our model. As tweets typically are not more than 140 characters mostly they consist of one or two sentences. Moreover, for tweets, as they are devoid of external factors such as body language, voice modulation, facial expression, text is the only means for conveying sentiment. We used the twitter API provided for obtaining tweets containing certain strings, we used #*sarcasm* to collect sarcastic tweets, and collected around 6K sarcastic tweets, out of which 3K were non-sarcastic (general tweets). This set did not include re-tweets for so that set is as versatile as possible. We employed tweepy library to collect tweets, although it only gave tweets from 2 weeks back and would not let tweets be collected after a certain amount, as a security measure.

For historical tweet extraction:

1. We generated a list of users whose tweets were present in the corpus. There were around 4K unique users.
2. Then, we extracted the historical tweets using username as search key (maximum five for each user) and saved their sentiment (using VADER sentiment analysis), calculated the average and stored the average and number of tweets in a dictionary.
3. When came across a tweet, we first checked if historical tweets were present, then overtness was calculated using formula mentioned in next section, the new average was then calculated using following formula:

$$S_{new} = S_{old} * n_{old} + S_{new}$$
$$n_{new} = n_{old} + 1$$

4. Storing the new values in the dictionary, the whole algorithm is then repeated for all tweets in the 6K tweets.

After collecting the tweets, preprocessing was done as following:

1. Other hashtags, URLs (links) and mentions were removed, (mentions were replaced by nouns).
2. Added space before and after punctuation marks for better word and sentence detection.
3. Replaced contractions (e.g. don't) with their expansion using dictionary available, this was to improve tokenization and get better sentiment analysis.
4. Replaced social media slangs and abbreviations with their full forms (e.g. lol, tbh, btw, etc.)
5. Spelling check was then done on all of the words after performing the above steps.

Following were the shortcomings of the dataset:

1. Tweepy: Gave error 429 (frequency of twitter extraction is too high) Tweepy does not give data more than two weeks old.
2. With time, people used more and more hashtags and less words. Sometimes Mentions are used as nouns (twitter handles) or adjectives (e.g. #awesome).
3. Sometimes, people may not add '#sarcasm' and sometimes, may add unnecessarily
4. Also, our model requires tweets to be extracted in one go (that is, we cannot combine the tweets extracted from weeks that are not consecutive using tweepy

(which allows a maximum of two weeks tweets to be extracted only)) as the score of overtness cannot be calculated if there are no previous tweets from some person

The corpus still may contain a lot of noise, like incorrect tagging of a sentence as sarcastic or absence of the hashtag while the sentence might be sarcastic, absence of historical data of a user, or all of the historical tweets by the user being sarcastic, therefore misleading the overtness feature.

6 Feature Description

Our model makes use of the following features:

The features are inspired from Detection of sarcasm in tweets: a rough set based approach by Bajpai et al. [9]

a. **Overtness**: As described earlier, overtness is a measure of how obvious the lie (untruth) is. So, this feature is a measure of how much overt or covert the statement is. If we are dealing with twitter data, we quantify this score by comparing the average sentiment of the tweets the same user has made in the past with the tweet that we are considering. The reason that we quantify this feature like this is because let us say a user is habitual of making positive (or negative) sentiment tweets and he suddenly makes a negative (or positive) tweet, then there are high chances that he is being sarcastic. To calculate the sentiment score of the tweet, we have made use of the NLTK library package and used the VADER implementation in it. Mathematically, the score is given by:

$$d = s - s_o$$

Here,

d is the overtness score
s is the sentiment of the tweet in consideration
s_o is the average sentiment o f he previous tweets by the same user

A nice observation that can be made here is that both s and s_o lie between the range $[-1, 1]$, where -1 signifies totally negative sentiment and $+1$ indicates totally positive sentiment. Hence, d lies in the range $[-2, 2]$

b. **Exaggeration**: As described earlier, exaggeration is a measure of how exaggerated (understated or overstated) the given statement or tweet is as compared to reality. To quantify this feature, we first observed and studied various exaggerated sentences and we came to a conclusion that adverbs, adjectives and verbs are the only figures of speech that would make a contribution to the exaggeration of a sentence or tweet. So, the next logical step that followed was to identify these figures of speech from the entire sentence or tweet. This was done with the help of POS-tagging using NLTK's library implementation of Penn Treebank tagging.

(POS-tagging assigns parts-of-speech tags to the constituent words of a sentence or tweet). Now, once we have identified the figures of speech of the constituent words of a sentence, we identified another problem, there are words in English like *"bass"* which can have different meaning depending upon the context of the sentence in which they are used. Following are the different meanings of the word *"bass"*:

1. a type of fish
2. tones of low frequency
3. A type of instrument.

It is quite evident that depending upon the meaning of the word in the actual sentence can have an impact on the sentiment score of the sentence. There are many algorithms that help in word-sense disambiguation (capturing the actual meaning of the word out of its different meanings). One such algorithm is the Lesk's algorithm [10] for word-sense disambiguation. It was introduced by Michael E. Lesk in 1986. The basic idea of the Lesk's algorithm is that we can find the actual sense of a word by looking at the overlap between the Dictionary definitions of the word and the context (neighbourhood) of the word in the sentence. The Lesk algorithm is:

(i) for every meaning (sense) of the word, count the number of words in its context (neighbourhood) that are present both in the dictionary meaning as well as the neighbourhood of the word
(ii) the algorithm returns that meaning (sense) of the word which gets the maximum overlap of words with the neighbourhood of the word

So, for word-sense disambiguation, we use an implementation of Lesk's algorithm in NLTK library. Once we are done with this, we go about finding the sentiment scores of the constituent words of the sentence using Sentiwordnet [11]. Mathematically,

$$S_c = p_s + n_s$$

where

S_c is the sentiment score (S-score) of the word
p_s is the positive sentiment score associated with the word
n_s is the negative sentiment score associated with the word.

When we provide a word as an input to the SentiWordNet, we get three scores associated with the word, namely, positivity score (the positive sentiment associated with the word), negativity score (the negative sentiment associated with the word) and an objectivity score (the neutral or objective sentiment associated with the word). All the three scores lie in the range [0, 1] and sum to 1. Hence, a straightforward observation that can be made here is that $S_c \leq 1$.

Now, we also observed that many a times, sentences or tweets contain degree modifiers, which are words like *"very"*, *"quite"*, etc., which basically increase or decrease the intensity of the word that follows it. Thus, there are two types of degree modifiers:

(a) **Degree Intensifying Modifiers**: These include words like "*very*", "*greatly*", etc., which increase the intensity of the words that follow them. When we provide degree intensifier words as an input to the SentiWordNet, we get $p_s > n_s$

(b) **Degree De-intensifying Modifiers**: These include words like "*rarely*", "*barely*", etc. which decrease the intensity of the words that follow them. When we provide degree de-intensifier words as an input to the SentiWordNet, we get $p_s < n_s$

Now, we had earlier observed that the sentiment score of the word S_c is less than 1. Therefore, we include the effect of degree modifiers in our model as follows: (Here, assume that w is the word that follows the degree modifier d)

$$S_c(\omega) = \begin{array}{ll} \sqrt{S_c(\omega)} & \text{if } a \text{ is a degree intensifier} \\ S_c(\omega) & \text{if } p_s(d) = n_s(d) \\ S_c(\omega)^2 & \text{if } d \text{ is a degree de - intensifier} \end{array}$$

The exaggeration score of the sentence or tweet is the average of the scores of all the constituent words in the tweet, that is

$$e = \left(\sum_{i=1}^{n} S_c(w_i) \right)/n$$

where, n is the total number of words in the sentence

(c) **Acceptability**: Acceptability is a measure of socially acceptable a statement or tweet is. The most logical way of quantifying this feature is measuring the number of acceptable or unacceptable words. To calculate the number of unacceptable words, we make use of a slang dictionary and perform string matching with the words of the dictionary to get the count. Then, mathematically the acceptability score is given by:

$$a = 1 - \frac{n_a}{n}$$

Here, n_a is the number of unacceptable words and n is the total number of words in the sentence or tweet. The lexicon of negative words was available online [12]

d. **Comparison**: This score measures if there is a comparison being made in the sentence or tweet (either similar or dissimilar objects are being compared). To mathematically quantify this score, we find the similarity score between the words being compared in the tweet. Now, comparison is a very frequent in sarcastic tweets or sentences; hence, comparison is an important score. The first step to find this score is to find the words being compared (these words can be adjectives, nouns, verbs, etc.). This is done using context-free grammar (CFG) rules to parse phrases of a sentence or tweet as follows:

S ⟹ NP "like" NP | ADJ "as" "NP" | ADJ "as" ADJ | ADJ "as" V | NP "like" V
NP ⟹ N | ADJ N | N N | "NNS" N
N ⟹ "NNP" | "NN"
V ⟹ "VBD" | "VB" | "VBG"
ADJ ⟹ "JJ" | "RB" | "RBR" | "VBG"

An important point to be noted here is that the above CFG is the one that we used in our work. It is not necessarily exhaustive. All the POS tags not present in the CFG will not be used in any further step to calculate this score and hence will be removed. An interesting observation from the above CFG is that a focus phrase cannot be of a length greater than 5. Now, once we have extracted the two words, next step is found out their similarity. This is calculated using Wu–Palmer similarity which returns the similarity score of two word senses based on the depth of the two senses and their least common subsumer (or least common ancestor).

Mathematically, the Wu–Palmer similarity is given by:

$$\text{sim}_{wup} = \frac{2 * \text{depth}(\text{lcs}(w_1, w_2))}{\text{depth}(w_1) + \text{depth}(w_2)}$$

Here, w_1 and w_2 are the words being compared and $lcs(w_1, w_2)$ is the least common subsumer of them.

The similarity score is calculated using the Wu–Palmer Similarity measure implementation available in NKTK package of python after applying Lesk's algorithm for word-sense disambiguation.

Two cases arise here,

Case 1 If an adjective or adverb is found before and after *"as"* or *"like"*, then similarity score between the words in target is calculated.

Eg. "He is as active as snoring kid"

Case 2 If an adjective or adverb is found before *"as"* or *"like"*, then that adjective is compared to all the words (maximum 2) following *"as"* or *"like"* and the similarity score is calculated of all words on the one side to the adjective or adverb on the other side.

Eg. "Alice is fast like a snail"

Case 3 If the phrase that we extract using context-free grammar has no adjective, then all the words coming before *"as"* or *"like"* are considered being compared to all that come after *"as"* or *"like"* and

Eg. "My boss is as human as a Neanderthal"

The final comparison score of the sentence or tweet is given by:

$$w = \left(\sum_p WP\text{sim}(p)\right)/n$$

Here, p is the number of word pairs found, n is the total number of words, $WP\text{sim}(p)$ is the Wu–Palmer similarity score for the pair p.

Apart from the above-mentioned four features, as discussed by Joshi et al., we also take consideration some other features which are based on incongruity and some lexical features, which are as follows:

1. ***Explicit Incongruity-Based Features***: As has been discussed in the previous sections, incongruity arises when two words or phrases of opposite sentiment occur together in a sentence or tweet. This feature is a measure of inherent incongruity in a sentence or tweet. We quantified this feature with a help of a number of sub features which individually capture things observed when an incongruity occurs. These are:

   a. ***Positive Word Count***: This is the overall count of words carrying a positive sentiment in the sentence or tweet in consideration. The sentiment of the words is identified using SentiWordNet.

   b. ***Negative Word Count***: This is the overall count of words carrying a negative sentiment in the sentence or tweet in consideration. The sentiment of the words is identified using SentiWordNet.

   c. ***Number of contextual incongruities***: This is perhaps the most important sub feature to capture explicit incongruity. As the definition of explicit incongruity suggests that incongruity arises when words (or phrases) of contrasting sentiments appear together in the sentence. This feature measures exactly the same thing. It is a count of how many times a positive sentiment word is followed by a negative sentiment word and vice versa. In short, it is count of sentiment switches occurring in the sentence or tweet in consideration.

   d. ***Longest sequence of positive or negative sentiment of words***: This feature measures the length of the longest sequence of words in the sentence that carry the same sentiment (the sentiment can be positive or negative).

   e. ***Overall sentiment of the sentence***: This is the overall sentiment of the sentence. It is calculated using VADER.

2. ***Lexical Features***: Theses are a set of features that capture important information about the structure of the focus sentence. It includes:

   a. ***Tf-Idf Values***: We also provide as input to our model, the Tf-Idf matrix of unigrams, bigrams and trigrams. The Tf-Idf value is a score that assigns importance to words based on how frequently they appear. For the purpose of our work, we used only top 3000 most frequent unigrams, bigrams and trigrams. The number of frequent words considered can be changed depending upon individual requirements.

   b. ***Capitalized Word Count***: As the name suggests, this is the count of words that are capitalized. We take this feature into consideration because it was observed that people tend to use capitalized words in tweets when they are being sarcastic to lay extra emphasis on the word that is capitalized.

   c. ***Number of smileys and emojis***: This is the count of number of smileys or emojis being used in the tweet.

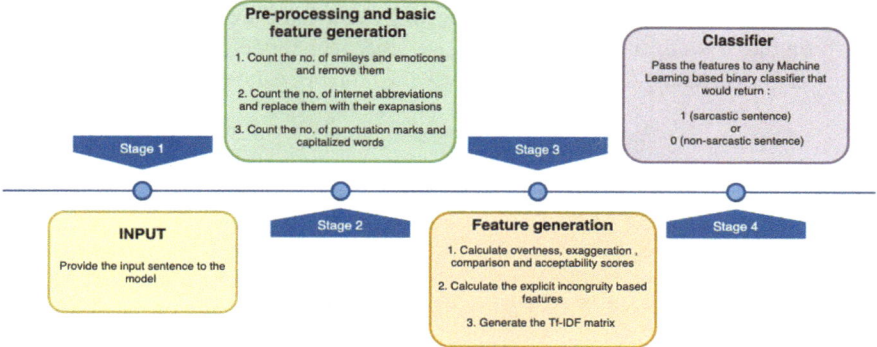

Fig. 1 Process outline

d. ***Number of Internet abbreviations***: Nowadays, people tend to use a lot of abbreviations instead of writing the entire text on social media. These abbreviations usually add to the nature of the sentence being considered. We used a dictionary of common Internet abbreviations and their full forms to get this count.

e. ***Number of punctuation marks***: This is the count of punctuation marks (',', '!', '?') in the focus tweet.

7 Process Outline

Figure 1 shows the process outline.

8 Models Used

We trained a number of machine-learning models so that we could compare them and draw further inferences. The models that we used are:

1. ***Decision Trees***: Decision trees are flow-chart like structure where at each node we make a test on some particular feature and depending upon the result, we go to the child node corresponding to the result.
2. ***Random Forest Classifier***: This fits a number of decision trees of given height on various subsets of input dataset and uses averaging to decrease over-fitting and increase accuracy.
3. ***Support Vector Machine***: Classifier which tries to learn the separating hyperplane between the instances by using support vectors instead of whole dataset, and may also mimic higher dimension by using kernel-based inner products.

4. ***Gradient Boosted Trees:*** Gradient boosting algorithms are those which generate ensemble of weak classifiers in such a way that each model generated next moves towards decreasing the error, i.e. moves towards opposite direction of gradient.

9 Experiments and Results

We obtained the following ROC curves for the different machine-learning models that we experimented with (using all the features mentioned previously as input to the model):

Some insights from the graph are:

1. Avg. AUC in 5 split cross validation = 0.90833
2. Avg. AUC in 10 split cross validation = 0.9316

This tells, if we have more data, the machine-learning model can perform even better (Fig. 2).

Table 2 compares the accuracies of various models having a different combination of features. (both F-score and AUC are used for comparison purposes):

Table 2 shows the comparison of results using different combinations of features.

While Model 3 has better Precision and F-score, Model 5 has better Recall and AUC. This tells that Model 3 was better at telling which is sarcastic, Model 5 is better at telling which is not.

Note: All the results in the above table are of random forest classifier as it has the highest score among the trained models.

Fig. 2 AUC curve of different methods, x-axis is false positive and y-axis is true positive

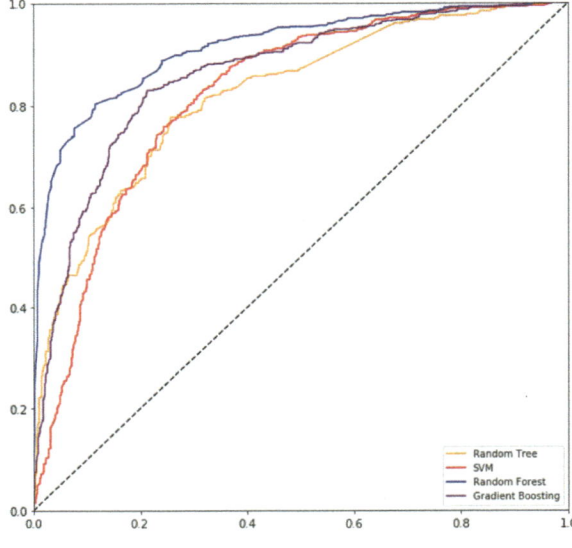

Table 2 Here, *L* lexical features, *EI* explicit incongruity-based features, *A* acceptability, *E* exaggeration, *C* comparison, *O* overtness

#	Features	Precision	Recall	F-score	AUC
1	L + EI	0.815154	0.883084	0.847761	0.902518
2	L + EI + A	0.836104	0.875621	0.855407	0.905609
3	L + EI + A + E	**0.883963**	0.843283	**0.863144**	0.9117731
4	L + EI + A + E + C	0.855361	0.853233	0.854296	0.912210
5	L + EI + A + E + C + O	0.820895	**0.889303**	0.853731	**0.914726**

10 Future Scope

1. Joshi et al. also mentions use of another feature, namely, implicit incongruity which can be added to the set of features used in the model.
2. A more rule-based algorithm can be used and its results can be compared with the regression-based models used.
3. The sentiment of the emojis and smileys used in the text can also be considered in the set of features used.
4. More historical data can be collected to calculate value of overtness.
5. More grammatical rules can be used to calculate the value of comparison and exaggeration.

References

1. Camp, Elisabeth. 2012. Sarcasm, pretense, and the semantics/pragmatics distinction. *Noûs* 46 (4): 587–634.
2. Wu, Z., and M. Palmer. 1994. Verb semantics and lexical selection. In *Proceedings of the 32nd Annual Meeting of the Associations for Computational Linguist*ics, 133–138.
3. Davidov, Dmitry, Oren Tsur and Ari Rappoport. 2010. Semi-supervised recognition of sarcasm in twitter and amazon. In *CoNLL*.
4. Riloff, Ellen, Ashequl Qadir, Prafulla Surve, Lalindra De Silva, Nathan Gilbert and Ruihong Huang. Sarcasm as contrast between a positive sentiment and negative situation.
5. Joshi, Aditya, Vinita Sharma, and Pushpak Bhattacharyya. 2015. Harnessing context incongruity for sarcasm detection. In *Proceedings of the 53rd Annual Meeting of the Association for Computational Linguistics and the 7th International Joint Conference on Natural Language Processing* (Volume 2: Short Papers), Vol. 2.
6. Zhang, Meishan, Yue Zhang and Guohong Fu. 2016. Tweet sarcasm detection using deep neural network. In *COLING*.
7. Nair, R.B. 1986. Telling lies: some literary and other violations of Grice's Maxim of quality. *Nottingham Linguistic Circular (Special Issue on Pragmatics)* 14: 53–71.
8. Grice, H.Paul. 1957. Meaning. *Philosophical Review* 66 (3): 377–388.
9. Bajpai, Anurag, Vaibhav Khandelwal and Niladri Chatterjee. Detection of sarcasm in tweets: a rough set based approach.

10. Lesk, M. 1986. Automatic sense disambiguation using machine readable dictionaries: how to tell a pine cone from an ice cream cone. In *SIGDOC'86: Proceedings of the 5th Annual International Conference on SYSTEMS Documentation, ACM*, 24–26, New York, USA.
11. Esuli, Andrea, and Fabrizio Sebastiani. 2006. Sentiwordnet: a publicly available lexical resource for opinion mining. In *LREC*, vol. 6.
12. https://github.com/shekhargulati/sentiment-analysis-python/blob/master/opinion-lexicon-English/negative-words.txt.

Deep Learning Approaches for Speech Emotion Recognition

Anjali Bhavan, Mohit Sharma, Mehak Piplani, Pankaj Chauhan, Hitkul and Rajiv Ratn Shah

Abstract In recent times, the rise of several multimodal (audio, video, etc.) content-sharing sites like Soundcloud and Dubsmash have made development of sentiment analytical techniques for these imperative. Particularly, there is much to explore when it comes to audio data, which has proliferated rapidly. Of all the various aspects of audio sentiment studies, emotion recognition in speech signals has gained momentum and attention in recent times. Recognizing specific emotions inherent in spoken language could go a long way in healthcare, information sciences, human–computer interaction, etc. This chapter examines the process of delineating sentiments from speech, and the impact of various deep learning techniques on the same. Factors like extracting relevant features and the performances of several deep learning architectures on such datasets are analyzed. Performances using various classical and deep learning approaches are presented as well. Finally, some conclusions and suggestions on the way forward are discussed.

Keywords Deep learning · Sentiment analysis · Emotion recognition · Multimodal data · Data · Audio data

A. Bhavan · M. Sharma · M. Piplani · P. Chauhan · Hitkul · R. R. Shah (✉)
MIDAS, IIIT-Delhi, New Delhi, India
e-mail: rajivratn@iiitd.ac.in

A. Bhavan
e-mail: anjalibhavan98@gmail.com

M. Sharma
e-mail: mohit.sharma.15cse@bml.edu.in

M. Piplani
e-mail: 201551072@iiitvadodara.ac.in

P. Chauhan
e-mail: cpankajr@gmail.com

Hitkul
e-mail: hitkuli@iiitd.ac.in

© Springer Nature Singapore Pte Ltd. 2020
B. Agarwal et al. (eds.), *Deep Learning-Based Approaches
for Sentiment Analysis*, Algorithms for Intelligent Systems,
https://doi.org/10.1007/978-981-15-1216-2_10

1   Introduction

With the passage of time, the quantity of data being created and shared is accumulating by the million, and also quite rapidly. Millions of new users are being added to the Web every minute, which is further giving rise to more data to process and analyze. Moreover, since social media is one of the principal ways for people to voice their opinions on a variety of issues, it is fast proving to be a valuable source of knowledge on emotions, sentiments and opinions. Sentiment and emotion analyses on data obtained from social media sites can come up with conclusions and insights on several products, ideas, etc., and can even help institutions gauge public opinion to come up with better products and applications. Recently, influencing public opinion by appealing to their sentiments and sharing, a lot of directed content has even led to massive upheaval on the political level in several countries, which gives an indication of the urgent need and potential of this field.

The rise of audio-based websites and apps like Soundcloud and Audible have, in particular, paved the way for gaining insights and analysis in newer, different ways. This provides for an immense amount of data for studying sentiments and opinions, and indeed, sentiment analysis of audio has been picking up pace in several research avenues.

Apart from classifying speech in terms of sentiment, one further step could be taken: recognizing particular emotions. Instead of just deciding if the given audio sample has a good or bad sentiment, could it be inferred that the person speaking is angry, happy or bored? This has been a topic of research attracting limited interest in the past few years. Since speech is the principal method of communication among people that also leads to valuable insights about a person's mood, it thus becomes a challenging problem to come up with models and metrics for the same. This chapter aims to explore recent approaches toward emotion recognition from speech, and also presents baselines and discussions on factors like feature engineering and evaluation metrics for the same.

In particular, deep learning has emerged in recent years as the lingua franca of the machine-learning research community. It has made inroads in a large number of fields like traffic monitoring, translation, speech recognition, Natural language processing [1–6], bio-medical imaging [7–9], etc. Deep learning models from numerous works are explored in this chapter as well, with a thorough analysis of their procedures, results and principles.

Sections 1, 2 and 3 present an introduction and literature study on the various features which can be analyzed in speech for identifying emotions as well as feature selection methods. Section 4 discusses the classical models for speech emotion recognition, and Sect. 5 discusses deep learning-based methods. Next, a model for ensemble learning for emotion recognition is proposed, and deep learning architectures from various research studies are compared with the same. Section 9 presents the final conclusions.

2   Feature Extraction

Various feature sets have been extracted from speech data for further analysis in several works. Which features are the best for any generalized speech sample, though, continues to be a topic for discussion and research.

The first and most commonly extracted features from speech are utterance-level and prosodic in nature. Prosodic features are concerned with how something is uttered, not with the properties of the constituent phonemes. For instance—an angry voice might have high energy and pitch; something sad might be low in energy. One of these prosodic features of sound is its pitch, which relates to how frequency is perceived by us. Pitch information has been used in several earlier works to develop emotion recognition systems with considerable success. Dellaert et al. [10] presented experiments with simple pitch information extracted from utterances, and improved on it by smoothing the pitch contour using cubic splines. Vogt and Andre [11] stated that a feature set made of single pitch values would not be of much use, and proposed the usage of statistical terms like maxima, minima, etc., instead to receive information spread over a period of time. These are termed as prosodic features with global utterance-level statistics, since they are calculated over entire speech samples.

Some other prosodic features of speech include vocal energy, intonation, zero crossing rate and intensity [12–14]. These features have been used both standalone and in combination with other kinds of features in several systems [15–17].

The majority of research on the feature extraction initially focused on prosodic features, with limited attention to spectral features. Spectral features rely on the power spectrogram of sound. From a speech signal's power spectrum, a discrete Fourier transform is first performed to map it to the spectral domain. Such spectra can provide more information about emotional content than mere prosodic features, but it is often limited to higher frequency ranges. This has been explored in [18, 19] as well.

One very important class of features is linear predictive coefficients (LPC) [20]. The motivation behind linear predictive coefficients is that any particular speech sample at a period of time can be approximated linearly by previous samples at different points of time. Morrison et al. [21] used these coefficients for emotion recognition in call-center speech samples. Razak et al. [22] used 18 features based on LPCs, and reported a recognition rate of approximately 60%. Many such applications and experiments continue to analyze LPCs for further advances in speech processing.

LPCs, however, can suffer from high variance, which does not make them very reliable when it comes to emotion recognition [23]. This motivated the need for better coefficients; thus, linear predictive cepstral coefficients (LPCCs) were next proposed. LPCCs are derived from LPCs themselves, by using an inverse Fourier transform to map the LPC coefficients to the cepstral domain [24].

For a long time, LPCCs continued to be the most frequently used features in speech emotion recognition [25]. Davis and Mermelstein [26] developed the concept of Mel-frequency cepstral coefficients (MFCCs), and showed that they outperform both LPCs and LPCCs.

MFCCs are a better way of representing sound as heard by the human ear. They are calculated by taking a discrete Fourier transform of the frames obtained from a speech sample first, deriving the Mel-filterbank, and calculating the inverse Fourier transform of the logarithm of the filterbank. They continue to be used and explored in multiple works [27]. LPCCs and MFCCs have also been demonstrated to show better performance than LPCs [28], and are currently the most common feature sets used in speech emotion recognition.

Molau et al. [29] demonstrated that it is possible to calculate MFCCs without the traditional filterbank technique; using this did not affect the performance of their speech recognition system. It can be argued that a similar approach can be used for speech emotion recognition as well, although to our knowledge, such experiments are yet to be undertaken.

Apart from these, many new features have been proposed as well; these include modulation spectral features [30], gammatone-frequency cepstral coefficients [31] and log-frequency power coefficients [32]. Recent trends indicate that a combination of various features tends to produce excellent results. This has been investigated in many recent works: Seehapoch and Wongthanavasu [33] compared various feature sets and demonstrated a combination of fundamental frequency, energy and MFCCs as the best performing compared to given baselines, while Schuller et al. [34] combined acoustic and linguistic features like pitch, energy, emotional key-phrases, etc., in a hybrid architecture for emotion classification. Kwon et al. [35] studied the effects of combining pitch, log energy, formant, mel-band energies and MFCCs on the classification rate (Table 1).

Feature	Description
Pitch	Perceived fundamental frequency of sound
ZCR	Number of times speech signal amplitude becomes zero
LPC	Coefficients estimating forthcoming output sample given previous samples
LPCC	LPC coefficients mapped to the cepstral domain using inverse Fourier transform
MFCC	Short-term signal power spectrum based on linear cosine transform of log power spectrum

Table 1 Various feature types

3 Feature Selection

Two kinds of feature selection methods are used for eliminating features: filter-based [36, 37] and wrapper-based [38]. Filter-based methods are independent of the classifier, and usually consist of statistical tests like Fisher criterion, variance-based selection, etc. Wrapper-based methods eliminate features based on the performance of the feature set; a typical approach is removing features recursively. Feature selection approaches have been explored only in limited numbers for speech emotion recognition, and a comprehensive list of such works is difficult to procure/create. There are a few, however, which we proceed to mention here.

Principal component analysis (PCA) is one of the important feature selection methods used for emotion recognition [39]. Sedaaghi et al. [40] hypothesized the usage of adaptive genetic algorithms for feature selection, combined with a Bayesian classifier. Similar genetic algorithms have also been used as feature generators for the same task [41]. Rong et al. [42] introduced a new feature selection method using decision trees and random forests called Ensemble Random Forest to Trees (ERFTrees). The final features were selected using a voting strategy. Petrushin [43] created several neural network models for classifying speech into happiness, sadness, anger, neutral and fear categories; the feature selection method used was RELIEF-F [44].

4 Classical Approaches

Many solutions, frameworks and approaches have been proposed for emotion recognition from speech. However, some aspects are frequently sidelined in such works:

1. The task of delineating emotions from speech is often an ambiguous one: someone angry might speak in a high-pitched, loud voice, but someone else might express anger in a quiet and more intense voice. This often varies according to a person's mood and personality, so it cannot be said that any system can determine a person's emotion based on certain aspects of their speech very reliably.
2. A significantly large proportion of speech emotion recognition (SER) research neglects data imbalance. If one emotion has more data samples corresponding to it than any other emotion, the model(s) will overfit that emotion and fail to reliably classify the others. While this may demonstrate an apparently high overall classification accuracy, the model in reality cannot be relied upon as a diverse emotion recognition system capable of recognizing a variety of emotions. Data resampling and augmentation techniques, therefore, must also be included while building SER systems (in case the concerned databases are imbalanced, that is). In case of such imbalances, using accuracy as the relevant evaluation metric will also fail to provide clear and accurate insight. Unfortunately, accuracy has been the primary evaluation metric in almost the entirety of research in SER.

Keeping these factors in mind, speech emotion recognition research has roughly branched out into two parts: speaker-independent and speaker-dependent. We proceed to discuss the current progress in both along with a historical analysis of proposed solutions. Further, more comprehensive analysis on SER research can be found in [45–47].

4.1 Speaker-Dependent SER

The primary research in SER tends to focus on speaker-independent analysis, for which the proposed frameworks and algorithms tend to be more reliable [47]. Speaker-dependent SER takes into consideration the person speaking: that is, a single emotion can be expressed differently by each speaker, which justifies the need for speaker-dependent emotion recognition.

Some frameworks have been developed for speaker-dependent analysis. Vogt et al. [48] created EmoVoice, a speech emotion classification framework which helps users train their own classifiers for recognizing emotions, essentially making this speaker-dependent. A user can use the framework for classifying four emotions: joy, anger, satisfaction and frustration. This has been successfully integrated with many applications involving emotion recognition as well. Kang et al. [49] created a corpus consisting of 4 female and 4 male speakers covering six emotions: happiness, fear, anger, sadness, neutral state and boredom. They extracted pitch and energy as features and experimented with three algorithms: nearest neighbor, maximum-likelihood Bayes and hidden Markov model. The reported classification accuracies were 69.3%, 68.9% and 89.1%, respectively. It was also observed that happiness and anger are confused by the classifier (possibly due to the commonly-observed high intensity in both), while sadness and boredom are confused as well, also perhaps due to the typical low energy in both. This was also observed by Kostoulas and Fakotakis [50] in their study: they noted that neutral emotion and sadness were confused by the classifier, as were anger, happiness and panic. They used a decision tree classifier for their framework. Their feature set comprised of pitch, energy, 13 MFCCs, harmonicity and signal formants, and they reported a mean accuracy of 80.46%.

Cen et al. [51] performed a canonical correlation analysis (CCA) to eliminate irrelevant features to enhance emotion classification accuracy on the LDC dataset. The features obtained were based on those used for speech recognition, and a probabilistic neural network was used as the classifier. Anagnostopoulos and Vovoli [52] performed a subset evaluation and chose pitch, energy, formants and MFCCs for their framework. They trained multi-layer perceptrons with one layer on the Emo-DB dataset, and observed that the emotions that were being identified with the highest accuracies were happiness and anger, both high energy, 'high-arousal' emotions. As shown in [52], Fig. 1 demonstrates the distribution of the emotions contained in Emo-DB database according to the valence-arousal factors, as shown by Lang [53]. Arousal is the extent to which a stimulus is calming or exciting, whereas valence is

Fig. 1 Characterization of Emo-DB based on valence-arousal factors [52]

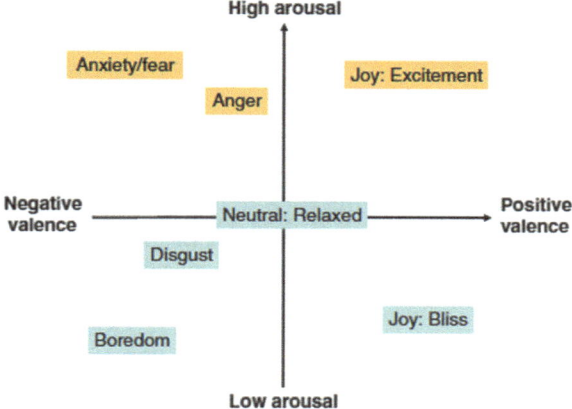

the extent to which a stimulus is negative or positive. The two together constitute the circumplex model of emotion classification, which was initially proposed by Russell [54].

4.2 Speaker-Independent SER

The majority of research in SER has focused on speaker-independent analysis, which consists of uniform analysis of emotions in speech irrespective of the nature of the person speaking. Some of the most frequently used machine-learning models for this include hidden Markov models (HMMs), support vector machines (SVMs) and Gaussian mixture models (GMMs). We discuss some models and frameworks that have been employed for SER: starting with single estimators and networks first, then going on to ensemble/multiple classifier methods.

Support Vector Machines (SVMs) SVMs [55] are based on the fact that a hyperplane (or decision boundary) such that all training examples are divided between their classes can be found. A SVM searches for the hyperplane with the largest margin, hence it is also known as a maximum margin classifier.

Consider a simple binary classification problem with N training examples, with each example consisting of a tuple $(\mathbf{x_i}, y)$. Let the class labels $y_i \in \{-1, 1\}$. Let $\mathbf{w}$ and b be the parameters of the learning problem. The learning methodology for a SVM can be written as the following optimization problem:

$$\min_{w} \frac{||w||^2}{2}$$
$$\text{subject to} \quad y_i (\mathbf{w}.x_i + b) \geq 1 \quad i = 1, 2, \ldots, N \tag{1}$$

This is a convex optimization problem since the loss function is quadratic and the constraints are linear in w and b. It can be solved using the Lagrange multiplier method. We can write the Lagrangian of this problem as follows:

$$L_P = \frac{1}{2}||w||^2 - \sum_{i=1}^{N} \lambda_i(y_i(\mathbf{w}.x_i + b) - 1) \tag{2}$$

where the parameters λ_i are called Lagrange multipliers. Solving this optimization problem will yield the following decision boundary:

$$(\sum_{i=1}^{N} \lambda_i y_i \mathbf{x_i}.\mathbf{x}) + b = 0 \tag{3}$$

b can be obtained by solving one of the constraints of the problem for the support vectors.

In case of data where separating it perfectly leads to a very small margin, it is possible to relax the boundary conditions slightly and accept some level of error to prevent over fitting. This is done by introducing slack variables, represented by ξ, into the constraints of the problem. Non-linear SVMs also work on a similar principle, except that the original coordinates $\mathbf{x}$ will be transformed into $\Phi(x)$, so they can be linearly separated. A comprehensive starting point for studying about SVMs can be found in [56].

Support vector machines have been used with immense success in diverse application areas [57–59]. SER research, in particular, has emphasized on the ability of SVMs to demonstrate excellent emotion classification performance. Chahvan et al. [60] performed gender-independent and gender-dependent analysis on the Emo-DB database and reported results using SVMs with Gaussian and polynomial kernels. They extracted two sets of features: MFCCs and Mel Energy spectrum Dynamic coefficients (MEDCs). Pan et al. [61] experimented with various feature set combinations for the Emo-DB database and a self-created Chinese database using SVM, and made an important observation: that energy factors may play an important role in Chinese speech emotion recognition. This would, perhaps, lead to the conclusion that different feature sets work on different languages—that is, language-dependent SER is an aspect that needs to be studied further. A justification for such analysis can be provided by the fact that the same emotion can be expressed differently in different languages. Similar work on the Chinese language has been conducted by Zhou et al. [62], who showed that the combination of spectral and prosodic features along with SVMs gave a high recognition rate and significantly lowered error rates compared to standard benchmarks. One of the more recent works on this is by Rajasekhar and Hota [63].

SVMs have been applied on Indian language databases as well. Ram and Ponnusamy [64] built a database for emotion recognition in Tamil speech consisting of 11 male and 7 female speakers covering anger, sad, neutral, fear and joy emotions.

They extracted MFCCs from the data, and reported results on application of SVMs to the feature set. It was observed that the emotion which was classified most easily was anger, while the least-performing one was fear. Sinith et al. [65] performed a similar analysis on a self-created Malayalam database, and obtained a classification accuracy of 95.83% with a feature set consisting of MFCCs, pitch and energy factors. Datasets and models on Telugu [66] and Hindi [67] have been introduced in recent works.

Hidden Markov Models (HMM) In hidden Markov Models (HMMs), we assume that the concerned system is a Markov process with unobserved states, i.e. the future states assumed by a system depend only on the current state, and not the past. This can be written mathematically as:

$$P[X_{n+1} = j | X_n = i, X_{n-1} = i_{n-1}, \ldots, X_0 = i_o] = P[X_{n+1} = j | X_n = i] = p_{ij} \tag{4}$$

where X_n is a stochastic process with discrete state and parameter spaces, $i, j = 0, 1, 2, \ldots$, and p_{ij} is the probability that if the system is in state i at time n, then it will next be in state j. These probabilities are called state transition probabilities, and they can be arranged in a matrix called the transition probability matrix.

Assuming the following notation for a HMM:

A = state transition probabilities (a_{ij})
B = observation probability matrix $(b_j(k))$
N = number of states in the model $\{1, 2, \ldots, N\}$ or the state at time t st
M = number of distinct observation symbols per state
$Q = \{q_0, q_1, \ldots, q_{N1}\}$ = distinct states of the Markov process
T = length of the observation sequence
$V = \{0, 1, \ldots, M1\}$ = set of possible observations
$O = (O_0, O_1, \ldots, O_{T1})$ = observation sequence
π = initial state distribution (π_i)

A discrete HMM can be stated as

$$\lambda = (A, B, \pi) \tag{5}$$

The task then comes down to three questions:

1. Evaluation: Given a sequence of observations $O = O_1 O_2 \ldots O_T$ and a model $\lambda = (A, B, \pi)$, how can $P(O|\lambda)$ be calculated?
2. Recognition: Given a sequence of observations $O = O_1 O_2 \ldots O_T$ and a model $\lambda = (A, B, \pi)$, what is the optimal hidden state sequence?
3. Learning/optimization: Given a sequence of observations $O = O_1 O_2 \ldots O_T$, what are the optimal model parameters such that $P(O|\lambda)$ can be maximized?

The solution for problem 1 can be constructed using the forward procedure:

$$P(O|\lambda) = \sum_{i=0}^{N-1} \alpha_{T-1}(i) \tag{6}$$

where α is the probability of observation and state sequence given model.

$$\alpha_t(i) = \left(\sum_{j=0}^{N-1} \alpha_{t-1}(j)\alpha_{ji} \right) b_i O_t \tag{7}$$

The solution for problem 2 can be constructed using the backward procedure:

$$\gamma_t(i) = \frac{\alpha_t(i)\beta_t(i)}{P(O|\lambda)} \tag{8}$$

where

$$\beta_t(i) = P(O_{t+1}, O_{t+2} \ldots O_{T-1}|x_t = q_i, \lambda) \tag{9}$$

There is no known way of arriving at globally optimal parameters for maximizing the probability of a certain state sequence, but methods for locally optimal solutions exist. The Baum–Welch algorithm [68] is one such popular and efficient solution.

Hidden Markov models have numerous application areas like protein sequence analysis [69], information retrieval [70] and part-of-speech tagging [71]. They are also used frequently for speech emotion recognition. Schuller et al. [72] calculated various global statistical quantities like standard deviation, maxima, minima, etc., for pitch and energy features, and trained continuous HMMs for emotion recognition. Lin and Wei [73] examined the usage of HMMs and SVMs for emotion recognition in the Danish language. They extracted 39 features for the HMM approach, and reported an accuracy of 99.5% for gender-independent analysis, 98.9% for female speakers and 100% for male speakers. Nwe et al. [32] trained four-state ergodic HMMs for each of the six emotions (Anger, Sadness, Surprise, Fear, Disgust and Joy) in their self-designed corpus. They observed that LPFCs (Log-frequency power coefficients) give better results on the model compared to MFCCs and LPCCs because they tend to better preserve information about fundamental frequency in the lower order filters— and besides, as noted above, different emotions may have similar characteristics of pitch, energy, etc.—which makes these quantities not very reliable for classifying emotions.

Many hybrid classifiers using HMMs have been proposed as well. Le and Provost [74] proposed an HMM-deep belief network hybrid architecture for classifying five emotions of the FAU Aibo dataset: anger, neutral, emphatic, rest and positive. The HMM was used for recording the temporal properties of emotion, while the deep belief network was used for calculating the emission probabilities. Huang and Ma [75] developed a conversation monitor for real-time speech emotion analysis. They modeled an HMM for the problem, and a Gaussian mixture model (GMM) for each state. The features extracted were zero crossing rate, pitch, energy and energy slope. **Gaussian Mixture Model (GMM)** A Gaussian mixture model is parameterized by two types of values, the component means and variances/covariances and the mixture component weights. For a Gaussian mixture model with K components, the kth component has a mean of μ_k and variance of σ_k for the univariate case and a mean of $\overrightarrow{\mu_k}$ and covariance matrix of Σ_k for the multivariate case. The mixture component weights are defined as ϕ_k for component C_k, with the constraint $\sum_{i=1}^{K} \phi_i = 1$ that so that the total probability distribution normalizes to 1. If the component weights are not learned, they can be viewed as an a-priori distribution over components such that p(x generated by component C_k) = ϕ_k. If they are instead learned, they are the a-posteriori estimates of the component probabilities given the data. Parameters for a GMM are estimated using the expectation–maximization (EM) algorithm.

GMMs are used in several fields for predictions and analyses [76, 77]. The motivation behind using GMMs for speech emotion recognition is their ability to model arbitrary densities, which makes them particularly successful in speaker identification [78]. Neiberg et al. [79] trained GMMs on a feature set comprising of standard MFCCs, MFCCs obtained from lower-ranged filterbanks and pitch. An initial root GMM was first trained using the EM algorithm, after which GMMs for each class were trained based on the root model using the maximum a-posteriori (MAP) criterion. Thapliyal and Amoli [80] extracted LPCs from a self-made dataset covering happy, angry, sad and neutral emotions, and used a GMM for emotion classification. They reported an overall classification rate of approximately 52–60%. Ververidis and Kotropoulos [81] performed experiments on the Danish emotional speech dataset and extracted about 87 features from pitch, energy, formants, etc., for their GMM algorithm. This work was further extended in [82] by using the sequential floating forward selection algorithm [83] to arrive at the best (and minimal) feature subset for the problem.

4.3   Other Models

The systems mentioned above are some of the more frequently used in the field of SER. Some other machine-learning models which are used for emotion recognition include:

1. K Nearest Neighbors: Classifiers that assign the label of the 'nearest neighbor' to a given sample. The nearest neighbor criterion can be calculated using various distance factors like Manhattan, Euclidean, etc. The value of K is significant to the performance of the classifier. Not much literature exists on KNN classifiers for SER; some can be found in [84, 85].
2. Trees: Trees are non-parametric learning methods that learn inference rules and patterns from the underlying data and categorize samples accordingly. Many decision tree implementations like ID3, CART, C4.5 and J48 are popularly used. Decision trees also serve as base estimators for ensembles like random forests or boosted trees. Some research on SER using trees can be found in [42, 86, 87].
3. Ensembles: Ensemble learning is the practice of combining multiple machine-learning models for arguably enhanced performances. Many techniques for creating ensembles exist, which can be loosely classified into bagging (bootstrap aggregating), boosting, voting or stacking (stacked generalization). Some works exploring such multiple classifier systems for SER can be found in [88, 89].

5 Deep Learning Approaches

Recently, a huge upsurge has been observed in the study of neural network architectures and end-to-end systems in several fields. Deep learning, in particular, has
emerged as one of the most applied learning techniques for problems ranging from
speech recognition to self-driving vehicles. The popularity of deep learning-based
solutions for numerous problem areas can be ascribed to the ability of such models
to perform feature learning on their own, instead of requiring efforts into developing
hand-crafted features.

Some of the main kinds of neural networks behind deep learning systems are
described below.

1. Deep Neural Networks (DNNs): A DNN is a type of artificial neural network
   with several hidden layers. The number of these layers can be significantly large,
   hence the term 'deep.' This depth allows learning representations and discriminative features for classification. DNNs have had extensive applications in a broad
   variety of areas, and are of significance in SER as well.
2. Convolutional Neural Networks (CNNs): CNNs are extensions of DNNs which
   work on data that comes in form of multiple arrays, especially images. Just like
   with signals which can be represented as a one-dimensional array, the input is
   convolved with filters which are also strided over it, after which they are pooled
   to reduce their dimensionality. This is done so as to capture the local statistics of
   the input and when repeated, they build up a hierarchy of features.
3. Long Short-Term Memory (LSTM): For sequential inputs, a recurrent neural network (RNN) is used. RNNs maintains a state vector which contains information
   about the sequential history of all the past elements. However, the gradients computed during RNN training tend to explode or vanish very quickly over many time
   steps, leading to poor long-term dependency capturing. Hence, to deal with long
   dependencies without underflow or overflow, long short-term memory(LSTM)
   networks are used, which have special hidden units(gates) to remember or forget
   inputs.
4. Auto-encoders: Auto-encoders are a special type of unsupervised DNNs which
   are used to reconstruct the input by first reducing it to a latent space representation
   (encoding) and then reconstructing from this latent space. Thus, this structure can
   be used not only for denoising but also dimensionality reduction.
5. Attention: An enhancement to the above methods is the concept of attention,
   which is loosely based on how humans focus on certain part of the input with
   high resolution and on all the other parts with a lower resolution. The attention
   mechanism allows us to "pay attention to" only the required parts of the input at
   each step of generating the output. Various networks, their structures and functions
   are described in detail in [90].

Out of these, the two earliest deep learning methods were DNNs [91] and DCNNs
[92], which consisting of more depth than CNN [93]. Deep learning techniques have
played the role of automatically learning high-level feature representations from raw

input data in fields of automatic speech recognition [94], classification of images [92] and object detection [95] as well.

It is, hence, a reasonable step to analyze deep learning approaches for SER. DNNs were some of the first ones to be studied; for instance, in [96], a neural network has been applied to learn high-level features from the low-level features extracted at an acoustic level for emotion classification. However, earlier works on this did not take proper feature extraction much into account, so the results tended to be unreliable.

The first study which brought about a turning point in this was [97], in which a DNN takes the features extracted at acoustic level and produces segment-level emotion state probability distributions. These features are then used to determine the class of emotion. The novelty to this model is brought about by adding a neural network having single-hidden-layer called extreme learning machine (ELM) [98] to perform emotion classification on utterance-level features. The training of ELM does not involve backpropagation of the weights. A large amount of training data is not required to train an ELM network because a considerable amount is already provided by the segment-level output.

Later works [99] have used CNNs for learning of features in case speech signals. Mao et al. [100] proposed a method to learn effective features for SER using CNNs using the fact that in a CNN, simple features are learned in the lower layers, and effective, discriminative features are obtained when depth increases. Their work brought into light the concept of two-phase learning by a CNN. First stage focuses on data which is not labeled to learn local features which are invariant with the help of a sparse auto-encoder (SAE) and the second phase involves using salient discriminative feature analysis (SDFA) as a feature extractor to learn affect-salient, discriminative features where the input to SDFA is the output from phase one. This was the first study which introduced feature learning to SER and demonstrated how CNNs can be effectively modeled to extract an optimal feature set.

Trigeorgis et al. [101] presented an end-to-end SER with a two-layer CNN and a long short-term memory (LSTM) network stacked on top of it. Bhargava and Rose [102] stated that intermediate representations learned by deep networks were not much different from hand-crafted features from speech. Sainath et al. [103] proposed a convolutional LSTM-DNN and showed that speech signals are temporally and contextually better modeled by their system in comparison with log Mel-filterbank energies. The above reasons motivate the use of an end-to-end pipeline and hence, features are obtained via convolutions and LSTMs are used for context dependency to set a common ground for comparison with other proposed architectures.

Another study which proposes using CNNs with LSTMs for SER is [104]. Till now, most of the techniques focused on extracting a good set of features and then feeding those into a dense classifier. There has been very little emphasis on capturing variation in features across time. The EmNet presented in [104] not only uses a standard feature set, but feeds this standard feature set into a CNN to extract local dependencies and then uses a global convolution layer to model higher-level features. Finally, the output of this layer is fed into an LSTM network to get a set of features which then get fed into a dense network.

Interestingly, all of the above-mentioned architectures focus on 1D frequency or time convolutions [99, 100], instead of 2D or 3D convolutions which was used in DCNN models [92]. They are also shallow models—that is, they have used one or two layers of CNNs, while the DCNN models are much more deeper. In fact, later research showed that deep multi-level networks comprising of convolutional and pooling layers performed better than the CNNs with lesser depth in field of vision [103, 104]. The reason behind such an observation is attributed to the fact that the DCNNs are able to preserve the hierarchical nature of information.

Motivated by the performance of models with more depth, [105] focuses on employing deep CNNs to develop an effective system for recognition of emotions. The technique proposed in this work used three levels of log Mel-spectrograms acquired from the one-dimensional utterances. A combination of temporal pyramid matching and optimal Lp-norm pooling for obtaining utterance-level features from segment-level features was also proposed.

In all the works discussed above, the fact that a DNN uses personalized features as input has not been taken into consideration. Hence, the models built cannot, arguably, be generalized enough because personalized features are affected by various parameters like way of speaking, the content, etc. We can define personalized features as the numerical values directly reflected in the features which embed in itself personal information, which is not an invariant quantity. Due to this reason, SER achieves promising results on speaker-dependent analysis due to the dependency of personal features.

Considering the good performance of Mel-frequency cepstral coefficients (MFCC) with deltas and double delta features in emotion recognition systems [106], it can be said that calculated deltas and delta-deltas are capable of reflecting the change of emotion and preserving emotional information while reducing the influence of not so relevant features. Another case in point is [105], which demonstrated that deltas and double deltas could be used as input to the convolutional recurrent network. Cheng et al. [107] analyzed RNNs with attention layer reported an enhancement of 11.26 and 111.26% for Emo-DB and IEMOCAP, respectively, compared to their baseline—the DNN-ELM model [97].

Rasmus et al. [108] introduced the concept of pure unsupervised networks which preserve enough information to reconstruct the input examples in conjunction with pure supervised networks, which only preserve relevant information for classification. This type of architecture which assumes both of these properties is a semi-supervised architecture known as ladder networks. When studied in the context of speech [109] and as shown in Fig. 2, a denoising auto-encoder (DAE) in which all hidden layers are injected with noise is used. Skip connections connect the noisy encoder–decoder pair. The encoder output acts as a feature input to SVMs. The loss function is formed by summing of cross entropy costs from the encoders with the reconstruction costs from decoders.

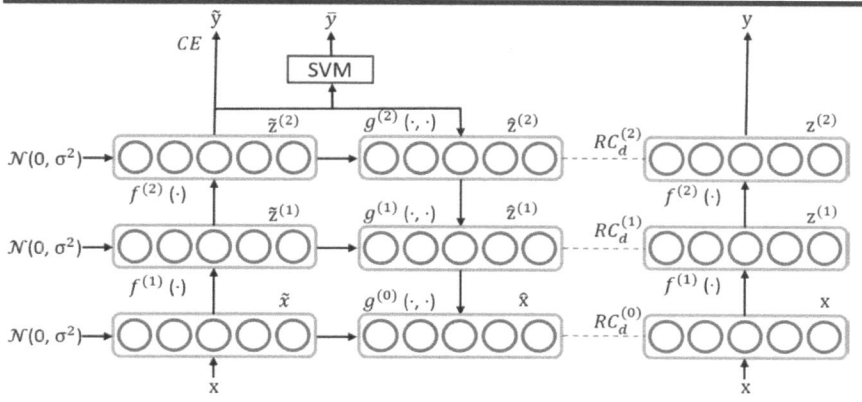

Fig. 2 Architecture of Ladder Networks [109] with noisy encoder, decoder and skip connections

6   System Overview

This section describes the classical and deep learning approaches explored for emotion recognition from speech, and the model overview and experimental setup in each case.

6.1   Classical Approach for SER

This section will cover the classical system overview—that is, the nature and quantity of features extracted and the structure and design of the model used.

Feature Extraction Keeping in mind that extraction of the right set of features is critical to the performance of a SER system, we extracted the following features:

1. Mel-frequency cepstral coefficients (MFCCs)
2. Delta and delta-delta MFCCs
3. Spectral sub-band centroids

Each audio file (signal) was first divided into frames of length 25ms each, with frame step 10ms. The above coefficients were calculated for each frame. Since the length of the audio files varies, these coefficients alone cannot give us a uniform feature vector—because the number of frames varies due to the varying audio file lengths. In order to get a proper feature vector from the above features, we calculated seven values for each audio file based on the values of each frame constituting the file: the mean, variance, maximum value, minimum value, skewness, kurtosis and inter-quartile range. These values were calculated for each audio file over all the frames and for each coefficient, which gave us a feature vector of size $(13 + 13 + 13 + 26) * 7 = 455$.

Data preprocessing The dataset was next preprocessed to make it appropriate for analysis using scaling and normalizing. Boruta [109], a wrapper-based all-relevant feature selection method was applied on the data in order to reduce the size of the feature vector.

Model description We use a bagging ensemble method as our model for the data. Bagging, short for bootstrap aggregating, consists of training samples (drawn at random, hence called bootstrap samples) fed into the various base estimators of the ensemble, then combining and deciding on the final predictions by using a majority voting rule.

Our base estimator was a support vector machine with a Gaussian kernel, penalty term 100 and kernel coefficient 0.1. We combined 20 of these in a bagging ensemble, and prepared it so samples were drawn from the training set as subsets of the feature set as well as the training examples. This took care of the correlation factor that could arise when similar estimators are trained on samples drawn with replacement.

Model training We trained and evaluated on the dataset using tenfold cross-validation, with accuracy chosen as the cross-validation metric (as the datasets are now balanced). A small holdout set (10%) was kept aside untouched, and the cross-validation procedure was performed on the remaining data. That is, the training data comprised of 90% of the dataset, and the test data comprised of the remaining 10%. The model was trained in the one versus rest fashion.

6.2 Deep Learning Approaches for SER

Adieu Features? End-to-end Speech Emotion Recognition Using A Deep Convolutional Recurrent Network With deep learning approaches, there has always been an attempt to automate the feature engineering part of the pipeline—that is, to go for end-to-end approaches which first receive the raw input signal. The network then learns the intermediate representations which not only eliminates the need to handcraft features, but also sometime lead to improved performance. This has been evident in [111], where the authors used a restricted Boltzmann machine (RBM) to get intermediate representations which resembled the bandpass behavior in the inner ear.

As shown in Fig. 3, 1D convolutions are performed on discrete-time waveforms $h(k)$ with 40 finite impulse filters with a window size of 5ms. Since convolutions increase the dimensionality of the output vectors, ax pooling across time which

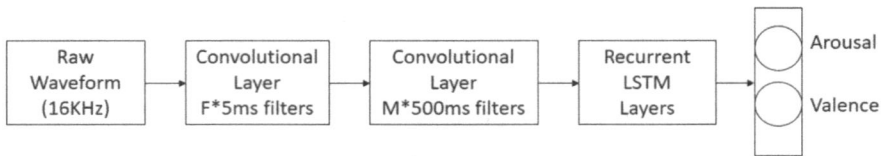

Fig. 3 Space time impulse filter convolutions on raw signals with recurrent LSTM layers

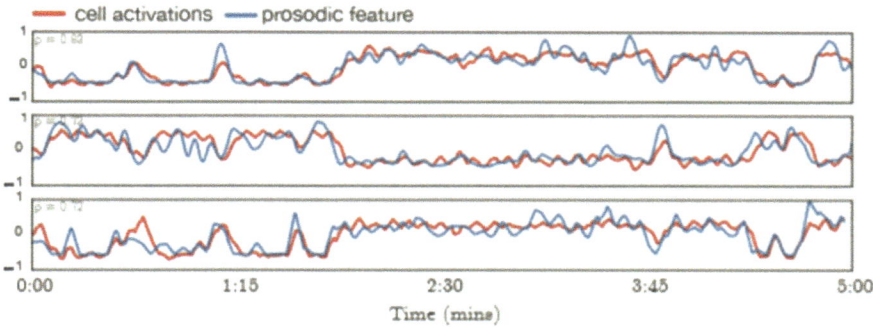

Fig. 4 Some cell activations closely resemble the variations in prosodic features [101]

resembles human cochlear transduction is done with a pool size of 2. Another convolution with 40 finite impulse filters is done with a 500ms window size for capturing
global variations and "roughness" of the signal. This is followed by pooling with a
pool size of 20. Next, this output is fed into a bidirectional LSTM network with
two layers and 128 cells each to model temporal dependencies and to compare with
existing approaches which use hand-crafted features with LSTMs.

The loss function makes use of a concordance correlation coefficient [112], for
evaluating reproducibility. In our case, we maximize this coefficient which in turn
minimizes out cost function. To train, the last-layer weights are backpropagated with
respect to L_c as shown in Eq. 10.

$$L_c = 1 - \rho_c = 1 - \frac{2\sigma_{xy}^2}{\sigma_x^2 + \sigma_y^2 + (\mu_x - \mu_y)^2} = 1 - 2\sigma_{xy}^2 \Psi^{-1} \qquad (10)$$

where $\Psi = \rho_x^2 + \rho_y^2 + (\mu_x - \mu_y)^2$, $\mu_x = E(x)$, $\mu_y = E(y)$, $\sigma_y^2 = var(y)$, $\sigma_x^2 = var(x)$, $\sigma_{xy}^2 = cov(x, y)$. Backpropagation training occurs by computing the derivative as shown in Eq. 11.

$$\frac{\partial L_c}{\partial x} \propto 2\frac{\sigma_{xy}^2(x - \mu_y)}{\Psi^2} + \frac{\mu_y - y}{\Psi} \qquad (11)$$

Results for this network were obtained on the RECOLA database [113], which is
a collection of 46 French speakers, each contributing 5 min worth of audio. On raw
signals, the network obtained a fairly good accuracy of 68.6% on arousal and 26.1%
on valence. Further, Fig. 4 shows that some cells closely resemble the variation of
acoustic and prosodic features which are strong indicators of affective states.

**3D Convolutional Recurrent Neural Networks With Attention Model for Speech
Emotion Recognition** In this study, an emphasis on two factors was observed: reducing the influence of dependencies on personal features which were dependent on
the speakers, and the removal of silent and irrelevant frames. In order to resolve the

second issue of the silent and/or emotionally irrelevant frames, the usage of attention mechanism was proposed. This comes from the fact that the attention mechanism has had considerable success in recognizing features suited for specific works. In addition, RNNs with attention layer learn correlation between the input and output sequences structures [114]. RNNs with attention layer are able to model the SER tasks well due to the following reasons:

- Speech is a type of sequential data having varied sequence lengths.
- Mostly, emotion labels are available at the level of the different utterances but these utterance-level segments contain a lot of silent frames. There are also cases where the inherent emotion is only associated with a few words. Hence, selection of emotionally relevant frames is an important step while modeling SER. Several studies [115, 116] have showed the success of attention layer in identifying the regions containing information related to emotion in order to achieve promising results on modeling utterance-level features.

The authors of this study have proposed a new 3D CRNN model which captures the relationship between time and frequency of the log-Mels. Chan et al. [117] discovered that 2D convolutions perform better than 1D convolutions when the amount of data is less. Also time-domain convolutions are equally important as frequency-domain convolutions. The researchers have hence used 3D convolutions since 3D convolutions can capture relevant information with better outcomes for SER as compared to 2D convolutions. The training is performed in the following way: First, CNN constituting three dimensions are modeled on the log-Mels. Then, the features of CNNs are given to LSTMs for obtaining temporal characteristics. Next step involves use of attention layer to produce utterance-level features. Finally, connected by fully connected layer.

The LSTM output at time step t, $h_t = [h_{tf}, h_{tb}]$, is passed through a softmax operation which computes normalized weights α_t using 12

$$\alpha_t = \frac{exp(W.h_t)}{\sum_{t=1}^{T} exp(W.h_t)} \tag{12}$$

The last step involves calculation of utterance-level features c using 13

$$c = \sum_{t=1}^{T} \alpha_t h_t \tag{13}$$

Speech Emotion Recognition Using Deep Convolutional Neural Network and Discriminant Temporal Pyramid Matching DCNNs had achieved great success in computer vision and text analysis-related tasks [92, 93] which prompted researchers to analyze its performance on SER. When the researchers tried to model DCNNs for SER, a few issues came up. Firstly, a proper representation of the speech signal was to be designed which could be fed as input to the DCNN. Previous works [99, 100] had used 1D convolutions in CNNs with 1D speech signal as input. However, as theorized above, 2D convolutions have more number of parameters and hence are capable of capturing the details of the correlations between temporal and frequency. Thus, there was a need to convert 1D features to 2D features that could be fed as input to the DCNN. The second issue that arose was the limited number of training examples in the existing datasets. Apart from this, another issue was that the speech signals had different durations but DCNN required an input of fixed size. Therefore, a feature pooling had to be designed which could formulate features at a global level called the utterance-level feature using the output of DCNNs. The training involved the following steps:

1. Generation of DCNN Input: The researchers found that 1D convolution operation done taking into account only the frequency axis was not able to capture sufficient temporal information required for recognizing the class of emotion. They also discovered that the frame length used for speech recognition, i.e. about 165 ms was not sufficient enough to differentiate between the emotions. References [119, 120] show that 250 ms is the least amount of length required of a speech signal to provide information for identifying the class of emotion. In order to resolve the above issues, overlapping Mel-spectrogram segments (Mel_{SS}) were extracted from the one-dimensional signals and were given input to the DCNN

$$Mel_{SS} \in \mathcal{R}^{F*C*T} \tag{14}$$

where F represents Mel-filter banks number, T shows the length of the segment in accordance to the frame number in a context window and C is indicative of channels of a spectrogram.
   The important point to be noted here is that $C = 1$ indicates the original form of the spectrogram, and $C = 3$ comprises of the normal, delta and double delta coefficients of a spectrogram.

2. DCNN Architecture and Training: The DCNN model (Fig. 5) includes five convolutional layers, the first three have a max-pooling layer after them, followed by two fully connected layers. For the model training, stochastic gradient descent (SGD) is employed using the following hyper-parameters batch size, learning rate, the momentum value and the decay rate for the weight. In this case, the weight w_i is updated by

$$v_{i+1} = 0.9.v_i.\eta.w_i - \eta.\langle \frac{\partial L}{\partial w} | w_i \rangle_{D_i}$$
$$w_{i+1} = w_i + v_{i+1} \tag{15}$$

More details of DCNN's training can be found in [89].
While training the DCNN, the network is first initialized with the parameters of AlexNet and then fine-tuned.
Given N overlapping segments of a spectogram which are inputs to the DCNN model, we get in result a feature representation at segment level $\in R^{dN}$. These

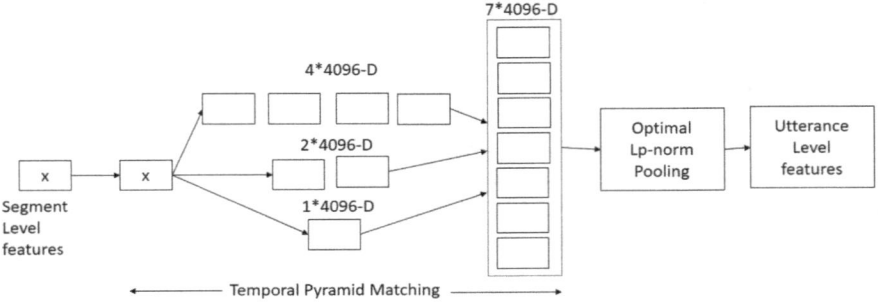

Fig. 5 The framework of discriminant temporal pyramid matching (DTPM)

features are then used as the input to DTPM algorithm which provides features at a global level called utterance-level features for identifying class of emotions.

3. DTPM for Utterance-Level Representations: A feature pooling is introduced wherein we need to convert the low-level features to the high-level features keeping in the mind that the dimensions should be same. Since the duration of speech utterances is not fixed, a segment-level feature set can have a varied number of segments. This process has been used to convert low-level features to higher-level features.

The DTPM approach proposed in the work has found inspiration from Spatial Pyramid Matching (SPM) [121] Finally, the concatenated feature contains integrated temporal clues at different scales. The formulation of feature pooling is

$$f^p(X_m) = (\frac{1}{n} \sum_{j=1}^{n} |x_j|^p)^{\frac{1}{p}} \tag{16}$$

where $f^p(X)$ denotes the acquired features after pooling operation, N denotes the segment features number and p controls the pooling strategy. It was discovered in [52], [53] indicative of the fact that p value has an important role to play in image classification accuracy.

Emotion Recognition from Human Speech Using Temporal Information and Deep Learning EmNet, the architecture proposed in this study aims to capture temporal information while using hand-crafted features as their input. There exists several feature extraction techniques and feature parameter sets like the geneva minimalistic acoustic parameter set (GeMAPS) [118]. One of the main problem with all of these approaches is that these extraction methods do not model the variations of the features over time, whose necessity has been supported by psychological studies [122]. Methods like computing standard deviations and explicitly appending temporal variations have their limitations in modeling a true temporal system.

DNNs has seen a big success in extracting higher-level features and have also shown good performance [100], but have failed to use the expert features and also need a lot of training data.

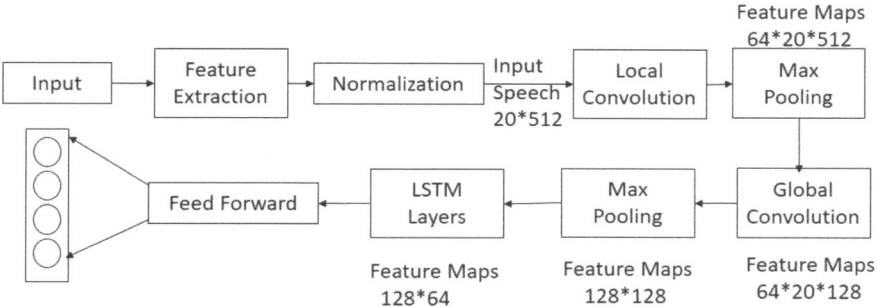

Fig. 6 EMNET architecture

The EmNet architecture is shown in Fig. 6. Based on their usefulness of their temporal variations, 20 of the 88 features in GeMAPS feature set are used. The feature vectors are then normalized and are then zero padded along the time axis to make their dimensions as 20×512.

These features then go as an input to a convolutional neural network (CNN) with local convolutions across the time axis, with 64 filters and ReLU activation. The output of these local convolutions, after max pooling, are fed to a global convolution layer with 128 filters to extract higher-level features. Finally, this output is fed into an LSTM network, two layers deep with 48 cells each.

This architecture was tested on EMO-DB dataset [123], by leaving one speaker out for validation and ADAM optimizer [124]. Out of the 98 trials of the number of free parameters in the EmNet, 11 surpassed the SOTA performance of 86% of the ComParE challenge, INTERSPEECH 2013. A t-SNE visualization of the LSTM output vector shows a close clustering between arousal emotions (happiness, anger and fear), while difficulty in separation for valence states.

6.3 Critical Comparision

Classical approaches depend a lot on two things: Feature extraction method and the underlying assumptions while choosing the correct algorithm type (non-linear kernel vs a linear kernel for SVM-based models). While in deep learning too we assume about the underlying architecture, we still keep the feature extraction part automated, like in CNNs. This reduces the hierarchy of assumptions we build choosing the correct learning algorithm. Here, we use ensemble model as our classical method, which would give us a good model with little data, but it is very much prone to overfitting due to the fact that ensembles try to model every possible variation in the training dataset. Deep Learning should perform better provided we have a good and large dataset with proper methods to keep overfitting in check like Batch normalization and regularization. As shown in case of Adieu features model, the prosodic variation actually becomes close to the cell activations, which further gives testament that deep learning is the way to go for an end-to-end solution.

7 Evaluation

7.1 Dataset Description

The interactive emotional dyadic motion capture (IEMOCAP) [110] dataset is a multimodal and multi-speaker corpus. The entire dataset contained various files for speech, song and other audio–visual data, out of which the speech part was chosen for our analysis. The corpus was created with several hours of recording sessions for capturing elicited emotions, and is one of the most widely used benchmark datasets for studying emotion recognition from speech.

7.2 Original Results

Adieu features? End-to-end speech emotion recognition using a deep convolutional recurrent network

Table 2 mentions the results obtained for each method. In all of the experiments, the [101] model outperforms the designed features in terms of ρ_c as explained earlier. The table shows results for baseline and proposed models on both raw signals and eGeMAPS feature set.

3D Convolutional Recurrent Neural Networks With Attention Model for Speech Emotion Recognition

Table 3 from [99] shows the comparison of author's method with the baseline model which is the state-of-the-art DNN-ELM method [98]. The result shows that

Table 2 Experimental results of SER accuracy from [101] on RECOLA dataset

Method (objective)	Feature type	Arousal result	Valence result
SVR (MSR)	eGeMAPS	.318	.169
SVR (MSR)	ComParE	.366	.180
BLSTM (MSR)	eGeMAPS	.300	.192
BLSTM (MSR)	ComParE	.132	.117
Proposed (MSR)	raw signal	.684	.249
BLSTM (Concor.)	eGeMAPS	.316	.195
BLSTM (Concor.)	ComParE	.382	.187
Proposed (Concor.)	raw signal	.686	.261

Table 3 Experimental results of SER accuracy from [107]

Method	IMEOCAP (%)	Emo-DB (%)
DNN-ELM [98]	51.24	71.56
2-D ARCNN	62.40	79.38
3-D ARCNN	64.74	82.82

author's method perform better than DNN-ELM and obtain improvement of about 13.5% in case of IEMOCAP and 11.26% in case of Emo-DB. The 3D ACRNN outperforms by 2.34% for IEMOCAP and 3.44% for Emo-DB. The reason behind effectiveness of 3D ACRNN as compared to 2D ACRNN indicates that the deltas and delta-deltas calculated retain the essential information for classifying accurately.

Speech Emotion Recognition Using Deep Convolutional Neural Network and Discriminant Temporal Pyramid Matching

Table 4 shows the results of the fine-tuned AlexNet. It can be observed that the fine-tuning procedure has significantly boosted the discriminative power of the extracted features. After using the fine-tuning technique, the performance of DCNN-DTPM has improved to 87.31%, 69.70%, 76.56%, and 44.61%, respectively, for the four datasets. We can also notice that accuracy on spontaneous BAUM-1s dataset is much lower than the other three datasets. This brings into light the fact that the spontaneous emotions can be more difficult to identify.

Emotion Recognition from Human Speech Using Temporal Information and Deep Learning

Table 5 compares the performance of the baseline model with the author's model. The baseline SVM system focusing on only 20 features achieved an accuracy of 77.3%. The author's proposed EmNet model achieved an accuracy of 88.9% with the same 20 features.

Table 4 Experimental results of SER accuracy from [105]

Method	Emo-DB (%)	RML (%)	eNTERFACE05 (%)	BAUM-1s (%)
DCNN-Average	82.65	66.17	72.80	42.26
DCNN-DPTM	87.31	69.70	76.56	44.61

Table 5 Experimental results of SER accuracy from [104]

Method	Result on Emo-DB
Baseline SVM	77.3
Proposed EmNet	88.9

Table 6 Experimental observations

Model	Test set accuracy (%)
Classical	76.13
3-D ARCNN	64.74
DCNN-DTPM	70.28
Adieu (1D Conv-LSTM)	25

7.3 Results Obtained

After doing a survey on various techniques, we have performed experiments of these architectures on a common dataset, IEMOCAP to establish a comparison between classical and deep approaches. Table 6 shows the resultant accuracies in both the approaches. 1D convolutions-LSTM architecture in [101] performed poorly as in the original experiment, it was trained on two classes (arousal and valence emotions), whereas IEMOCAP has 9 classes and very little training data. With more data, the performance should increase.

Although [101] showed us that the variations of prosodic features and cell activations match, a lot of training data is required in order to achieve this resonance, inferring from our results and the complexity of the proposed network. Given this abundant data problem is resolved, this architecture is end-to-end, which is always beneficial. Also, the current architecture uses 1D convolutions, but as shown in [107], higher-order convolutions perform much better. Chen et al. [107] brought out certain novel ideas like the usage of 3D convolutions to extract features which are able to preserve the emotional information of the audio signal. This study also introduced the concept of attention to reduce the impact of irrelevant frames. It also justified the importance of learning discriminative salient features for classification.

Zhang et al. [105] showed how transfer learning could be useful when the amount of training data is less. The authors used the AlexNet DCNN model pretrained on the large ImageNet dataset. This study also introduced the concept of discriminant pyramid matching which combines temporal pyramid matching and optimal Lp-norm pooling to extract high-level utterance features more accurate for emotion classification.

8 Comparison of Existing Approaches

In terms of classical models, the existing approaches to SER focus on estimators like SVMs and GMMs. Up till now, emotion recognition systems built using these have been prevalent in the field, but these come with their limitations. Recent approaches involving neural networks and deep learning have shown immense promise, and research in SER has been slowly shifting focus toward such deep learning-based methods—since one of the main advantages of deep learning models is that they could be engineered to learn feature representations on their own, thus being significantly more efficient than conventional classical learning-based approaches, where feature sets are required to be extracted and constructed by hand.

9 Conclusions

The chapter provides as a thorough introduction to ideas and research in speech emotion recognition, and analyzed many prevalent models for the same. An introduction to the subject is given, followed by an analysis of the various features extracted from speech. This is followed by feature selection approaches, and a study of the several classical and deep learning-based models used for SER. Finally, we provided our own experimentation, using an ensemble of classifiers with MFCC features and spectral centroids for emotion classification, and compared it with deep learning models proposed in several other works.

The classical model performed better on the IEMOCAP dataset than the deep learning models. This, however, came at an increase in computational requirements owing to the training of multiple models on subsets of the data. The trade-off of computational complexity and performance can be assessed depending on the problem size in general situations.

With a small dataset, classical models tend to perform better as evident by our results.

References

1. Sawhney, R., P. Manchanda, P. Mathur, R. Shah, and R. Singh. 2018, October. Exploring and learning suicidal ideation connotations on social media with deep learning. In *Proceedings of the 9th Workshop on Computational Approaches to Subjectivity, Sentiment and Social Media Analysis*, 167–175
2. Mathur, P., R. Shah, R. Sawhney, and D. Mahata. 2018, July. Detecting offensive tweets in Hindi-English code-switched language. In *Proceedings of the Sixth International Workshop on Natural Language Processing for Social Media*, 18–26
3. Mathur, P., R. Sawhney, M. Ayyar, and R. Shah. 2018, October. Did you offend me? Classification of offensive tweets in hinglish language. In *Proceedings of the 2nd Workshop on Abusive Language Online (ALW2)*, 138–148
4. Sawhney, R., P. Manchanda, R. Singh, and S. Aggarwal. 2018, July. A computational approach to feature extraction for identification of suicidal ideation in tweets. In *Proceedings of ACL 2018, Student Research Workshop*, 91–98
5. Mishra, R., P. Sinha, R. Sawhney, D. Mahata, P. Mathur, and R. Shah. 2019, June. SNAP-BATNET: Cascading author profiling and social network graphs for suicide ideation detection on social media. In *Proceedings of the 2019 Conference of the North American Chapter of the Association for Computational Linguistics: Student Research Workshop*

6. Chowdhury, A., R. Sawhney, P. Mathur, D. Mahata, and R. Shah. 2019, June. Speak up, Fight Back! Detection of social media disclosures of sexual harassment. In *Proceedings of the 2019 Conference of the North American Chapter of the Association for Computational Linguistics: Student Research Workshop*

7. Sawhney, R., P. Mathur, and R. Shankar. 2018, May. A firefly algorithm based wrapper-penalty feature selection method for cancer diagnosis. In *International Conference on Computational Science and Its Applications*, 438–449. Cham: Springer

8. Jain, R., R. Sawhney, and P. Mathur. 2018, March. Feature selection for cryotherapy and immunotherapy treatment methods based on gravitational search algorithm. In *2018 International Conference on Current Trends towards Converging Technologies (ICCTCT)*, 1–7. IEEE

9. Sawhney, R., and R. Jain. 2018, February. Modified Binary Dragonfly algorithm for Feature Selection in Human Papillomavirus-Mediated disease treatment. In *2018 International Conference on Communication, Computing and Internet of Things (IC3IoT)*

10. Dellaert, F., T. Polzin, and A. Waibel. 1996, October. Recognizing emotion in speech. In *Proceeding of Fourth International Conference on Spoken Language Processing. ICSLP'96*, vol. 3, 1970–1973. IEEE

11. Vogt, T., and E. Andr. 2005, July. Comparing feature sets for acted and spontaneous speech in view of automatic emotion recognition. In *2005 IEEE International Conference on Multimedia and Expo*, 474–477. IEEE

12. Petrushin, V.A. 2000. Emotion recognition in speech signal: Experimental study, development, and application

13. Banse, R., Scherer, K.R.: Acoustic profiles in vocal emotion expression. Journal of Personality and Social Psychology **70**(3), 614 (1996)

14. Rabiner, L.R., Sambur, M.R.: An algorithm for determining the endpoints of isolated utterances. Bell System Technical Journal **54**(2), 297–315 (1975)

15. Zhu, A., and Q. Luo. 2007. Study on speech emotion recognition system in E-learning. In *Human Computer Interaction, Part III, HCII*, ed. J. Jacko, 544–552. LNCS. Berlin: Springer

16. Zhang, S. 2008. Emotion recognition in Chinese natural speech by combining prosody and voice quality features. In *Advances in Neural Networks*, ed. Sun et al. 457–464. Lecture Notes in Computer Science. Berlin: Springer

17. Origlia, A., V. Galatá, and B. Ludusan. 2010. Automatic classification of emotions via global and local prosodic features on a multilingual emotional database. In *Proceeding of the 2010 Speech Prosody*. Chicago

18. Hu, H., M.X. Xu, and W. Wu. 2007, April. GMM supervector based SVM with spectral features for speech emotion recognition. In *2007 IEEE International Conference on Acoustics, Speech and Signal Processing-ICASSP'07*, vol. 4, IV-413. IEEE

19. Bitouk, D., Verma, R., Nenkova, A.: Class-level spectral features for emotion recognition. Speech Communication **52**(7–8), 613–625 (2010)

20. Atal, B.S.: Effectiveness of linear prediction characteristics of the speech wave for automatic speaker identification and verification. The Journal of the Acoustical Society of America **55**(6), 1304–1312 (1974)

21. Morrison, D., Wang, R., De Silva, L.C.: Ensemble methods for spoken emotion recognition in call-centres. Speech Communication **49**(2), 98–112 (2007)

22. Razak, A.A., R. Komiya, M. Izani, and Z. Abidin. 2005, July. Comparison between fuzzy and NN method for speech emotion recognition. In *Third International Conference on Information Technology and Applications (ICITA'05)*, vol. 1, 297–302. IEEE

23. Shrawankar, U., and M. Thakare. 2013. Techniques for feature extraction in speech recognition system: A comparative study. arXiv preprint arXiv:1305.1145

24. Gupta, H., and D. Gupta. 2016, January. LPC and LPCC method of feature extraction in Speech Recognition System. In *2016 6th International Conference-Cloud System and Big Data Engineering (Confluence)*, 498–502. IEEE

25. Mao, X., Chen, L., Zhang, B.: Mandarin speech emotion recognition based on a hybrid of HMM/ANN. International Journal of Computers **1**(4), 321–324 (2007)

26. Davis, S., Mermelstein, P.: Comparison of parametric representations for monosyllabic word recognition in continuously spoken sentences. IEEE Transactions on Acoustics, Speech, and Signal Processing **28**(4), 357–366 (1980)
27. Sato, N., Obuchi, Y.: Emotion recognition using mel-frequency cepstral coefficients. Information and Media Technologies **2**(3), 835–848 (2007)
28. Bou-Ghazale, S.E., Hansen, J.H.: A comparative study of traditional and newly proposed features for recognition of speech under stress. IEEE Transactions on Speech and Audio Processing **8**(4), 429–442 (2000)
29. Molau, S., M. Pitz, R. Schluter, and H. Ney. 2001. Computing mel-frequency cepstral coefficients on the power spectrum. In *2001 IEEE International Conference on Acoustics, Speech, and Signal Processing. Proceedings (Cat. No. 01CH37221)*, vol. 1, 73–76. IEEE
30. Wu, S., Falk, T.H., Chan, W.Y.: Automatic speech emotion recognition using modulation spectral features. Speech Communication **53**(5), 768–785 (2011)
31. Liu, G.K. 2018. Evaluating Gammatone Frequency Cepstral Coefficients with neural networks for emotion recognition from speech. arXiv preprint arXiv:1806.09010
32. Nwe, T.L., Foo, S.W., De Silva, L.C.: Speech emotion recognition using hidden Markov models. Speech Communication **41**(4), 603–623 (2003)
33. Seehapoch, T., and S. Wongthanavasu. 2013, January. Speech emotion recognition using support vector machines. In *2013 5th International Conference on Knowledge and Smart Technology (KST)*, 86–91. IEEE
34. Schuller, B., G. Rigoll, and M. Lang. 2004, May. Speech emotion recognition combining acoustic features and linguistic information in a hybrid support vector machine-belief network architecture. In *2004 IEEE International Conference on Acoustics, Speech, and Signal Processing*, vol. 1, I-577. IEEE
35. Kwon, O.W., K. Chan, J. Hao, and T.W. Lee. 2003. Emotion recognition by speech signals. In *Eighth European Conference on Speech Communication and Technology*
36. Guyon, I., Elisseeff, A.: An introduction to variable and feature selection. Journal of Machine Learning Research **3**, 1157–1182 (2003)
37. Ladha, L., Deepa, T.: Feature selection methods and algorithms. International Journal on Computer Science and Engineering **3**(5), 1787–1797 (2011)
38. Chrysostomou, K. 2009. Wrapper feature selection. In *Encyclopedia of Data Warehousing and Mining*, 2nd ed, 2103–2108. IGI Global
39. Lee, C.M., S. Narayanan, and R. Pieraccini. 2001. Recognition of negative emotions from the speech signal. In *IEEE Workshop on Automatic Speech Recognition and Understanding, 2001. ASRU'01*, 240–243. IEEE
40. Sedaaghi, M.H., C. Kotropoulos, and D. Ververidis. 2007, October. Using adaptive genetic algorithms to improve speech emotion recognition. In *2007 IEEE 9th Workshop on Multimedia Signal Processing*, 461–464. IEEE
41. Schuller, B., D. Arsi, F. Wallhoff, M. Land, and G. Rigoll. 2005. Bioanalogacoustic emotion recognition by genetic feature generation based on low-level-descriptors. In *Proceedings of the International Conference on Computer as Tool (EUROCON)*, 1292–1295
42. Rong, J., Li, G., Chen, Y.P.P.: Acoustic feature selection for automatic emotion recognition from speech. Information Processing & Management **45**(3), 315–328 (2009)
43. Petrushin, V. 1999, November. Emotion in speech: Recognition and application to call centers. In *Proceedings of Artificial Neural Networks in Engineering*, vol. 710, 22
44. Kononenko, I. 1994, April. Estimating attributes: Analysis and extensions of RELIEF. In *European Conference on Machine Learning*, 171–182. Berlin: Springer
45. El Ayadi, M., Kamel, M.S., Karray, F.: Survey on speech emotion recognition: Features, classification schemes, and databases. Pattern Recognition **44**(3), 572–587 (2011)
46. Koolagudi, S.G., Rao, K.S.: Emotion recognition from speech: A review. International Journal of Speech Technology **15**(2), 99–117 (2012)
47. Anagnostopoulos, C.N., Iliou, T., Giannoukos, I.: Features and classifiers for emotion recognition from speech: A survey from 2000 to 2011. Artificial Intelligence Review **43**(2), 155–177 (2015)

48. Vogt, T., E. Andr, and N. Bee. 2008, June. EmoVoiceA framework for online recognition of emotions from voice. In *International Tutorial and Research Workshop on Perception and Interactive Technologies for Speech-Based Systems*, 188–199. Berlin: Springer
49. Kang, B.S., C.H. Han, S.T. Lee, D.H. Youn, and C. Lee. 2000. Speaker dependent emotion recognition using speech signals
50. Kostoulas, T.P., and N. Fakotakis. 2006, July. A speaker dependent emotion recognition framework. In *Proceedings of 5th International Symposium, Communication Systems, Networks and Digital Signal Processing (CSNDSP)*, 305–309. University of Patras
51. Cen, L., W. Ser, and Z.L. Yu. 2008, December. Speech emotion recognition using canonical correlation analysis and probabilistic neural network. In *2008 Seventh International Conference on Machine Learning and Applications*, 859–862. IEEE
52. Anagnostopoulos, C.N., and E. Vovoli. 2009. Sound processing features for speaker-dependent and phrase-independent emotion recognition in Berlin Database. In *Information Systems Development*, 413–421. Boston, MA: Springer
53. Lang, P.J.: The emotion probe: Studies of motivation and attention. American Psychologist **50**(5), 372 (1995)
54. Russell, J.A.: A circumplex model of affect. Journal of Personality and Social Psychology **39**(6), 1161 (1980)
55. Cortes, C., Vapnik, V.: Support-vector networks. Machine Learning **20**(3), 273–297 (1995)
56. Wang, L. (ed.). 2005. *Support Vector Machines: Theory and Applications*, vol. 177. Springer Science & Business Media
57. Laptev, I., and B. Caputo. 2004, August. Recognizing human actions: A local SVM approach. In *Proceedings of the 17th International Conference on Pattern Recognition, 2004. ICPR 2004*, 32–36. IEEE
58. Joachims, T. 1998, April. Text categorization with support vector machines: Learning with many relevant features. In *European Conference on Machine Learning*, 137–142. Berlin: Springer
59. Furey, T.S., Cristianini, N., Duffy, N., Bednarski, D.W., Schummer, M., Haussler, D.: Support vector machine classification and validation of cancer tissue samples using microarray expression data. Bioinformatics **16**(10), 906–914 (2000)
60. Chavhan, Y., Dhore, M.L., Yesaware, P.: Speech emotion recognition using support vector machine. International Journal of Computer Applications **1**(20), 6–9 (2010)
61. Pan, Y., Shen, P., Shen, L.: Speech emotion recognition using support vector machine. International Journal of Smart Home **6**(2), 101–108 (2012)
62. Zhou, Y., Y. Sun, J. Zhang, and Y. Yan. 2009. Speech emotion recognition using both spectral and prosodic features. In *International Conference on Information Engineering and Computer Science, ICIECS, Wuhan*, December 1920, 1–4. New York: IEEE Press
63. Rajasekhar, A., and M.K. Hota. 2018, April. A study of speech, speaker and emotion recognition using mel frequency cepstrum coefficients and support vector machines. In *2018 International Conference on Communication and Signal Processing (ICCSP)*, 0114–0118. IEEE
64. Ram, C.S., and R. Ponnusamy. 2014, February. An effective automatic speech emotion recognition for Tamil language using Support Vector Machine. In *2014 International Conference on Issues and Challenges in Intelligent Computing Techniques (ICICT)*, 19–23. IEEE
65. Sinith, M.S., E. Aswathi, T.M. Deepa, C.P. Shameema, and S. Rajan. 2015, December. Emotion recognition from audio signals using Support Vector Machine. In *2015 IEEE Recent Advances in Intelligent Computational Systems (RAICS)*, 139–144. IEEE
66. Koolagudi, S.G., S. Maity, V.A. Kumar, S. Chakrabarti, and K.S. Rao. 2009, August. IITKGP-SESC: Speech database for emotion analysis. In *International Conference on Contemporary Computing*, 485–492. Berlin: Springer
67. Koolagudi, S.G., R. Reddy, J. Yadav, and K.S. Rao. 2011, February. IITKGP-SEHSC: Hindi speech corpus for emotion analysis. In *2011 International Conference on Devices and Communications (ICDeCom)*, 1–5. IEEE
68. Baum, L.E., Petrie, T., Soules, G., Weiss, N.: A maximization technique occurring in the statistical analysis of probabilistic functions of Markov chains. The Annals of Mathematical Statistics **41**(1), 164–171 (1970)

69. Sonnhammer, E.L., Von Heijne, G., Krogh, A.: July. A hidden Markov model for predicting transmembrane helices in protein sequences. Ismb **6**, 175–182 (1998)
70. Miller, D.R., Leek, T., Schwartz, R.M.: August. A hidden Markov model information retrieval system. SIGIR **99**, 214–221 (1999)
71. Kupiec, J.: Robust part-of-speech tagging using a hidden Markov model. Computer Speech & Language **6**(3), 225–242 (1992)
72. Schuller, B., G. Rigoll, and M. Lang. 2003, April. Hidden Markov model-based speech emotion recognition. In *2003 IEEE International Conference on Acoustics, Speech, and Signal Processing, 2003. Proceedings. (ICASSP'03)*, vol. 2, II-1. IEEE
73. Lin, Y.L., and G. Wei. 2005, August. Speech emotion recognition based on HMM and SVM. In *2005 International Conference on Machine Learning and Cybernetics*, vol. 8, 4898–4901. IEEE
74. Le, D., and E.M. Provost. 2013, December. Emotion recognition from spontaneous speech using hidden Markov models with deep belief networks. In *2013 IEEE Workshop on Automatic Speech Recognition and Understanding*, 216–221. IEEE
75. Huang, R., and C. Ma. 2006, August. Toward a speaker-independent real-time affect detection system. In *18th International Conference on Pattern Recognition (ICPR'06)*, vol. 1, 1204–1207. IEEE
76. Yang, M.H., and N. Ahuja. 1998, December. Gaussian mixture model for human skin color and its applications in image and video databases. In *Storage and Retrieval for Image and Video Databases VII*, vol. 3656, 458–467. International Society for Optics and Photonics
77. Huang, Y., Englehart, K.B., Hudgins, B., Chan, A.D.: A Gaussian mixture model based classification scheme for myoelectric control of powered upper limb prostheses. IEEE Transactions on Biomedical Engineering **52**(11), 1801–1811 (2005)
78. Reynolds, D.A., Rose, R.C.: Robust text-independent speaker identification using Gaussian mixture speaker models. IEEE Transactions on Speech and Audio Processing **3**(1), 72–83 (1995)
79. Neiberg, D., K. Elenius, K. Laskowski. (2006). Emotion recognition in spontaneous speech using GMMs
80. Thapliyal, N., Amoli, G.: Speech based emotion recognition with gaussian mixture model. International Journal of Advanced Research in Computer Engineering & Technology **1**(5), 2278–1323 (2012)
81. Ververidis, D., and C. Kotropoulos. 2005, May. Emotional speech classification using Gaussian mixture models. In *2005 IEEE International Symposium on Circuits and Systems*, 2871–2874. IEEE
82. Ververidis, D., and C. Kotropoulos. 2005, July. Emotional speech classification using Gaussian mixture models and the sequential floating forward selection algorithm. In *2005 IEEE International Conference on Multimedia and Expo*, 1500–1503. IEEE
83. Pudil, P., F.J. Ferri, J. Novovicova, and J. Kittler. 1994, October. Floating search methods for feature selection with non-monotonic criterion functions. In *Proceedings of the 12th IAPR International Conference on Pattern Recognition, Vol. 3-Conference C: Signal Processing (Cat. No. 94CH3440-5)*, vol. 2, 279–283. IEEE
84. Bombatkar, A., G. Bhoyar, K. Morjani, S. Gautam, and V. Gupta. 2014. Emotion recognition using Speech Processing Using k-nearest neighbor algorithm. *International Journal of Engineering Research and Applications (IJERA)*. ISSN, 2248-9622
85. Khan, M., Goskula, T., Nasiruddin, M., Quazi, R.: Comparison between K-NN and SVM method for speech emotion recognition. International Journal on Computer Science and Engineering **3**(2), 607–611 (2011)
86. Lee, C.C., Mower, E., Busso, C., Lee, S., Narayanan, S.: Emotion recognition using a hierarchical binary decision tree approach. Speech Communication **53**(9–10), 1162–1171 (2011)
87. Cichosz, J., and K. Slot. 2007. Emotion recognition in speech signal using emotion-extracting binary decision trees. In *Proceedings of Affective Computing and Intelligent Interaction*
88. Schuller, B., S. Reiter, R. Muller, M. Al-Hames, M. Lang, and G. Rigoll. 2005, July. Speaker independent speech emotion recognition by ensemble classification. In *2005 IEEE International Conference on Multimedia and Expo*, 864–867. IEEE

89. Schuller, B., M. Lang, and G. Rigoll. 2005. Robust acoustic speech emotion recognition by ensembles of classifiers. In *Tagungsband Fortschritte der Akustik-DAGA 05*, München
90. LeCun, Y., Bengio, Y., Hinton, Geoffrey: Deep learning. Nature **521**(7553), 436–444 (2015)
91. Hinton, G.E., Salakhutdinov, R.R.: Reducing the dimensionality of data with neural networks. Science **313**(5786), 504–507 (2006)
92. Krizhevsky, A., I. Sutskever, and G.E. Hinton. 2012. Imagenet classification with deep convolutional neural networks. In *Proceedings of Advances in Neural Information Processing Systems*, 1097–1105
93. LeCun, Y., Bottou, L., Bengio, Y., Haffner, P.: Gradient-based learning applied to document recognition. Proceedings of the IEEE **86**(11), 2278–2324 (1998)
94. Hinton, G., et al.: Deep neural networks for acoustic modeling in speech recognition: The shared views of four research groups. IEEE Signal Processing Magazine **29**(6), 82–97 (2012)
95. He, K., X. Zhang, S. Ren, and J. Sun. 2014. Spatial pyramid pooling in deep convolutional networks for visual recognition. In *Proceedings of the 13th European Conference on Computer Vision*, 346–3610. New York, NY, USA: Springer
96. Stuhlsatz, A., et al. 2011. Deep neural networks for acoustic emotion recognition: Raising the benchmarks. In *Proceedings of IEEE International Conference on Acoustics, Speech and Signal Processing*, 5688–5691
97. Han, K., D. Yu, and I. Tashev. 2014. Speech emotion recognition using deep neural network and extreme learning machine. In *Proceedings of Interspeech*, 223–227
98. Huang, G.-B., Zhu, Q.-Y., Siew, C.-K.: Extreme learning machine: Theory and applications. Neurocomputing **70**(1), 489–501 (2006)
99. Huang, Z., M. Dong, Q. Mao, Y. Zhan. 2014. Speech emotion recognition using CNN. In *Proceedings of ACM International Conference on Multimedia, New York, NY, USA*, 801–804
100. Mao, Q., Dong, M., Huang, Z., Zhan, Y.: Learning salient features for speech emotion recognition using convolutional neural networks. IEEE Transactions on Multimedia **16**(8), 2203–2213 (2014)
101. Trigeorgis, G., et al. 2016. Adieu features? End-to-end speech emotion recognition using a deep convolutional recurrent network. In *Proceedings of the 41st IEEE International Conference on Acoustics, Speech, and Signal Processing, Shanghai, China*, 5200–5204
102. Bhargava, M., and R. Rose. 2015. Architectures for deep neural network based acoustic models defined over windowed speech waveforms. In *Sixteenth Annual Conference of the International Speech Communication Association*
103. Sainath, T.N., R.J. Weiss, A. Senior, K.W. Wilson, and O. Vinyals. 2015. Learning the speech front-end with raw waveform CLDNNs. In *Sixteenth Annual Conference of the International Speech Communication Association*
104. Kim, J., and R.A. Saurous. 2018. Emotion recognition from human speech using temporal information and deep learning. In *Proceedings of Interspeech*, 937–940
105. Zhang, S., Zhang, S., Huang, T., Gao, W.: Speech emotion recognition using deep convolutional neural network and discriminant temporal pyramid matching. IEEE Transactions on Multimedia **20**(6), 1576–1590 (2018)
106. G. Alex, et al. 2005. Bidirectional LSTM networks for improved phoneme classification and recognition. In *International Conference on Artificial Neural Networks*, September 11–15, 799–804
107. Chen, M., He, X., Yang, J., Zhang, H.: 3-D convolutional recurrent neural networks with attention model for speech emotion recognition. IEEE Signal Processing Letters **25**(10), 1440–1444 (2018)
108. Rasmus, A., M. Berglund, M. Honkala, H. Valpola, and T. Raiko. 2015. Semi-supervised learning with ladder networks. In *Advances in Neural Information Processing Systems*, 3546–3554
109. Huang, J., Y. Li, J. Tao, Z. Lian, M. Niu, and J. Yi. 2018, May. Speech emotion recognition using semi-supervised learning with ladder networks. In *2018 First Asian Conference on Affective Computing and Intelligent Interaction (ACII Asia)*, 1–5. IEEE

110. Kursa, M.B., Rudnicki, W.R.: Feature selection with the Boruta package. Journal of Statistical Software **36**(11), 1–13 (2010)
111. Jaitly, N., and G. Hinton. 2011, May. Learning a better representation of speech soundwaves using restricted boltzmann machines. In *2011 IEEE International Conference on Acoustics, Speech and Signal Processing (ICASSP)*, 5884–5887. IEEE
112. Lawrence, I., Lin, K.: A concordance correlation coefficient to evaluate reproducibility. Biometrics **45**, 255–268 (1989)
113. Ringeval, F., A. Sonderegger, J. Sauer, and D. Lalanne. 2013, April. Introducing the RECOLA multimodal corpus of remote collaborative and affective interactions. In *2013 10th IEEE International Conference and Workshops on Automatic Face and Gesture Recognition (FG)*, 1–8. IEEE
114. Wei, H.C., and S.S. Narayanan. 2016. Attention assisted discovery of sub-utterance structure in speech emotion recognition. In *Proceedings of Interspeech*, 1387–1391
115. Huang, C.W., S.S. Narayanan. 2017. Deep convolutional recurrent neural network with attention mechanism for robust speech emotion recognition. In *Proceedings of IEEE International Conference on Multimedia and Expo, Hong Kong*, 583–588
116. Michael, N., and N.T. Vu. 2017. Attentive convolutional neural network based speech emotion recognition: A study on the impact of input features, signal length, and acted speech. In *Proceedings of Interspeech*, 1263–1267
117. Chan, W., and I. Lane. 2015. Deep convolutional neural networks for acoustic modeling in low resource languages. In *Proceedings of the IEEE International Conference on Acoustics, Speech and Signal Processing*, 2056–2060
118. Eyben, F., et al. 2016, April-June. The Geneva minimalistic acoustic parameter set (GeMAPS) for voice research and affective computing. *IEEE Transactions on Affective Computing* 7 (2): 190–202
119. Provost, E.M. 2013. Identifying salient sub-utterance emotion dynamics using flexible units and estimates of affective ow. In *Proceedings of the IEEE International Conference on Acoustics, Speech and Signal Processing*, 3682–3686
120. Wollmer, M., Kaiser, M., Eyben, F., Schuller, B., Rigoll, G.: LSTM modeling of continuous emotions in an audiovisual affect recognition framework. Image and Vision Computing **31**(2), 153–163 (2013)
121. Lazebnik, S., Schmid, C., Ponce, J.: Beyond bags of features: Spatial pyramid matching for recognizing natural scene categories. Proceedings of the IEEE Conference on Computer Vision and Pattern Recognition **2**, 2169–2178 (2006)
122. Murray, I., Arnott, J.: Toward the simulation of emotion in synthetic speech: A review of the literature on human vocal emotion. The Journal of the Acoustical Society of America **32**(2), 1097–1108 (1993)
123. Chaspari, T., D. Dimitriadis, and P. Maragos, Emotion classification of speech using modulation features. In *Proceedings of the European Signal Processing Conference (EUSIPCO)*
124. Kingma, D., and J. Ba. 2015. Adam: A method for stochastic optimization. In *Proceedings of ICLR, San Diego, USA*

Bidirectional Long Short-Term Memory-Based Spatio-Temporal in Community Question Answering

Nivid Limbasiya and Prateek Agrawal

Abstract Community-based question answering (CQA) is an online-based crowd-sourcing service that enables users to share and exchange information in the field of natural language processing. A major challenge of CQA service is to determine the high-quality answer with respect to the given question. The existing methods perform semantic matches between a single pair of a question and its relevant answer. In this paper, a Spatio-Temporal bidirectional Long Short-Term Memory (ST-BiLSTM) method is proposed to predict the semantic representation between the question–answer and answer–answer. ST-BiLSTM has two LSTM network instead of one LSTM network (i.e., forward and backward LSTM). The forward LSTM controls the spatial relationship and backward LSTM for examining the temporal interactions for accurate answer prediction. Hence, it captures both the past and future context by using two networks for accurate answer prediction based on the user query. Initially, preprocessing is carried out by name-entity recognition (NER), dependency parsing, tokenization, part of speech (POS) tagging, lemmatization, stemming, syntactic parsing, and stop word removal techniques to filter out the useless information. Then, a par2vec is applied to transform the distributed representation of question and answer into a fixed vector representation. Next, ST-BiLSTM cell learns the semantic relationship between question–answer and answer–answer to determine the relevant answer set for the given user question. The experiment performed on SemEval 2016 and Baidu Zhidao datasets shows that our proposed method outperforms than other state-of-the-art approaches.

N. Limbasiya · P. Agrawal (✉)
Lovely Professional University, Jalandhar, Punjab, India
e-mail: prateek061186@gmail.com

N. Limbasiya
e-mail: nlimbasiya24@gmail.com

N. Limbasiya
VVP Engineering College, Rajkot, Gujarat, India

P. Agrawal
University of Klagenfurt, Klagenfurt, Austria

© Springer Nature Singapore Pte Ltd. 2020
B. Agarwal et al. (eds.), *Deep Learning-Based Approaches for Sentiment Analysis*, Algorithms for Intelligent Systems,
https://doi.org/10.1007/978-981-15-1216-2_11

291

Keywords Answer quality prediction · BiLSTM · Community question answering · Deep learning · Par2vec · Spatio-Temporal

1   Introduction

With the tremendous use of the Internet, more peoples are connected to the Internet to discuss their problems, ask questions, share their opinion, and obtain some advice via social question-answering sites and social media. CQA [1, 2] is a familiar online service which is used to share and exchange information for the users need. Some of the familiar CQA sites are stack exchange [3], news sharing, Yahoo response, Baidu Zhidao, Yahoo! Answers, Quora, etc. [4, 5] This service allows users to share their opinions without any restrictions, ask questions, and suggest a solution to the problem with the entire world [6]. This comprises millions of questions and their related answers from the previous decade. For most of the questions, the users receive the answers within a short period of time, but the quality of the answers is not guaranteed. Some of the answers are really good and others of poor quality. The answer selection from the CQA aims to extract the best answer from the multiple answer pair set. The main challenge arises when establishing the semantic gap between the QA set. The users are also allowed to vote the answer to determine the best answer for the question. Here, the count of up votes and down votes helps to estimate the top score for the question posted on the service. This reflects the positive and negative attitudes toward the particular statement and the early detection of quality prediction based on its voting score. But, there exists a certain correlation between the voting of the low-quality answers toward the user's statement [7].

In the health sector, the physicians share health information and the remedies to certain diseases [8]. However, the quality answers are not fully guaranteed as it provides different solutions shared by various physicians. In additions, many irrelevant answers are added and or the physicians post answers for the advertisements of the hospital. These low-quality answers minimize the user's experience for health consumers in the HQA service. Also, some answers are not answered by the expertise in that particular area of health field [9]. These factors affect the quality of question–answer health information extraction in the CQA systems.

In CQA, the user can find out a similar question and answer for their query [10]. CQA has attracted a lot of attention to the area of NLP and information retrieval research [11]. CQA is able to tacit knowledge of answering the enormous questions posted on a daily life event and its associated incidents. Basically, CQA has two advantages in information retrieval (IR): First, it uses natural language rather than keyword as a query. Second, instead of giving the ranked list, it will give several possible answers. Some of the major challenges we are facing today in CQA are to identify the quality of answers perfectly based on the given question. The question quality also determines the answer quality estimation to the user. Some of the factors include the high-quality questions need the best answer from the service and low-quality questions determine the irrelevant answer to the task [12].

Various methods are introduced to solve these issues; in Ref. [13] a new tab method has been introduced to minimize the Matthew effect in CQA. The goal of this tab was to provide a quality answer based on the user's need and to minimize the Matthew effect, but this minimization causes a lexical gap problem. To mitigate these problems, the multi-scale matching model [14] was introduced, which solves the lexical gap problem and retrieves the reasonable answer to the user's question. Moreover, this method observes the correlation between word and n-grams (word to n-grams) for various levels of granularity. A novel framework named Coupled Semi-Supervised Mutual Reinforcement-based Label Propagation (CSMRLP) [15] was adapted to reduce the data sparsity problem, which was used to rank the quality of answer based on the given query. For the answer selection task [16], the recurrent convolutional neural network (RCNN) model could be used to integrate both CNN and LSTM. This determines the semantic relationship between QA pair in the CQA community. For this purpose, first CNN was used to learn the joint representation of QA pair and then joint representation is given to the LSTM to perform the matching quality of each answer.

KABLSTM [17] method is a knowledge-aware architecture that uses the knowledge graphs (KG) for a deeper representation of ranking the question and answer pair. For the answer classification task, HITSZ-ICRC [18] team participates in three tasks, i.e., Arabic task, English subtask A, and English subtask B. Ensemble method and hierarchical classification were the two multi-classifier approaches, which perform the tasks in SemEval 2015. To forecast the best answers for the new question, Ref. [19] uses Latent Dirichlet Allocation (LDA) with the collaborative voting mechanism. From the above analysis, we still found some limitations, so the existing techniques are inconsistent with various methods and datasets. This will encourage us to develop a new method called spatial influence and temporal interaction (ST-biLSTM), which controls and updates the spatial influence and temporal interaction. This bidirectional LSTM are used for various task such as text classification [20], sequence classification [21], sentence classification [22], and sentiment classification [23]. Initially, w e first perform preprocessing to filter out the useless information, and then the transformation of the word to word vector via pre-trained model par2vec is performed for vector representation. Then, the spatial influence and temporal interaction biLSTM (ST-biLSTM) model learns the semantic relationship between the Q&A and estimates the quality answer for the input query. Here, the ST-biLSTM model captures both the past and future context by using two LSTM networks. For the experimental purpose, we use two datasets, i.e., SemEval 2016 and Baidu Zhidao, which are more suitable for answer selection task.

The contribution of this paper is described below:

1. We developed an ST-biLSTM architecture that effectively enhances both the spatial influence between question–answer and temporal interaction between answer–answer.
2. ST-biLSTM model captures both the past and future context by using two LSTM networks (i.e., forward and backward model).

3. The experimental analysis conducted on SemEval 2016 and Baidu Zhidao datasets shows that the proposed method outperforms well in answer selection task.

The outline of this work is modeled as follows: In Sect. 2, we first reviewed the related works and its research gaps of the QA platform are mentioned. Section 3 describes the preprocessing methods and the ST-biLSTM architecture in detail. Section 4 discusses the dataset details and the performance evaluation in terms of different metrics are made and compared with existing works. Finally, we conclude the paper based on the results in Sect. 5.

2 Related Works

In question-answering platform, answer selection is one of the crucial research areas in CQA community. Prior studies for answer selection can be divided into two tasks, namely answer classification and answer ranking. Answer classification is to classify the quality of answers based on the given question but ranking aims to select the relevant answer among a large number of candidate answer based on the posted question. Recent works focused on CQA techniques are discussed as follows.

Elalfy et al. [24] presented a hybrid method to determine the best answer to the question in CQA. The answer prediction mainly depends on two features, i.e., content and non-content module. In this, three types of features were present in a content module named answers content feature, answer–answer feature, and question–answer feature. In the second module, a reputation score function is used to estimate the matching answer from the sentence set. At the final stage, both the features from both the module are combined and used for the prediction stage. Initially, preprocessing is carried out to process the question and answer to extract the tokens suitable for further processing. Then, the relationship between the question–answer features is extracted, and finally, the classifier is trained to get the classified result into two classes as best and not the best answer to the input question.

Hu et al. [25] analyzed the physician's quality answer in health question–answer (HQA) service based on multimodal deep belief network-based learning framework. In this, both the textual and non-textual models are determined to obtain the semantic representation of answers. The learning framework consists of three phases: feature learning, fusion, and supervised learning. In the first phase, the high-level features are learned from both the textual and non-textual modalities by means of three-layered deep belief network (DBN). Then, the fusion process is carried out by Gaussian RBM that determines the nonlinear relationship from different modalities. Then, the joint semantic representation is performed, and learned representation is fed as an input of RBM classifier to determine the quality of answers. This framework learns the high-level semantic features from both the features to predict the quality answer set. It overcomes the data sparsity problem, which occurred due to the short text answers.

Zhou et al. [26] introduced a novel recurrent convolutional neural network (RCNN) for the CQA answer selection issues. RCNN is a hybrid technique of both recurrent neural network (RNN) and convolutional neural network (CNN) learning model. Initially, the semantic matching patterns of question and answer are determined separately via CNN. Then, the fully connected network learns the fixed length representation of the QA pair. Then, the output is fed to RNN that estimates the correlation between the sequence of answers. This hybrid model learns the context-dependent valuable representation of QA pairs. Finally, the softmax classifier identifies the semantic relevance between the given questions and answers based on the prediction requirement in answer selection. However, the result obtained is not satisfactory due to less unlabeled data and still needs an improvement in the answer selection task in the question–answer community.

Hu et al. [27] introduced a collaborative decision CNN (CDCNN)-based deep learning framework that estimates the quality of a physician's answer in the HQA service. In the first phase, CNN and dependency sensitive (DSCNN) learn the semantic knowledge independently from the pre-trained word embedding model. Next, the learned features are fused for the mixed representation. In the third phase, joint representation is performed by fusing the semantic and temporal features by multimodel learning component, and then the quality score is estimated at the final stage. The quality score of an answer could be obtained by using factorization machines (FM) to estimate the best score of the answer. This model is used to determine the nonlinear semantic feature embedded in the data. However, not a guaranteed answer set is extracted from the HQA service.

Fang et al. [28] introduced a novel framework called heterogeneous social network learning to determine the quality of answer related to the question. In this, the random walk method is employed to tackle the data sparsity problem. Here, the question–answers and the users are displayed simultaneously to estimate the social relationship and their textual contents in the prediction stage. In addition, the graph regularized structure utilized in this network will be extended to large social networks. In the training stage, the sliding window approach is used to extract the textual content data from the question–answer set. For training the network, we use the LSTM-based network that learns the semantic embedding relationship between the content of question and answer. This method is applied for large-scale social networks and calculates the identical score between the questions and answers. However, the deep walk method performs only on the basis of the graph knowledge-based structure of social information extraction in CQA tasks.

Xiang et al. [29] introduced an attentive deep neural network (NN) architecture that learns the deterministic information for predicting the answer named A-ARC. This model consists of three variants to attain answer selection in natural language processing include CNN, attention LSTM and conditional random field (CRF). Initially, CNN first encodes the features and compresses to the fixed length vector. Then, the attentive LSTM further encodes the features and determines the dependencies of the answer set. Here, it uses the bidirectional LSTM network followed by soft attention layer that estimates the correlation of the whole sequences. The integration is carried out in linear form to obtain the global perspective of the sequences. At the final stage, the CRF estimates the predictions based on the encoded features and their label transitions. The attention networks address the information loss while handling large sequences. However, the word embedding task extends the processing features, and learning the ability of the deep learning model is low.

Liu et al. [15] proposed a Coupled Semi-Supervised Mutual Reinforcement-Based Label Propagation algorithm (CSMRLP) which predicts the quality of questions (QQ) and quality (AQ) for user information. The question function affects QQ related to user data and solves the latency in AQ, but the correlation has been neglected. Therefore, the CSMRLP method is used to handle this problem. First, the additional QQ is used in the AQ measurement is used to evaluate the correlation. The value of the correlation is measured between the lists of complete features such as the related functionalities, the answer, and the quality of the questions. Secondly, this method is used to predict the probability value of each application to obtain a high AQ. Here, the analysis process is performed in Yahoo! to list the confirmed response, which is used to detect the influence of the function on AQ. Finally, statistical analysis is used to verify the QQ survey analysis from Yahoo! From this, they extract various types of functionality that include asker, relating to the answers and functions of questions, categories. This correlation analysis is also used to consider interactions between multi-class features.

Roy et al. [11] introduced a new tabulation method to provide a good quality response to all users based on their usefulness. The three different tabulation methods are used to answer the list of questions, such as the oldest, active, and voting tabs. In the oldest tab, the answer is organized according to age; active tab, the answer is provided based on recent recurrence, and in the voting form, the answer is organized based on their total response vote obtained from afar. This whole method is used to find the best answer from a set of answers, but this whole method does not minimize the Mathew effect on the community question answering (CQA) site as a way to predict a better answer. The proposed method improves the quality of the CQA site, and the high-level answer is well recognized using the machine learning algorithm and reducing the Mathew effect.

Liu and Jansen [30] proposed a predictive model that identifies the knowledge share behavior of the user based on the non-Q&A features. In this, four different aspects are considered which include profile, posting style, language, and social behaviors. Then, we build a binary classifier that automatically distinguishes the knowledge shared from the non-share that tends to increase the response rate. This model is designed based on the individual aspect of answering the questions in community-based question-answering services. However, this model requires more time in answering strange questions.

The limitations of the existing studies calculate only the past information leads to the lower prediction rate. In addition, they use answer ranking and classification separately which affects the performance and accuracy measure in answer prediction. To overcome those complications, we use both the spatial and temporal features that effectively determine the best quality answer for the given query.

3 Methodology

For quality answer prediction, we introduce a new technique named Spatio-Temporal bidirectional Long Short-Term Memory (ST-biLSTM) which captures both the past and future dependencies. The existing LSTM method captures only the past information leads to lower performance prediction of the answer for the given query. To improve the performance, we use ST-biLSTM, which has two LSTM network instead of one LSTM network (i.e., forward and backward LSTM). In this, the forward LSTM is used to hold the question–answer influence, and the backward LSTM captures the answer–answer interactions. Initially, preprocessing is carried out to filter out the useless information. Then, par2vec is applied to transform the distributed representation of the question and answer into the fixed vector representation. Next, ST-biLSTM cell learns the semantic relationship between question–answer and answer–answer to determine the relevant answer set for the given user question.

3.1 Preprocessing Steps

In CQA, the posted question must have grammar and spelling errors. These errors can affect the performance and calculations of similarity in the field of information retrieval (IR). To solve this, we use preprocessing steps to improve the quality of data in terms of accuracy, redundancy, and consistency. Various steps have been used to preprocess the question and answers are discussed as follows:

(a) **Name-Entity Recognition**

NER [31] otherwise known as entity extraction, entity identification, and entity chunking. This algorithm takes a sentence or paragraph (a string of text) as an input and identifies related nouns indicated in the sentence or paragraph. This is a subtask of information extraction (IE) from the natural language document. This is mainly used to identify the organization, name of the person, location, quantities, expression of time, etc.

For example, Eugene sold much products from inivos in 2015
Name: Eugene
Organization: inivos
Time: 2015

(b) **Dependency Parsing**

Parsing solves the structural problem in a formal way. There are two types of parsing used in NLP: dependency parsing and phrase structure parsing. Here, we use dependency parsing which mainly focuses on the relationship between the words in the sentences.

(c) **Tokenization**

It is a process of splitting the sequence of words into pieces such as keywords, phrases, words, symbols, and other elements called tokens. A program that performs tokenization can be represented as tokenizer, lexer, or scanner.

(d) **POS-Tagging**

It is the process of labeling (tagging) the words as a proper part of speech. The part of speech comprises the noun, article, verb, adjective, pronoun, preposition, adverb, interjection, and conjunction. POS-tagging will work after the tokenization process.

- For example, Word: hung, Tag: noun
- Word: sit, Tag: verb
- Word: happy, Tag: adjective

(e) **Lemmatization/Stemming**

The objective of both lemmatization and stemming is to eliminate the inflectional forms in the search query. Lemmatization is a process of transforming word of a sentence into its dictionary form while stemming changes the word of a sentence into its non-changing portions. Several algorithms can be used in the stemming process, but in English, Porter stemmer is widely used. These two algorithms not only reduce the noise but also improve the accuracy of information retrieval.

For example, national and nationalize are lemmatized into nation.

(f) **Syntactic Parsing**

It is used to find the structural equivalence between the words in the sentence. Knowledge of syntax is useful for QA, parsing, generation, translation, and information extraction.

(g) **Stop Word Removal**

The main concern of the preprocessing is to filter out the useless data. Some of the commonly used stop words include are, a, an, in, the, etc. In the field of IR, stop word removal approach helps to eliminate/dropout the useless words or data present in NLP.

3.2 Best Answer Prediction

In the existing methods, various techniques have been introduced to solve the best answer prediction but still have some limitations in terms of prediction accuracy, performance, etc. To overcome these, ST-biLSTM is used for best answer predictions, which uses two LSTM network, i.e., forward and backward. The forward LSTM is estimating the spatial impact on examining the appropriate answer for the given question and backward LSTM for determining the temporal features within the answers.

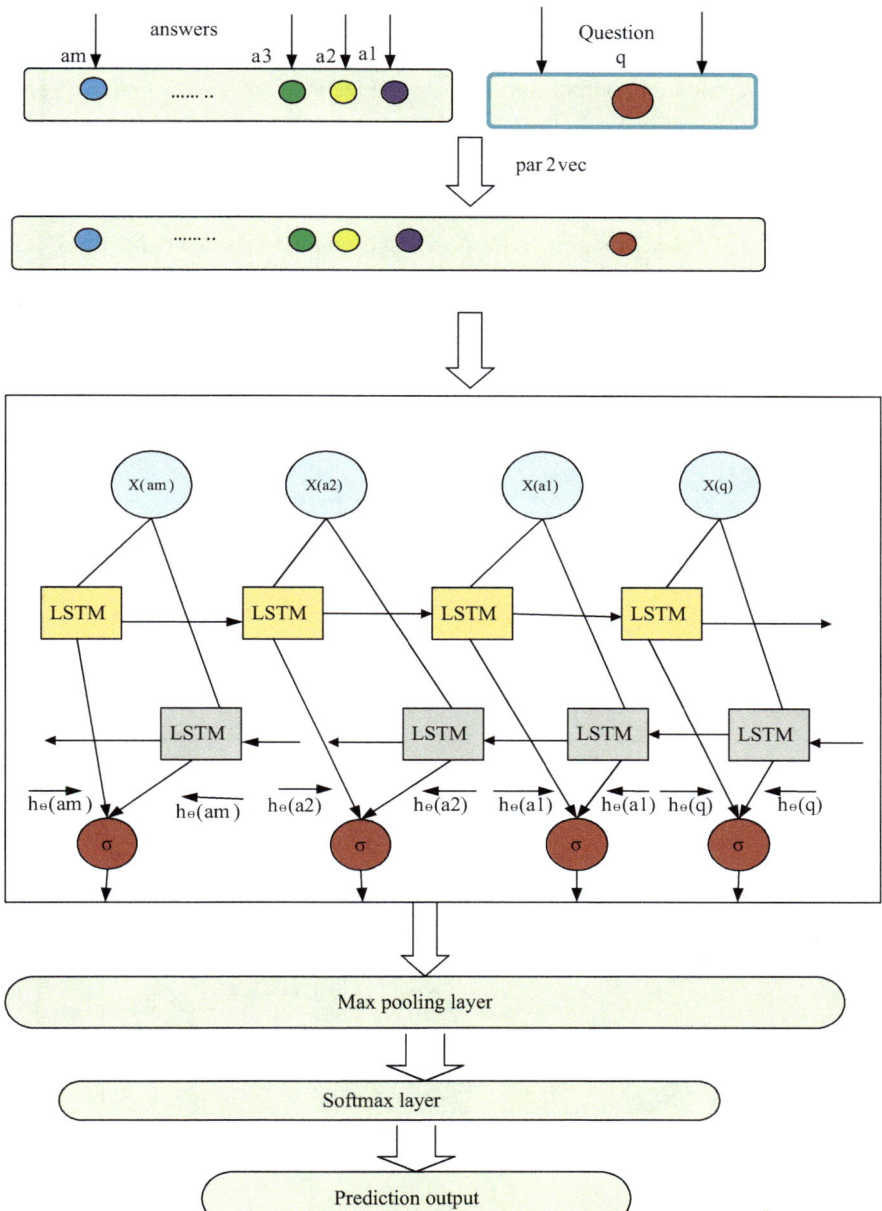

Fig. 1 General review of our proposed ST-biLSTM cell architecture

Figure 1 illustrates the architecture of our proposed ST-biLSTM technique. In this first, we read the input question q and its answer set A = [a₁, a₂, a₃ ... aₘ] to obtain the corresponding output for the answer selection task. In this first, we apply a par2vec model to transform the sentences of every question and answers into the fixed vector representation. This vector output is then fed into ST-biLSTM network for answer selection process which considers both the spatial and temporal features between the sentences. The spatial features estimate the relevant answer for the query but the temporal-based features perform the high-quality answer prediction. In the previous works, they consider either spatial or temporal features but we consider both question–answer and answer–answer interactions. ST-biLSTM is used to enhance the prediction performance and accuracy. In this, we use the maximum pooling strategy to downsample the hidden layer output and reduce the dimensionality of the parameter. Finally, the softmax layer computes a probability for each possible class, and thereby, predicts the accepted (quality) answer for the posted question.

(a) **ST-LSTM**

Figure 2 elucidates the details of ST-LSTM model. In this, $\Omega(t-1)$ is the state gate output at the time interval of $t-1$ where $h_e(1)$, $h_e(2)$, ..., $h_e(t-1)$ are the hidden layer output at $t = 1, 2, 3, ..., t-1$, respectively. In this, $d(t)$ is the input question or answer in the distributed manner, tanh and σ denote the activation functions.

$n(t)$ and $a(t)$ denote the input gate and candidate gate which change the incoming signal of the ST cell, and the output gate $u(t)$ can allow the cell to have an effect on other neurons. These gates are computed in Eqs. (1–3) as follows:

$$n(t) = \sigma(w_{nd}(t)d(t) + w_{nv}(t)v(t) + b_n) \qquad (1)$$

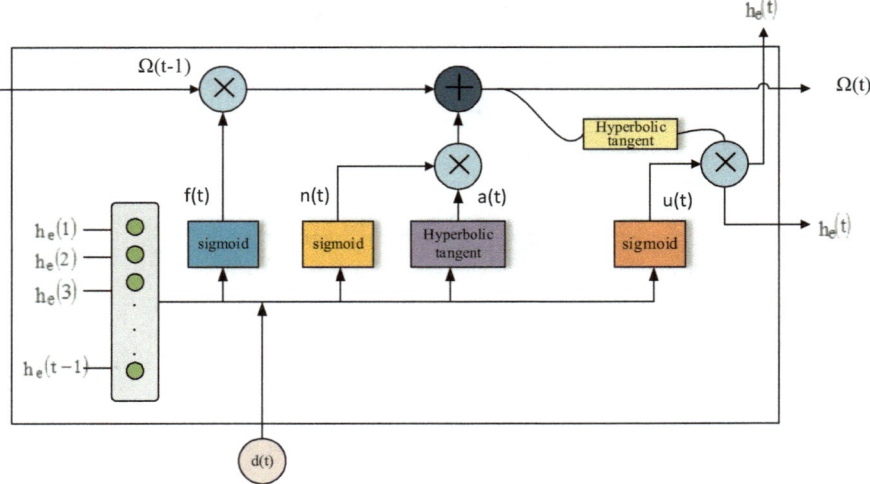

Fig. 2 Structure of ST-LSTM

$$u(t) = \sigma(w_{ud}(t)d(t) + w_{uv}(t)v(t) + b_u) \tag{2}$$

$$a(t) = \tanh[w_{ad}(t)d(t) + w_{av}(t)v(t) + b_a] \tag{3}$$

where $v(t)$ denotes the uniform vector. Moreover, forget gate $f(t)$ is used to forget or discard the useless prior information computed in Eq. (4).

$$f(t) = \sigma[w_{fd}(t)d(t) + w_{fv}(t)v(t) + b_f] \tag{4}$$

Depending on the result of the above four equations, the state gate $\Omega(t)$ and $h_e(t)$ is calculated as follows:

$$\Omega(t) = a(t) \cdot n(t) + \Omega(t-1) \cdot f(t) \tag{5}$$

$$h_e(t) = u(t) \cdot \tanh(\Omega(t)) \tag{6}$$

where σ represents the function of logistic sigmoid, W_{*d}, W_{*v} are the weight matrices, and b is the bias vector for the input gate.

(b) **ST-biLSTM**

The ST-LSTM network conserves only the past dependencies and does not consider the future analysis of the data in the sentences set. Hence, to predict the high-quality answer, we introduce ST-biLSTM that preserves both the past and future dependencies between the question–answer. Compared with ST-LSTM, the prediction and accuracy performance of ST-biLSTM are improved in predicting the high-quality answer for the given query. This is originated from bidirectional RNN, which executes or processes the sequence of data in both directions, i.e., forward and backward direction with two individual hidden layers. These two hidden layers are linked to the same output layer.

Initially, the questions and answers are given to the forward layer ST-biLSTM network. In the forward layer, it checks the answer from the training set and matches the correct answer related to the required question. Thus, high-quality answers are predicted for the input query.

Then, the backward ST-biLSTM model predicts the temporal features between the answers. In general, temporal interaction obtains the best answer from the previous answers. This layer checks the appropriate answer by considering the future and previous dependencies for the question. Thus, it predicts more subjective answers based on the integration of previous answers. Finally, the top answer is obtained by using the temporal interactions between answers–answer estimation. In the ST-biLSTM network, where σ function combines the two output answers from both layer, and the summation or a multiplication function is used to combine both outputs and obtain the best answer for the related questions. Thus, the bidirectional LSTM model is considerably better than the standard LSTM network as it considers both the past and future context features for the quality answer prediction.

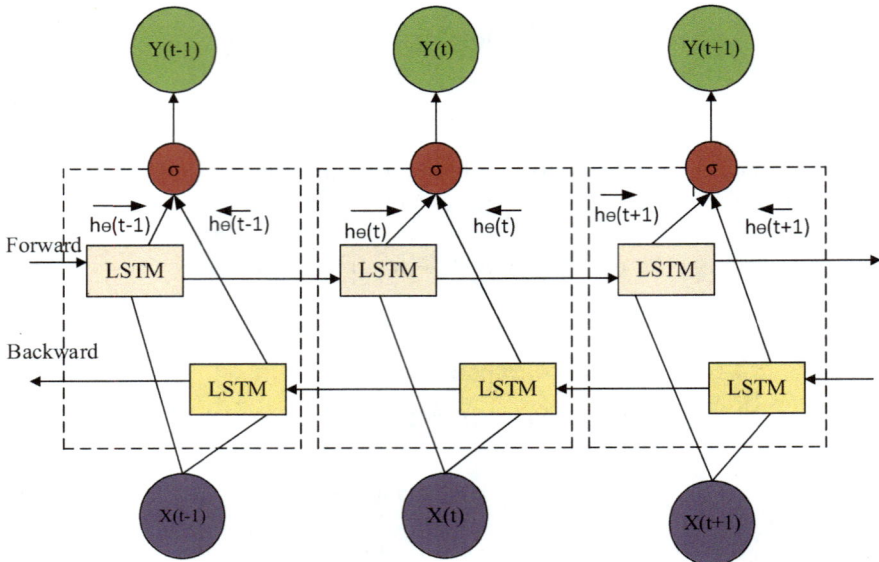

Fig. 3 Structure of ST-biLSTM cell

Figure 3 illustrates the structure of ST-biLSTM layer, which holds both the forward LSTM and backward LSTM layer. BiLSTM combines two LSTM to process both the forward and backward directions. Thus, forward and backward contexts are considered simultaneously to improve the prediction accuracy and performances in the question-answering system. The forward and backward biLSTM are calculated based on the following equations:

$$\overrightarrow{h_\theta}(t) = \text{LSTM}(\overrightarrow{h_{t-1}}, Q(w_{\text{tot}})) \tag{7}$$

$$\overleftarrow{h_\theta}(t) = \text{LSTM}(\overleftarrow{h_{t-1}}, Q(w_{\text{tot}})) \tag{8}$$

Concatenation of both hidden forwarding and backward state is represented as $h_t = \overrightarrow{h_\theta}(t); \overleftarrow{h_\theta}(t)$. Standard LSTM updating Eqs. (3–8) have been used to calculate the output of both backward and forward layers. BiLSTM layer can produce an output vector y_t, in which every element is computed by incorporating the following equation:

$$Y_t = \sigma(\overrightarrow{h_\theta}(t); \overleftarrow{h_\theta}(t)) \tag{9}$$

From Eq. (11), two output sequences are integrated using the function σ. This can be a summing function, concatenation function, multiplication function, or an average function.

Maximum Pooling

From the above equation, the output vector $Y_t = \sigma(\overrightarrow{h_\theta}(t); \overleftarrow{h_\theta}(t))$ is passed through the maximum pooling layer. This reads the information and computes the maximum pooling value. It is calculated as follows:

$$Y_{\text{max pool}} = \max(\overrightarrow{h_\theta}(t) . \overleftarrow{h_\theta}(t)) \tag{10}$$

Softmax Layer

Next, the maximum pooling output is passed to the softmax and fully connected layer for high-quality answer selection and is calculated as follows:

$$p(y = i|x) = \frac{e^{x^T \phi_i}}{\sum_{m=1}^{M} e^{x^T \phi_m}} \tag{11}$$

where ϕ_m is the weight vector of the mth class. This layer is a fully connected softmax layer, which acquires $Y_{\text{max pool}}$ as a feature and generates the predicted probability distribution. It is mainly used for a binary classification problem. This can be trained to minimize the cross-entropy cost function.

$$L_{\text{lossfn}} = \sum_{j=1}^{n} y_i \log p_j + (1 - y_j) \log(1 - p_j) + \lambda \|\Theta\|_2^2 \tag{12}$$

where p represents the softmax layer output. Θ includes all the parameter in the network, and $\gamma \|\Theta\|_2^2$ is the L2 regularization. Finally, the hidden output is used to predict the quality answer with respect to the given question.

4 Experimental Setup: Answer Classification

Datasets

We regulate our analysis on a public dataset of answer classification problem in SemEval 2016 CQA dataset. Here, the data is bifurcated into three subsets: a training, a development, and testing set. This dataset consists of 4439 questions with 21,818 answers as shown in Table 1. The answer is categorized into three classes: relevant, potential, and irrelevant. Relevant means the appropriate answers for the given question; potential defines some valuable information about one question. Irrelevant class determines the answer which is not suitable for the given question.

Table 1 Statistic for
SemEval 2016 dataset

Dataset	Question	Answer
Training	3600	17,451
Development	400	2189
Test	439	2178
All	4439	21,818

Para2vec

Par2vec is used to transform the distributive representation for every question and every answer into a fixed vector representation. We pre-train this par2vec using Wikipedia English corpus to eliminate the stop words. Here, we consider every single question/answer as a paragraph, and thus, it obtains 23,391 paragraphs from the dataset. As a result, for all input question and answer for this method contain the 300-dimensional vector.

Performance Metrics

The proposed method has evaluated three performance metrics for an answer classification task that is precision (P), recall (R), and accuracy (ACC). These three performance metrics are explained as follows:

Precision

Precision is otherwise known as positive predictive value. It is the proportion of the true positive to the summation of a true positive T_{+ive} and false positive F_{+ive} result. It is computed as follows:

$$P = \frac{T_{+ive}}{T_{+ive} + F_{+ive}} \tag{13}$$

Recall

The recall is otherwise known as sensitivity. It is the proportion of the true positive versus the true and positive classes and is calculated by Eq. (14).

$$R = \frac{T_{+ive}}{T_{+ive} + F_{-ive}} \tag{14}$$

Accuracy

Accuracy is the fraction of both true positive outcomes to the entire number of classes examined.

$$ACC = \frac{T_{+ive} + F_{-ive}}{T_{+ive} + F_{+ive} + F_{-ive} + T_{-ive}} \tag{15}$$

$F1$ Score

F-measure is a harmonic average of recall and precision, where $F1$ measure attains 1 for best value (absolute precision and recall) and 0 for worse.

$$F1_{measure} = 2 \cdot \frac{P \cdot R}{P + R} \qquad (16)$$

where P is denoted as precision, and R represents recall.

Baseline

In Ref. [16], RCNN performs lower semantic matching quality for every answer, and it is improved by using biLSTM, so we leverage the prediction performance in our proposed method. In Ref. [18], they introduced ensemble learning and hierarchical classification approach for the answer prediction task. The hybrid approach [24] with content and non-content features was used to predict the best answer with respect to the user's needs. The accuracy of the hybrid method [26] is increased but influenced by the reliability problem between Q&A. Hence, we use ST-biLSTM to improve the reliability among the question–answer pair, and the proposed method achieves a higher value in terms of F-score and accuracy than the previous methods. CDCNN [27] model does not predict the quality answer set due to the lack of semantic relationship estimation between the answer sets. In the proposed technique, we use spatial and temporal interaction to obtain quality answers. Attentive deep neural network [29] architecture for answer selection learns the deterministic information but lacks in the classification of the relevant answer for the question. Deep belief network (DBN) [25] forecasts the high-quality answer relevant to the given question. Existing methods use either answer classification or answer ranking leads to inaccurate answer prediction from the sentence set. But our proposed method integrates both answer classification and answer ranking, and thereby, improves the prediction performance and accuracy. The result shows that our proposed method outperforms than the other existing approaches in terms of precision, recall, accuracy, and F-score metrics and mentioned in Table 2.

Table 3 shows the $F1$ score for answer classification task, and its macro-averaged result for answer classification task is shown in Table 4. From the estimation, we observe that the deep neural architectures obtain better result because it uses the simplex features than the handcrafted features. Also, the RCNN architecture uses the joint representation between the QA which captures the richer matching features than the existing techniques. But the proposed method achieves a better result especially in relevant answer estimation because of the temporal interaction performed by our backward LSTM structure that enhances the accurate answer prediction.

Table 2 Performance measures based on precision, recall, and accuracy (high scores are boldfaced)

Model	Precision	Recall	F1 measure	Accuracy
Zhou et al. [16], R-CNN	56.41	56.16	56.14	72.32
Hou et al. [18], HITSZ	57.83	56.82	56.41	68.67
Elalfy et al. [24] Hybrid model (content + non-content features)	88.7	88.7	–	88.65
Hu et al. [25] deep belief network (multimodality)	97.50	97.7	97.8	97.8
Zhou et al. [26] recurrent convolutional neural network	59.41	58.84	58.77	–
Hu et al. [27] CDCNN	87	51.20	64.40	74.60
Xiang et al. [29] attentive deep neural network architecture	60.12	58.41	58.35	77.18
ST-biLSTM (answer classification)	**99.34**	**98.02**	**97.11**	**98.03**

Table 3 F1 score for answer classification task (high scores are highlighted)

Methods	Relevant	Potential	Irrelevant
Zhou et al. [16], R-CNN	77.31	15.22	**75.88**
Hou et al. [18], HITSZ	76.52	18.41	74.32
Zhou et al. [26] recurrent convolutional neural network	78.81	**79.58**	17.92
Xiang et al. [29] attentive deep neural network architecture	81.28	81.65	13.77
ST-biLSTM (answer classification)	**87.15**	20.11	72.22

Table 4 Macro-averaged scores for an answer classification task

For answer classification task	
Macro-averaged result	
Precision	99.34
Recall	98.02
F1-measure	97.11
Accuracy	98.03
F1 score	
Relevant	87.15
Potential	20.11
Irrelevant	72.22

5 Experiment II: Answer Ranking

Dataset
The dataset we use for answer ranking is Baidu Zhidao, and its statistical details are mentioned in Table 5. This dataset is subdivided into training, development, and testing set. It comprises 323,092 answers and 114,437 questions. Here, each answer has a score value assumed based on the thump up and down count. The high score defines the high-quality answers that can cover the user's information needs. SemEval 2017 dataset can also be used for ranking purpose, but it has lower QA data than the Baidu Zhidao dataset.

Para2vec
Distributive characterization for every question and every answer is initialized by Baidu Zhidao dataset. Setup for this dataset is similar to that of the previous experiment. In this, pre-train the par2vec using a Wikipedia Chinese corpus to filter out the stop words which contain 300-dimensional vector.

Baseline
For the ranking purpose, we perform a comparison of the proposed method with familiar supervised and unsupervised techniques. The supervised methods achieve a better result when sufficient training samples are available and matching scores in the QA task. On the other hand, the unsupervised methods are more efficient to the training size variance. Tables 6 and 7 represent the evaluation result of precision@1 and accuracy measures. Para2vec [26] is an unsupervised algorithm, which uses the LSTM network that encodes the distributed representation of question and answers in a low-dimensional feature space but its performance and semantic roles in matching are poor. To overcome certain disadvantages, we use a bidirectional LSTM network. On comparing with RNN and LSTM network, the MemNN model performs quite well, but it has some semantic matching problem. To avoid this, we use both spatial and temporal interactions in the biLSTM architecture. Also, the RCNN and T-LSTM show deteriorative performance compared with the proposed method, because it utilizes only the temporal features which cause low prediction. Hence, we use both spatial influence and temporal interactions in ST-biLSTM model that enhances the prediction accuracy in answer ranking,

Table 5 Statistic for Baidu Zhidao dataset

Dataset	Question	Answer
Training	100,398	308,725
Development	7000	7189
Test	7039	7178
All	114,437	323,092

Table 6 Comparison of training samples at different proportions on precision@1 measure

Training	25%	50%	75%	100%
Zhou et al. [16], Recurrent–CNN	29.57	38.28	48.63	60
Le et al. [32], Para2vec	31.44	33.61	34.59	35.36
T-LSTM	39.77	39.54	50.12	62.94
Weston et al. [33], MemNN	32.69	40.4	58	70.43
ST-LSTM	32.12	45.67	58.76	72.01

Table 7 Comparison on accuracy with training samples at different proportions

Training	25%	50%	75%	100%
Zhou et al. [16], Recurrent–CNN	40.62	50.63	61.27	69.60
Le et al. [32], Para2vec	44/73	46.58	47.17	47.61
T-LSTM	39.76	51	64.52	73.52
Weston MemNN [33]	44.06	55.19	71	78.90
ST-LSTM	51.01	56.12	69.12	79.33

6 Conclusion

In this work, we present a deep learning-based Spatio-Temporal biLSTM network (ST-biLSTM), which effectively enhances both the answer selection and answer ranking task. In particular, ST-biLSTM is used to capture both the previous and future context by using two LSTM networks. The forward LSTM controls the spatial relationship and backward LSTM for examining the temporal interactions for accurate answer prediction. The experiment conducted using two datasets shows that ST-biLSTM effectively determines the best quality of the answer from the answer sequence and performs better than the existing methods in the quality answer selection task.

References

1. Wang, X., Huang, C., Yao, L., Benatallah, B., Dong, M.: A survey on expert recommendation in community question answering. Journal of Computer Science and Technology **33**(4), 625–653 (2018)
2. Zhao, Z., Zhang, L., He, X., Ng, W.: Expert finding for question answering via graph regularized matrix completion. IEEE Transactions on Knowledge and Data Engineering **27**(4), 993–1004 (2015)
3. Lal, S., D. Correa, and A. Sureka. 2014. Migrated question prediction on StackExchange. In *2014 21st Asia-Pacific Software Engineering Conference*, Jeju, pp. 35–38
4. Sahu, T.P., Nagwani, N.K., Verma, S.: Selecting best answer: an empirical analysis on community question answering sites. IEEE Access **4**, 4797–4808 (2016)

5. Toba, H., Ming, Z.Y., Adriani, M., Chua, T.S.: Discovering high quality answers in community question answering archives using a hierarchy of classifiers. Information Sciences **10**(261), 101–115 (2014)
6. Palanisamy, R., Foshay, N.: Impact of user's internal flexibility and participation on usage and information systems flexibility. Global Journal of Flexible Systems Management **14**(4), 195–209 (2013)
7. Yao, Y., Tong, H., Xie, T., Akoglu, L., Xu, F., Lu, J.: Detecting high-quality posts in community question answering sites. Information Sciences **302**, 70–82 (2015)
8. Beloborodov, A., P. Braslavski, and M. Driker. 2014. Towards automatic evaluation of health-related CQA data. In *International Conference of the Cross-Language Evaluation Forum for European Languages 2014 Sept 15*, pp. 7–18. Cham: Springer
9. Liu, F., Antieau, L.D., Yu, H.: Toward automated consumer question answering: automatically separating consumer questions from professional questions in the healthcare domain. Journal of Biomedical Informatics **44**(6), 1032–1038 (2011)
10. Fensel, D.: Ontology-based knowledge management. Computer **35**(11), 56–59 (2002)
11. Roy, P.K., Ahmad, Z., Singh, J.P., Alryalat, M.A., Rana, N.P., Dwivedi, Y.K.: Finding and ranking high-quality answers in community question answering sites. Global Journal of Flexible Systems Management **19**(1), 53–68 (2018)
12. Agichtein, E., C. Castillo, D. Donato, A. Gionis, G. Mishne. (2008). Finding high-quality content in social media. In *Proceedings of the 2008 International Conference on Web Search and Data Mining*, ACM, pp 183–194
13. Molino, P., Aiello, L.M., Lops, P.: Social question answering: textual, user, and network features for best answer prediction. ACM Transactions on Information Systems (TOIS) **35**(1), 4 (2016)
14. Yang, X., M. Wang, W. Wang, M. Khabsa, and A. Awadallah. (2018). Adversarial Training for Community Question Answer Selection Based on Multi-scale Matching. arXiv preprint arXiv: 1804.08058
15. Liu, J., Shen, H., Yu, L.: Question quality analysis and prediction in community question answering services with coupled mutual reinforcement. IEEE Transactions on Services Computing **10**(2), 286–301 (2017)
16. Zhou, X., B. Hu, Q. Chen, B. Tang, and X. Wang. Answer Sequence Learning with Neural Networks for Answer Selection in Community Question Answering. arXiv preprint arXiv: 1506.06490.201
17. Shen, Y., Y. Deng, M. Yang, Y. Li, N. Du, W. Fan, K. Lei. 2018. Knowledge-aware attentive neural network for ranking question answer pairs. In *The 41st International ACM SIGIR Conference on Research & Development in Information Retrieval ACM*, 901–904
18. Hou, Y., C. Tan, X. Wang, Y. Zhang, J. Xu, Q. Chen. 2015. HITSZ-ICRC: exploiting classification approach for answer selection in community question answering. In *Proceedings of the 9th International Workshop on Semantic Evaluation (SemEval 2015)*, pp. 196–202
19. Dong, H., J. Wang, H. Lin, B. Xu, Z. Yang. 2015. Predicting best answerers for new questions: an approach leveraging distributed representations of words in community question answering. In *2015 Ninth International Conference on Frontier of Computer Science and Technology (FCST) IEEE*, pp. 13–18
20. Xia, W., Zhu, W., Liao, B., Chen, M., Cai, L.N., Huang, L.: Novel architecture for long short-term memory used in question classification. Neurocomputing **299**, 20–31 (2018)
21. Jurgovsky, J., Granitzer, M., Ziegler, K., Calabretto, S., Portier, P.E., He-Guelton, L., Caelen, O.: Sequence classification for credit-card fraud detection. Expert Systems with Applications **100**, 234–245 (2018)
22. Ouyang, X., K. Gu and P. Zhou. 2018. Spatial pyramid pooling mechanism in 3D convolutional network for sentence-level classification. *IEEE/ACM Transactions on Audio, Speech, and Language Processing*
23. Lee, G., Jeong, J., Seo, S., Kim, C., Kang, P.: Sentiment classification with word localization based on weakly supervised learning with a convolutional neural network. Knowledge-Based Systems **152**, 70–82 (2018)

24. Elalfy, D., Gad, W., Ismail, R.: A hybrid model to predict best answers in question answering communities. Egyptian Informatics Journal **19**(1), 21–31 (2018)
25. Hu, Z., Zhang, Z., Yang, H., Chen, Q., Zuo, D.: A deep learning approach for predicting the quality of online health expert question-answering services. Journal of Biomedical Informatics **71**, 241–253 (2017)
26. Zhou, X., B. Hu, Q. Chen and X. Wang. 2018. Recurrent convolutional neural network for answer selection in community question answering. *Neurocomputing* **274**, 8–18
27. Hu, Z., Z. Zhang, H. Yang, Q. Chen, R. Zhu, and D. Zuo. Predicting the quality of online health expert question-answering services with temporal features in a deep learning framework. *Neurocomputing* **275**, 2769–2782
28. Fang, H., F. Wu, Z. Zhao, X. Duan, Y. Zhuang and M. Ester. 2016. Community-based question answering via heterogeneous social network learning. In *Thirtieth AAAI Conference on Artificial Intelligence*
29. Xiang, Y., Chen, Q., Wang, X., Qin, Y.: Answer selection in community question answering via attentive neural networks. IEEE Signal Processing Letters **24**(4), 505–509 (2017)
30. Liu, Z., Jansen, B.J.: Identifying and predicting the desire to help in social question and answering. Information Processing & Management **53**(2), 490–504 (2017)
31. Freihat, A.A., G. Bella, H. Mubarak, F. Giunchiglia. 2018. A single-model approach for Arabic segmentation, POS tagging, and named entity recognition. In *2018 2nd International Conference on IEEE Natural Language and Speech Processing (ICNLSP)*, pp. 1–8 (2018)
32. Le, Q., and T. Mikolov. 2014. Distributed representations of sentences and documents. In *International Conference on Machine Learning*, pp. 1188–1196
33. Weston, J., S. Chopra, and A. Bordes. 2014. Memory networks. CoRR, vol.abs/1410.3916

Comparing Deep Neural Networks to Traditional Models for Sentiment Analysis in Turkish Language

Savaş Yildirim

Abstract Traditional bag-of-words (BOW) draws advantage from *distributional theory* to represent document. The drawback of BOW is high dimensionality. However, this disadvantage has been solved by various dimensionality reduction techniques such as principal component analysis (PCA) or singular value decomposition (SVD). On the other hand, neural network-based approaches do not suffer from dimensionality problem. They can represent documents or words with shorter vectors. Especially, recurrent neural network (RNN) architectures have gained big attractions for short sequence representation. In this study, we compared traditional representation (BOW) with RNN-based architecture in terms of capability of solving sentiment problem. Traditional methods represent text with BOW approach and produce one-hot encoding. Further well-known linear machine learning algorithms such as logistic regression and Naive Bayes classifier could learn the decisive boundary in the data points. On the other hand, RNN-based models take text as a sequence of words and transform the sequence using hidden and recurrent states. At the end, the transformation finally represents input text with dense and short vector. On top of it, a final neural layer maps this dense and short representation to a sentiment of a list. We discussed our findings by conducting several experiments in depth. We comprehensively compared traditional representation and deep learning models by using a sentiment benchmark dataset of five different topics such as books and kitchen in Turkish language.

Keywords Bow · Deep learning · RNN · Turkis language

S. Yildirim (✉)
Department of Computer Engineering, Faculty of Engineering and Natural Science,
Istanbul Bilgi University, Istanbul, Turkey
e-mail: savas.yildirim@bilgi.edu.tr

© Springer Nature Singapore Pte Ltd. 2020   311
B. Agarwal et al. (eds.), *Deep Learning-Based Approaches for Sentiment Analysis*, Algorithms for Intelligent Systems,
https://doi.org/10.1007/978-981-15-1216-2_12

1 Introduction

Sentiment analysis is a computational process identifying an author's opinion expressed in a text to detect whether the attitude toward specific topics such as book and film is positive, negative, or neutral. It is also called opinion mining and has been the most attractive topic in the field of natural language processing [1]. Because social media gets popular and user comments are easily accessible in real-time applications, many researchers have focused on sentiment analysis. The main characteristic of the social media is that users are likely to use causal expression, which is also called texting language. Another feature is that text has limited size. These two properties make the problem harder than other NLPs. Some previous studies applied lexicon-based approaches to the problem [2] and some focused on machine learning-based methods [3]. Lexicon-based approaches rely on high quality of emotion lexicon and word polarity. Machine learning approaches rely on supervised architecture and word features. Well-known features are bag-of-words (or n-gram) and lexicon-based polarity words. Traditional machine learning algorithms such as Naive Bayes and support vector machine have been successfully applied to the problem so far. It is also considered either *binary text classification* problem if there exist two sentiments (positive and negative) or *multi-class classification* when three categories of negative, positive, and neutral exist.

For document representation, formerly the traditional studies utilized BOW approach, which is also called distributional approach to represent text. The *distributional hypothesis* relies on the idea that similar documents share similar context and words. They do not use predefined sources such as polarity dictionary or linguistics rules. The distributional approaches have applied one-hot encoding, also called *bag-of-words* (BOW). A document is represented by a vector that keeps word count in it, a.k.a term-document matrix. However, such approaches raised the problems of high dimensionality and sparsity. Recently, the dimensionality curse has been solved by various neural network approaches [4–8]. This network-based study also improved the performance of word semantics. They showed that document and words could be represented in very short and dense vectors by mean of neural layers, namely *document or word embeddings*.

In this study, we assessed the capacity of deep learning (DL) approaches for the sentiment analysis. We utilized BOW approaches and traditional machine learning algorithm as baseline function. We evaluate deep learning representations by comparing the performances in solving the sentiment problem. Most recently, deep learning approaches such as RNN and long short-term memory (LSTM) have achieved good results for a variety of NLPs. The approach transforms a sequence of words, input text, by using recurrent layer and hidden state which favors the order of words. Therefore, it is notably better for some problems such as time series, predictive maintenance, weather forecasting problem, and so forth. Text is finally represented with a short and dense numeric vector. This final transformed vector is likely to be ready to be classified by any linear classifier. We compared several RNNs such as LSTM, gated recurrent unit (GRU), bidirectional RNN. And we also utilized the deep learning optimization methods to improve the performance. Other features such as loss function and epoch size have been examined and tuned to learn better sentiment classifier. Finally, we showed that deep learning architectures outperformed other traditional BOW models for sentiment analysis in Turkish language.

2 Methodology

Neural network approaches have gained a big attention by virtue of some driving technical forces, better hardware, huge amount of dataset, and algorithmic improvement. Algorithmic advances in the machine learning became possible once when large amount of data and computational power of cheap hardware have been available. The term *deep learning* has been coined as many layers efficiently have been added to the neural architecture. In last decades, some algorithmic improvement accelerates the studies in the field; different optimization algorithms, several activation functions, and better regularization such as dropout and early stopping have significantly improved the results. The most important achievement has been done for optimization phase. Efficient optimization algorithms such as Adam and RMSProp have been designed by varying classical *gradient descent*, momentum, redefining regularization, adaptive learning factors, and so on [9–12].

Loss function J (θ), a.k.a. *objective function*, is another important factor for getting better performance. The loss measure needs to be minimized during training. Each problem requires a specific loss function to be designed. Generally speaking while binary classification application needs *binary cross-entropy* loss function, multi-class or multi-label classification problem requires *categorical cross-entropy* loss function. But it depends on the problem and varies.

Deep learning studies offer a variety of architectures. Recently, Natural Language Problems have been successfully solved by recurrent neural networks (RNNs). Contrary to bag-of-words approaches, text is processed as a sequence of words in order. As the architecture spans the text from head to tail, it updates and produces a representation of the input text. It comes up with final representation at the end. This could be mapped to a class category of multi-classes (text classification) or translated to another representation (machine translation and summarization).

The main problem of RNNs is that it suffers from vanishing gradients. As text size gets longer, long-term dependencies such as co-reference relation between words or subject–verb agreement are impossible to learn. This phenomenon is due to the problem of vanishing gradient. As it keeps adding many layers to a deep network, training network weights eventually becomes impossible. However, other variants of RNN such as LSTM and GRU are capable of keeping the gradients from vanishing problem. Thus, recent studies showed that they have been successfully applied to the problem [13, 14].

Another possible variant of RNN is use of bidirectional representation. The bidirectional model aims at improving representational capacity of the layers. RNN is explicitly time and order dependent and processes sequence from head to tail without shuffling or random access. It finally extracts representations from the sequence in one direction. A bidirectional architecture aims at exploiting other direction as well and consists of two main RNNs that could be simple RNN, LSTM or GRU. Each RNN processes the input sequence in one direction left to right (chronologically) and right to left (antichronologically), and finally then concatenates these two representations. The bidirectional RNN can catch the patterns that might be ignored by a simple unidirectional RNN.

3 Experimental Setup and Results

3.1 Dataset

We used product and movie dataset provided by a study [15]. This dataset includes movie and product reviews. The products are book, DVD, electronics, and kitchen. The movie dataset is taken from a cinema Web page (www.beyazperde.com) with 5331 positive and 5331 negative sentences. Reviews in the Web page are marked in scale from 0 to 5 by the users who made the reviews. The study considered a review sentiment positive if the rating is equal to or bigger than 4, and negative if it is less or equal to 2. They also built Turkish product review dataset from an online retailer Web page. They constructed benchmark dataset consisting of reviews regarding some products (book, DVD, etc.). Likewise, reviews are marked in the range from 1 to 5, and majority class of reviews are 5. Each category has 700 positive and 700 negative reviews in which average rating of negative reviews is 2.27 and of positive reviews is 4.5 [15].

3.2 Traditional BOW Approach

A document could be represented in many ways. The most important way is to use words as dimension, vector space models (VSMs). VSM has been found very convenient for similarity and other tasks such as classification or clustering [16]. Salton [17] represented a word or a document by using co-occurrence statistics where the column size of the representation matrix is equal to the vocabulary. While clustering algorithms use cosine similarity to measure the proximity, the classification algorithm selects the most informative words to represent the document. It is also called the bag-of-words where the bag includes the words (or token) of a document by ignoring word order. The main disadvantage of this representation is using huge number of words. The well-known solution is feature selection in the preparation step, which discards non-informative words based on corpus statistics. The study [16] addressed that the frequent words are likely to be informative. Some selectional criteria such as IG, PMI, and chi-square (χ^2) are also found very useful to find informative terms [16]. Another important step is term weighting where the count

Table 1 Performance of traditional machine learning algorithm with BOW+ TF-IDF representation

	BOOKS	DVD	ELEC	KITC	FILM
SVM	77.9	72.8	73.9	71.6	84.9
LR	**78.9**	**73.6**	75.7	**74.3**	85.5
mNB	78.8	72.9	**77.6**	**73.9**	**86.5**
RF	76.3	73.4	75.4	68.7	82.1
XGB	74.7	71.1	73.3	66.2	77.5
VOC	*1854*	*1693*	*2179*	*1919*	*5000*
P-mean	*34.3*	*33.3*	*39.3*	*35.65*	*21.7*
P-med	*29.1*	*28*	*31*	*30*	*21*
N-mean	*40.3*	*38.44*	*44.3*	*37.2*	*22.5*
N-med	*33.1*	*31*	*33.5*	*31*	*21*

Numbers in bold represents summary statistics
Numbers in italics represents topic-wise best score

values in vector are normalized over documents and words where globally high frequent terms receive less weights. The widely used scheme is term frequency-inverse document frequency (TF-IDF).

In this study, we select most frequent K words as vocabulary for each subproblem. TF-IDF scheme is used for term-document matrix weighting. Next step is to apply machine learning to solve sentiment problem. Machine learning algorithms could be easily applied to term-document matrix where each document is labeled with sentiment. We trained well-known algorithms such as support vector machine (linear kernel), logistic regression, multinominal Naive Bayes (mNB) algorithm, random forest, XGBoost algorithm. The last two algorithms are an example of bagging and ensemble learning where the base learner is decision tree (DT). We select these algorithms since they have been found very successful for the text classification problem recently [18].

We applied the models to five different topics: book, DVD, electronics, kitchen, and film. All but film categories have 700 examples for positive and negative sentiments. Only film dataset has more instances: 5331 negative and positive instances. Table 1 shows machine learning algorithm performances across each product. The table also shows review length median/mean for positive and negative classes, respectively. The table and also our findings suggest that LR and mNB are the most successful algorithms at a glance. They outperformed SVM, RF, and XGBoost. Among products, books are easily classified. Other observation is that negative interviews are longer than positive ones in size. We also observed that TF-IDF schema notably improved the results.

Table 2 DL model performance for sentiment analysis

	BOOK	DVD	ELEC	KITC	FILM
MAX-LEN	*150*	*172*	*195*	*122*	*52*
VOC SIZE	*1812*	*1665*	*2099*	*1867*	*7727*
RNN	62.1	60.17	56	67.5	82.14
LSTM	77.5	77.5	81.43	**80.71**	88.42
LSTM drop	77.1	78.01	**82.14**	78.21	**89.26**
GRU	77.86	78.21	81.79	78.93	87.11
GRU drop	**78.5**	75.77	77.8	74.64	**88.8**
Bi-LSTM	77.5	78.9	80	**80.71**	86.92
Bi-LSTM drop	77.14	**80.1**	**83.5**	77.86	**88.7**
Bi-GRU	77.86	**78.93**	**83.2**	79.64	88
Bi-GRU drop	75.71	77.45	76.07	76.43	88.33

Numbers in bold represents summary statistics
Numbers in italics represents topic-wise best score

3.3 Deep Learning Architecture

Deep learning offers a variety of architectures. Simple RNN and its variants, LSTM, and GRU are the most suitable models for NLPs. We also applied bidirectional variant of LSTM and GRU architecture. In order to cope with overfitting problem, we applied dropout mechanism. Table 2 shows performance of DL models across product sentiments. The first row (MAX-LEN) represents the maximum number of words appeared in reviews. We keep this maximum number, max, to fit deep learning model so that each review is treated the way that its size is max. For example, each review in book category is represented as if its text size is 150. Shorter text is filled with dummy words since this process is necessary to standardize the input. Words need a short embedding representation to be learned. This is done by embedding layer of the models. In Keras library, we use embedding class to learn word embeddings. When training the model, it learns word embeddings by means of back-propagation. The embedding size is experimentally kept 64 for all setup. VOC SIZE is the number of unique words appeared in the documents once some low noisy words have been eliminated.

At a glance, LSTM and GRU are better than simple RNN. The bidirectional representation slightly outperformed than one directional design. The results show that even though dropout regularization does not apparently improve the result, it helps to reduce the variance of the model. These two tables show that DL architecture gets better results than traditional approaches. For example, DL models get 89.26 accuracy at their best, and the traditional models, however, can get 86.5 when evaluating film category. Likewise, for DVD categories the success rates are 80.1 and 73.4 at their best accordingly.

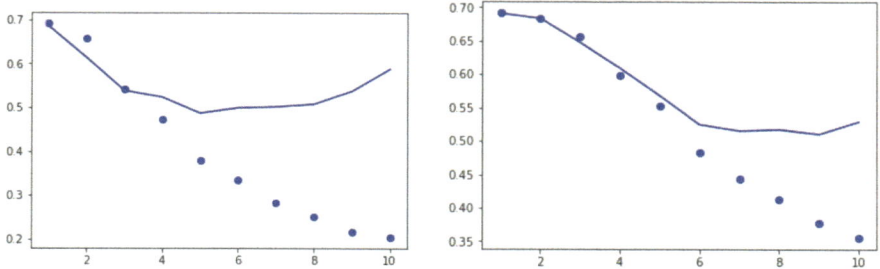

Fig. 1 Training and validation loss for LSTM and LSTM (+dropout)

One of the important problems for machine learning algorithm is overfitting. When the model keeps fitting the training data, it might not correctly classify the test data points. The overfitting problems can be solved by several regularization procedures. Figure 1 shows that dropout mechanism can cope with overfitting problem. It shows the loss of training set (solid line) and validation set (dotted line) over epoch size up to 10. The left figure is the output of normal LSTM design without dropout, and the right plot is regarding LSTM with dropout where drop rate is 50%. Dropout mechanism notably shows that the area between training and validation loss is smaller than the other. This leads to a less variance model and better classifier.

We also evaluated the performance of various optimization algorithms: stochastic gradient descent (SGD), NAG, Nadam, AdaGrad, RMSProp, and Adam optimizers. These optimization variants have successfully improved the results of many problems [9, 12, 19, 20]. We observed that RMSProp and Adam are slightly better than Nadam and AdaGrad. On the other hand, we found that simple SGD and NAG hardly converge the training data and require more time for training. They demand much more epoch step to fit the data, which make the fitting process slower. The most important issue in deep learning is to tune hyperparameters such as layer size, word embedding size, learning rate, the number of hidden unit, and so forth. Our experimental results showed that the optimum number of epoch was found 10, which maximizes validation accuracy. The word embedding size is kept 64, which helps to represent words in text sequence. We observe that as embedding size increases, the performance does not improve. We used binary cross-entropy loss function due to binary classification. Besides, we mostly applied default settings of other hyperparameters defined by Keras library. Our architectures have been implemented using *TensorFlow API*[1] *and Keras wrapper library.*[2] For example, weight initialization scheme was *random uniform* that generates unit weights with a uniform distribution within the range (−0.05 and 0.05). The validation set split is set to 0.2. The batch size during training phase is set to 128.

---

[1] https://www.tensorflow.org.

[2] www.keras.io.

4   Conclusion

Traditional BOW approach has dimensionality curse problem. On the other hand, neural network-based approaches do not suffer from dimensionality problem. They represent documents or words with low and dense vector. Moreover, recent studies showed that DL models comparatively get better performance than traditional model for the NLP. Especially, recurrent neural network (RNN) architectures and its variants have gained big attractions for shorter representation and improved results in many ways. In this study, we compared traditional representation (BOW) with deep learning architecture in terms of capability of solving sentiment problem in Turkish language. We used benchmark dataset collected from Turkish retailer Web pages regarding some product groups such as book, DVD, and film. Our experiments showed that DL models showed better performance than traditional BOW models. We also observed that bidirectional variant of LSTM and GRU is slightly better than one directional design. LSTM and GRU significantly outperformed simple RNN, because they do now suffer from vanishing gradient problem. Another important finding is that dropout mechanism can cope with overfitting and variance problem. We optimized drop rate of dropout as 50% that helps for less variance. In conclusion, the tables indicate that deep learning architectures get better results than traditional approaches on average. For example, for film category DL models get 89.26 accuracy at their best, and the traditional models, however, can get 86.5%. Likewise, we observe same results for other categories as shown in the paper.

References

1. Liu, Bing. 2012. *Sentiment Analysis and Opinion Mining*. Morgan & Claypool Publishers
2. Hu, Minqing and Bing Liu. 2004. Mining and summarizing customer reviews. In *Proceedings of the Tenth ACM SIGKDD International Conference on Knowledge Discovery and Data Mining (KDD '04)*, 168–177. New York: ACM
3. Pang, Bo, Lillian Lee, and Shivakumar Vaithyanathan. 2002. Thumbs up? Sentiment classification using machine learning techniques. In *Proceedings of the ACL-02 Conference on Empirical Methods in Natural Language Processing, vol. 10 (EMNLP '02)*, 79–86. Stroudsburg, PA: Association for Computational Linguistics
4. Yu, Z., H. Wang, X. Lin, and M. Wang. 2015. Learning term embeddings for hypernymy identification. In *Proceedings of the Twenty-Fourth International Joint Conference on Artificial Intelligence*, 1390–1397
5. Turian, J., L. Ratinov, and Y. Bengio. 2010. Word representations: A simple and general method for semi-supervised learning. In *Proceedings of the 48th Annual Meeting of the Association for Computational Linguistics, ACL '10*, 384–394. Stroudsburg, PA, USA
6. Pennington, J., R. Socher, C. Manning. 2014. Glove: Global vectors for word representation. In *Proceedings of the 2014 Conference on Empirical Methods in Natural Language Processing*, 1532–1543, Doha, Qatar
7. Mikolov, T., W. Yih, and G. Zweig. 2013. Linguistic regularities in continuous space word representations. In *Human Language Technologies: Conference of the North American Chapter of the Association of Computational Linguistics*, 746–751. Atlanta, Georgia, USA

8. Rothe, Sascha, Sebastian Ebert, Hinrich Schütze. 2016. Ultradense word embeddings by or-thogonal transformation. In *Proceedings of the 2016 Conference of the North American Chapter of the ACL*, San Diego, California

9. Duchi, John, Elad Hazan, and Yoram Singer. 2011. Adaptive subgradient methods for online learning and stochastic optimization. *The Journal of Machine Learning Research* 12: 2121–2159

10. Tieleman, Tijmen and Geoffrey Hinton. 2012. Lecture 6.5-RMSPROP: Divide the gradient by a running average of its recent magnitude. COURSERA: Neural Networks for Machine Learning, 4

11. Goodfellow, I., Y. Bengio, and Courville Aaron. 2016. *Deep Learning*. Cambridge: MIT Press

12. Sutskever, I., J. Martens, G. Dahl, G. Hinton. 2013. On the importance of initialization and momentum in deep learning

13. Hochreiter, Sepp and Jüurgen Schmidhuber. 1997. Long short-term memory. Neural Comput. 9(8): 1735–1780

14. Bengio, Y., P. Simard, and P. Frasconi. 1994. Learning long-term dependencies with gradient descent is difficult

15. Demirtas, Erkin and Mykola Pechenizkiy. 2013. Cross-lingual polarity detection with ma-chine translation. In *Proceedings of the Second International Workshop on Issues of Sentiment Discovery and Opinion Mining (WISDOM '13)*

16. Schütze, Hinrich, David A. Hull, and Jan O. Pedersen. 1995. A comparison of classifiers and document representations for the routing problem. In *Proceedings of the 18th Annual International ACM SIGIR Conference on Research and Development in Information Retrieval (SIGIR '95)*, ed. Edward A. Fox, Peter Ingwersen, and Raya Fidel, 229–237. New York, NY: ACM

17. Salton, G. 1971. *The SMART Retrieval System-Experiments in Automatic Document Processing*. Upper Saddle River, NJ: Prentice-Hall Inc

18. Bishop, Christopher M. 2011. *Pattern Recognition and Machine Learning*. Berlin: Springer

19. Dozat, T. 2016. Incorporating Nesterov Momentum into Adam

20. Kingma, Diederik and Jimmy Ba. 2014. Adam: A method for stochastic optimization. arXiv preprint arXiv:1412.6980